Large-Scale Inference

We live in a new age for statistical inference, where modern scientific technology such as microarrays and fMRI machines routinely produce thousands and sometimes millions of parallel data sets, each with its own estimation or testing problem. Doing thousands of problems at once involves more than repeated application of classical methods.

Taking an empirical Bayes approach, Bradley Efron, inventor of the bootstrap, shows how information accrues across problems in a way that combines Bayesian and frequentist ideas. Estimation, testing, and prediction blend in this framework, producing opportunities for new methodologies of increased power. New difficulties also arise, easily leading to flawed inferences. This book takes a careful look at both the promise and pitfalls of large-scale statistical inference, with particular attention to false discovery rates, the most successful of the new statistical techniques. Emphasis is on the inferential ideas underlying technical developments, illustrated using a large number of real examples.

BRADLEY EFRON is Max H. Stein Professor of Statistics and Biostatistics at the Stanford University School of Humanities and Sciences, and the Department of Health Research and Policy at the School of Medicine.

INSTITUTE OF MATHEMATICAL STATISTICS
MONOGRAPHS

IMS Monographs are concise research monographs of high quality on any branch of statistics or probability of sufficient interest to warrant publication as books. Some concern relatively traditional topics in need of up-to-date assessment. Others are on emerging themes. In all cases the objective is to provide a balanced view of the field.

Large-Scale Inference

Empirical Bayes Methods for Estimation, Testing, and Prediction

BRADLEY EFRON
Stanford University

CAMBRIDGE
UNIVERSITY PRESS

CAMBRIDGE
UNIVERSITY PRESS

University Printing House, Cambridge CB2 8BS, United Kingdom

Published in the United States of America by Cambridge University Press, New York

Cambridge University Press is part of the University of Cambridge.

It furthers the University's mission by disseminating knowledge in the pursuit of education, learning and research at the highest international levels of excellence.

www.cambridge.org
Information on this title: www.cambridge.org/9781107619678

First published 2010
First paperback edition 2013

A catalogue record for this publication is available from the British Library

ISBN 978-0-521-19249-1 Hardback
ISBN 978-1-107-61967-8 Paperback

Contents

Prologue

At the risk of drastic oversimplification, the history of statistics as a recognized discipline can be divided into three eras:

1 The age of Quetelet and his successors, in which huge census-level data sets were brought to bear on simple but important questions: Are there more male than female births? Is the rate of insanity rising?
2 The classical period of Pearson, Fisher, Neyman, Hotelling, and their successors, intellectual giants who developed a theory of optimal inference capable of wringing every drop of information out of a scientific experiment. The questions dealt with still tended to be simple — Is treatment A better than treatment B? — but the new methods were suited to the kinds of small data sets individual scientists might collect.
3 The era of scientific mass production, in which new technologies typified by the microarray allow a single team of scientists to produce data sets of a size Quetelet would envy. But now the flood of data is accompanied by a deluge of questions, perhaps thousands of estimates or hypothesis tests that the statistician is charged with answering together; not at all what the classical masters had in mind.

The response to this onslaught of data has been a tremendous burst of statistical methodology, impressively creative, showing an attractive ability to come to grips with changed circumstances, and at the same time highly speculative. There is plenty of methodology in what follows, but that is not the main theme of the book. My primary goal has been to ground the methodology in familiar principles of statistical inference.

This is where the "empirical Bayes" in my subtitle comes into consideration. By their nature, empirical Bayes arguments combine frequentist and Bayesian elements in analyzing problems of repeated structure. Repeated structures are just what scientific mass production excels at, e.g., expression levels comparing sick and healthy subjects for thousands of genes at the same time by means of microarrays. At their best, the new methodologies are successful from both Bayes and frequentist viewpoints, which is what my empirical Bayes arguments are intended to show.

False discovery rates, Benjamini and Hochberg's seminal contribution, is the great success story of the new methodology. Much of what follows is an attempt to explain that success in empirical Bayes terms. FDR, indeed, has strong credentials in both the Bayesian and frequentist camps, always a good sign that we are on the right track, as well as a suggestion of fruitful empirical Bayes explication.

The later chapters are at pains to show the limitations of current large-scale statistical practice: Which cases should be combined in a single analysis? How do we account for notions of relevance between cases? What is the correct null hypothesis? How do we handle correlations? Some helpful theory is provided in answer, but much of the argumentation is by example, with graphs and figures playing a major role. The examples are real ones, collected in a sometimes humbling decade of large-scale data analysis at the Stanford School of Medicine and Department of Statistics. (My examples here are mainly biomedical, but of course that has nothing to do with the basic ideas, which are presented with no prior medical or biological knowledge assumed.)

In moving beyond the confines of classical statistics, we are also moving outside its wall of protection. Fisher, Neyman et al. fashioned an almost perfect inferential machine for small-scale estimation and testing problems. It is hard to go wrong using maximum likelihood estimation or a t-test on a typical small data set. I have found it very easy to go wrong with huge data sets and thousands of questions to answer at once. Without claiming a cure, I hope the various examples at least help identify the symptoms.

The classical era of statistics can itself be divided into two periods: the first half of the 20th century, during which basic theory was developed, and then a great methodological expansion of that theory in the second half. Empirical Bayes stands as a striking exception. Emerging in the 1950s in two branches identified with Charles Stein and Herbert Robbins, it represented a genuinely new initiative in statistical theory. The Stein branch concerned normal estimation theory, while the Robbins branch was more general, being applicable to both estimation and hypothesis testing.

Typical large-scale applications have been more concerned with testing than estimation. If judged by chapter titles, the book seems to share this imbalance, but that is misleading. Empirical Bayes blurs the line between testing and estimation as well as between frequentism and Bayesianism. Much of what follows is an attempt to say how well we can estimate a testing procedure, for example how accurately can a null distribution be estimated? The false discovery rate procedure itself strays far from the spirit of classical hypothesis testing, as discussed in Chapter 4.

About this book: it is written for readers with at least a second course in statistics as background. The mathematical level is not daunting — mainly multidimensional calculus, probability theory, and linear algebra — though certain parts are more intricate, particularly in Chapters 3 and 7 (which can be scanned or skipped at first reading). There are almost no asymptotics. Exercises are interspersed in the text as they arise (rather than being lumped together at the end of chapters), where they mostly take the place of statements like "It is easy to see ..." or "It can be shown ...". Citations are concentrated in the **Notes** section at the end of each chapter. There are two brief appendices, one listing basic facts about exponential families, the second concerning access to some of the programs and data sets featured in the text.

I have perhaps abused the "mono" in monograph by featuring methods from my own work of the past decade. This is not a survey or a textbook, though I hope it can be used for a graduate-level lecture course. In fact, I am not trying to sell any particular methodology, my main interest as stated above being how the methods mesh with basic statistical theory.

There are at least three excellent books for readers who wish to see different points of view. Working backwards in time, Dudoit and van der Laan's 2009 *Multiple Testing Procedures with Applications to Genomics* emphasizes the control of Type I error. It is a successor to *Resampling-based Multiple Testing: Examples and Methods for p-Value Adjustment* (Westfall and Young, 1993), which now looks far ahead of its time. Miller's classic text, *Simultaneous Statistical Inference* (1981), beautifully describes the development of multiple testing before the era of large-scale data sets, when "multiple" meant somewhere between two and ten problems, not thousands.

I chose the adjective *large-scale* to describe massive data analysis problems rather than "multiple," "high-dimensional," or "simultaneous," because of its bland neutrality with regard to estimation, testing, or prediction, as well as its lack of identification with specific methodologies. My intention is not to have the last word here, and in fact I hope for and expect a healthy development of new ideas in dealing with the burgeoning statistical problems of the 21st century.

Acknowledgments

The Institute of Mathematical Statistics has begun an ambitious new monograph series in statistics, and I am grateful to the editors David Cox, Xiao-Li Meng, and Susan Holmes for their encouragement, and for letting me in on the ground floor. Diana Gillooly, the editor at Cambridge University Press (now in its fifth century!) has been supportive, encouraging, and gentle in correcting my literary abuses. My colleague Elizabeth Halloran has shown a sharp eye for faulty derivations and confused wording. Many of my Stanford colleagues and students have helped greatly in the book's final development, with Rob Tibshirani and Omkar Muralidharan deserving special thanks. Most of all, I am grateful to my associate Cindy Kirby for her tireless work in transforming my handwritten pages into the book you see here.

Bradley Efron
Department of Statistics
Stanford University

1

Empirical Bayes and the James–Stein Estimator

Charles Stein shocked the statistical world in 1955 with his proof that maximum likelihood estimation methods for Gaussian models, in common use for more than a century, were inadmissible beyond simple one- or two-dimensional situations. These methods are still in use, for good reasons, but Stein-type estimators have pointed the way toward a radically different *empirical Bayes* approach to high-dimensional statistical inference. We will be using empirical Bayes ideas for estimation, testing, and prediction, beginning here with their path-breaking appearance in the James–Stein formulation.

Although the connection was not immediately recognized, Stein's work was half of an energetic post-war empirical Bayes initiative. The other half, explicitly named "empirical Bayes" by its principal developer Herbert Robbins, was less shocking but more general in scope, aiming to show how frequentists could achieve full Bayesian efficiency in large-scale parallel studies. Large-scale parallel studies were rare in the 1950s, however, and Robbins' theory did not have the applied impact of Stein's shrinkage estimators, which are useful in much smaller data sets.

All of this has changed in the 21st century. New scientific technologies, epitomized by the microarray, routinely produce studies of thousands of parallel cases — we will see several such studies in what follows — well-suited for the Robbins point of view. That view predominates in the succeeding chapters, though not explicitly invoking Robbins' methodology until the very last section of the book.

Stein's theory concerns estimation, whereas the Robbins branch of empirical Bayes allows for hypothesis testing, that is, for situations where many or most of the true effects pile up at a specific point, usually called 0. Chapter 2 takes up large-scale hypothesis testing, where we will see, in Section 2.6, that the two branches are intertwined. Empirical Bayes theory blurs the distinction between estimation and testing as well as between frequentist and Bayesian methods. This becomes clear in Chapter 2, where we will undertake frequentist estimation of Bayesian hypothesis testing rules.

1.1 Bayes Rule and Multivariate Normal Estimation

This section provides a brief review of Bayes theorem as it applies to multivariate normal estimation. Bayes rule is one of those simple but profound ideas that underlie statistical thinking. We can state it clearly in terms of densities, though it applies just as well to discrete situations. An unknown parameter vector $\boldsymbol{\mu}$ with prior density $g(\boldsymbol{\mu})$ gives rise to an observable data vector $\boldsymbol{z}$ according to density $f_{\boldsymbol{\mu}}(\boldsymbol{z})$,

$$\boldsymbol{\mu} \sim g(\cdot) \quad \text{and} \quad \boldsymbol{z}|\boldsymbol{\mu} \sim f_{\boldsymbol{\mu}}(\boldsymbol{z}). \tag{1.1}$$

Bayes rule is a formula for the conditional density of $\boldsymbol{\mu}$ having observed $\boldsymbol{z}$ (its *posterior distribution*),

$$g(\boldsymbol{\mu}|\boldsymbol{z}) = g(\boldsymbol{\mu}) f_{\boldsymbol{\mu}}(\boldsymbol{z}) / f(\boldsymbol{z}) \tag{1.2}$$

where $f(\boldsymbol{z})$ is the *marginal distribution* of $\boldsymbol{z}$,

$$f(\boldsymbol{z}) = \int g(\boldsymbol{\mu}) f_{\boldsymbol{\mu}}(\boldsymbol{z})\, d\boldsymbol{\mu}, \tag{1.3}$$

the integral being over all values of $\boldsymbol{\mu}$.

The hardest part of (1.2), calculating $f(\boldsymbol{z})$, is usually the least necessary. Most often it is sufficient to note that the posterior density $g(\boldsymbol{\mu}|\boldsymbol{z})$ is proportional to $g(\boldsymbol{\mu}) f_{\boldsymbol{\mu}}(\boldsymbol{z})$, the product of the prior density $g(\boldsymbol{\mu})$ and the *likelihood* $f_{\boldsymbol{\mu}}(\boldsymbol{z})$ of $\boldsymbol{\mu}$ given $\boldsymbol{z}$. For any two possible parameter values $\boldsymbol{\mu}_1$ and $\boldsymbol{\mu}_2$, (1.2) gives

$$\frac{g(\boldsymbol{\mu}_1|\boldsymbol{z})}{g(\boldsymbol{\mu}_2|\boldsymbol{z})} = \frac{g(\boldsymbol{\mu}_1)}{g(\boldsymbol{\mu}_2)} \frac{f_{\boldsymbol{\mu}_1}(\boldsymbol{z})}{f_{\boldsymbol{\mu}_2}(\boldsymbol{z})}, \tag{1.4}$$

that is, the posterior odds ratio is the prior odds ratio times the likelihood ratio. Formula (1.2) is no more than a statement of the rule of conditional probability but, as we will see, Bayes rule can have subtle and surprising consequences.

Exercise 1.1 Suppose μ has a normal prior distribution with mean 0 and variance A, while z given μ is normal with mean μ and variance 1,

$$\mu \sim \mathcal{N}(0, A) \quad \text{and} \quad z|\mu \sim \mathcal{N}(\mu, 1). \tag{1.5}$$

Show that

$$\mu|z \sim \mathcal{N}(Bz, B) \qquad \text{where } B = A/(A+1). \tag{1.6}$$

Starting down the road to large-scale inference, suppose now we are dealing with many versions of (1.5),

$$\mu_i \sim \mathcal{N}(0, A) \quad \text{and} \quad z_i|\mu_i \sim \mathcal{N}(\mu_i, 1) \qquad [i = 1, 2, \ldots, N], \tag{1.7}$$

the (μ_i, z_i) pairs being independent of each other. Letting $\boldsymbol{\mu} = (\mu_1, \mu_2, \ldots, \mu_N)'$ and $\boldsymbol{z} = (z_1, z_2, \ldots, z_N)'$, we can write this compactly using standard notation for the N-dimensional normal distribution,

$$\boldsymbol{\mu} \sim \mathcal{N}_N(\mathbf{0}, AI) \tag{1.8}$$

and

$$\boldsymbol{z}|\boldsymbol{\mu} \sim \mathcal{N}_N(\boldsymbol{\mu}, I) \tag{1.9}$$

where I is the $N \times N$ identity matrix. Then Bayes rule gives posterior distribution

$$\boldsymbol{\mu}|\boldsymbol{z} \sim \mathcal{N}_N(B\boldsymbol{z}, BI) \qquad [B = A/(A+1)], \tag{1.10}$$

this being (1.6) applied component-wise.

Having observed $\boldsymbol{z}$ we wish to estimate $\boldsymbol{\mu}$ with some estimator $\hat{\boldsymbol{\mu}} = t(\boldsymbol{z})$,

$$\hat{\boldsymbol{\mu}} = (\hat{\mu}_1, \hat{\mu}_2, \ldots, \hat{\mu}_N)'. \tag{1.11}$$

We use total squared error loss to measure the error of estimating $\boldsymbol{\mu}$ by $\hat{\boldsymbol{\mu}}$,

$$L(\boldsymbol{\mu}, \hat{\boldsymbol{\mu}}) = \|\hat{\boldsymbol{\mu}} - \boldsymbol{\mu}\|^2 = \sum_{i=1}^{N} (\hat{\mu}_i - \mu_i)^2 \tag{1.12}$$

with the corresponding risk function being the expected value of $L(\boldsymbol{\mu}, \hat{\boldsymbol{\mu}})$ for a given $\boldsymbol{\mu}$,

$$R(\boldsymbol{\mu}) = E_{\boldsymbol{\mu}} \{L(\boldsymbol{\mu}, \hat{\boldsymbol{\mu}})\} = E_{\boldsymbol{\mu}} \left\{\|t(\boldsymbol{z}) - \boldsymbol{\mu}\|^2\right\}, \tag{1.13}$$

$E_{\boldsymbol{\mu}}$ indicating expectation with respect to $\boldsymbol{z} \sim \mathcal{N}_N(\boldsymbol{\mu}, I)$, $\boldsymbol{\mu}$ fixed.

The obvious estimator of $\boldsymbol{\mu}$, the one used implicitly in every regression and ANOVA application, is $\boldsymbol{z}$ itself,

$$\hat{\boldsymbol{\mu}}^{(\text{MLE})} = \boldsymbol{z}, \tag{1.14}$$

the maximum likelihood estimator (MLE) of $\boldsymbol{\mu}$ in model (1.9). This has risk

$$R^{(\text{MLE})}(\boldsymbol{\mu}) = N \tag{1.15}$$

for every choice of $\boldsymbol{\mu}$; every point in the parameter space is treated equally by $\hat{\boldsymbol{\mu}}^{(\text{MLE})}$, which seems reasonable for general estimation purposes.

Suppose though we have prior belief (1.8) which says that $\boldsymbol{\mu}$ lies more or less near the origin $\mathbf{0}$. According to (1.10), the Bayes estimator is

$$\hat{\boldsymbol{\mu}}^{(\text{Bayes})} = B\boldsymbol{z} = \left(1 - \frac{1}{A+1}\right)\boldsymbol{z}, \tag{1.16}$$

this being the choice that minimizes the expected squared error given $\boldsymbol{z}$. If $A = 1$, for instance, $\hat{\boldsymbol{\mu}}^{(\text{Bayes})}$ shrinks $\hat{\boldsymbol{\mu}}^{(\text{MLE})}$ halfway toward $\mathbf{0}$. It has risk

$$R^{(\text{Bayes})}(\boldsymbol{\mu}) = (1 - B)^2\|\boldsymbol{\mu}\|^2 + NB^2, \tag{1.17}$$

(1.13), and overall Bayes risk

$$R_A^{(\text{Bayes})} = E_A\left\{R^{(\text{Bayes})}(\boldsymbol{\mu})\right\} = N\frac{A}{A+1}, \tag{1.18}$$

E_A indicating expectation with respect to $\boldsymbol{\mu} \sim \mathcal{N}_N(\mathbf{0}, AI)$.

Exercise 1.2 Verify (1.17) and (1.18).

The corresponding Bayes risk for $\hat{\boldsymbol{\mu}}^{(\text{MLE})}$ is

$$R_A^{(\text{MLE})} = N$$

according to (1.15). If prior (1.8) is correct then $\hat{\boldsymbol{\mu}}^{(\text{Bayes})}$ offers substantial savings,

$$R_A^{(\text{MLE})} - R_A^{(\text{Bayes})} = N/(A+1); \tag{1.19}$$

with $A = 1$, $\hat{\boldsymbol{\mu}}^{(\text{Bayes})}$ removes half the risk of $\hat{\boldsymbol{\mu}}^{(\text{MLE})}$.

1.2 Empirical Bayes Estimation

Suppose model (1.8) is correct but we don't know the value of A so we can't use $\hat{\boldsymbol{\mu}}^{(\text{Bayes})}$. This is where empirical Bayes ideas make their appearance. Assumptions (1.8), (1.9) imply that the marginal distribution of $\boldsymbol{z}$ (integrating $\boldsymbol{z} \sim \mathcal{N}_N(\mu, I)$ over $\boldsymbol{\mu} \sim \mathcal{N}_N(0, A \cdot I)$) is

$$\boldsymbol{z} \sim \mathcal{N}_N\left(\mathbf{0}, (A+1)I\right). \tag{1.20}$$

The sum of squares $S = \|\boldsymbol{z}\|^2$ has a scaled chi-square distribution with N degrees of freedom,

$$S \sim (A+1)\chi_N^2, \tag{1.21}$$

so that

$$E\left\{\frac{N-2}{S}\right\} = \frac{1}{A+1}. \tag{1.22}$$

Exercise 1.3 Verify (1.22).

The *James–Stein estimator* is defined to be

$$\hat{\boldsymbol{\mu}}^{(\mathrm{JS})} = \left(1 - \frac{N-2}{S}\right)\boldsymbol{z}. \tag{1.23}$$

This is just $\hat{\boldsymbol{\mu}}^{(\mathrm{Bayes})}$ with an unbiased estimator $(N-2)/S$ substituting for the unknown term $1/(A+1)$ in (1.16). The name "empirical Bayes" is satisfyingly apt for $\hat{\boldsymbol{\mu}}^{(\mathrm{JS})}$: the Bayes estimator (1.16) is itself being empirically estimated from the data. This is only possible because we have N similar problems, $z_i \sim \mathcal{N}(\mu_i, 1)$ for $i = 1, 2, \ldots, N$, under simultaneous consideration.

It is not difficult to show that the overall Bayes risk of the James–Stein estimator is

$$R_A^{(\mathrm{JS})} = N\frac{A}{A+1} + \frac{2}{A+1}. \tag{1.24}$$

Of course this is bigger than the true Bayes risk (1.18), but the penalty is surprisingly modest,

$$R_A^{(\mathrm{JS})} \Big/ R_A^{(\mathrm{Bayes})} = 1 + \frac{2}{N \cdot A}. \tag{1.25}$$

For $N = 10$ and $A = 1$, $R_A^{(\mathrm{JS})}$ is only 20% greater than the true Bayes risk.

The shock the James–Stein estimator provided the statistical world didn't come from (1.24) or (1.25). These are based on the zero-centric Bayesian model (1.8), where the maximum likelihood estimator $\hat{\boldsymbol{\mu}}^{(0)} = \boldsymbol{z}$, which doesn't favor values of $\boldsymbol{\mu}$ near $\mathbf{0}$, might be expected to be bested. The rude surprise came from the theorem proved by James and Stein in 1961[1]:

Theorem *For $N \geq 3$, the James–Stein estimator everywhere dominates the MLE $\hat{\boldsymbol{\mu}}^{(0)}$ in terms of expected total squared error; that is,*

$$E_\mu\left\{\|\hat{\boldsymbol{\mu}}^{(\mathrm{JS})} - \boldsymbol{\mu}\|^2\right\} < E_\mu\left\{\|\hat{\boldsymbol{\mu}}^{(\mathrm{MLE})} - \boldsymbol{\mu}\|^2\right\} \tag{1.26}$$

for every *choice of $\boldsymbol{\mu}$.*

Result (1.26) is frequentist rather that Bayesian — it implies the superiority of $\hat{\boldsymbol{\mu}}^{(\mathrm{JS})}$ no matter what one's prior beliefs about $\boldsymbol{\mu}$ may be. Since versions of $\hat{\boldsymbol{\mu}}^{(\mathrm{MLE})}$ dominate popular statistical techniques such as linear regression, its apparent uniform inferiority was a cause for alarm. The fact that linear regression applications continue unabated reflects some virtues of $\hat{\boldsymbol{\mu}}^{(\mathrm{MLE})}$ discussed later.

[1] Stein demonstrated in 1956 that $\hat{\boldsymbol{\mu}}^{(0)}$ could be everywhere improved. The specific form (1.23) was developed with his student Willard James in 1961.

A quick proof of the theorem begins with the identity

$$(\hat{\mu}_i - \mu_i)^2 = (z_i - \hat{\mu}_i)^2 - (z_i - \mu_i)^2 + 2\,(\hat{\mu}_i - \mu_i)\,(z_i - \mu_i). \tag{1.27}$$

Summing (1.27) over $i = 1, 2, \ldots, N$ and taking expectations gives

$$E_{\boldsymbol{\mu}}\left\{\|\hat{\boldsymbol{\mu}} - \boldsymbol{\mu}\|^2\right\} = E_{\boldsymbol{\mu}}\left\{\|\boldsymbol{z} - \hat{\boldsymbol{\mu}}\|^2\right\} - N + 2\sum_{i=1}^{N} \operatorname{cov}_{\boldsymbol{\mu}}\left(\hat{\mu}_i, z_i\right), \tag{1.28}$$

where $\operatorname{cov}_{\boldsymbol{\mu}}$ indicates covariance under $\boldsymbol{z} \sim \mathcal{N}_N(\boldsymbol{\mu}, I)$. Integration by parts involving the multivariate normal density function $f_{\boldsymbol{\mu}}(z) = (2\pi)^{-N/2}\exp\{-\frac{1}{2}\sum(z_i - \mu_i)^2\}$ shows that

$$\operatorname{cov}_{\boldsymbol{\mu}}\left(\hat{\mu}_i, z_i\right) = E_{\boldsymbol{\mu}}\left\{\frac{\partial \hat{\mu}_i}{\partial z_i}\right\} \tag{1.29}$$

as long as $\hat{\mu}_i$ is continuously differentiable in $\boldsymbol{z}$. This reduces (1.28) to

$$E_{\boldsymbol{\mu}}\|\hat{\boldsymbol{\mu}} - \boldsymbol{\mu}\|^2 = E_{\boldsymbol{\mu}}\left\{\|\boldsymbol{z} - \hat{\boldsymbol{\mu}}\|^2\right\} - N + 2\sum_{i=1}^{N} E_{\boldsymbol{\mu}}\left\{\frac{\partial \hat{\mu}_i}{\partial z_i}\right\}. \tag{1.30}$$

Applying (1.30) to $\hat{\boldsymbol{\mu}}^{(\text{JS})}$ (1.23) gives

$$E_{\boldsymbol{\mu}}\left\{\left\|\hat{\boldsymbol{\mu}}^{(\text{JS})} - \boldsymbol{\mu}\right\|^2\right\} = N - E_{\boldsymbol{\mu}}\left\{\frac{(N-2)^2}{S}\right\} \tag{1.31}$$

with $S = \sum z_i^2$ as before. The last term in (1.31) is positive if N exceeds 2, proving the theorem.

Exercise 1.4 (a) Use (1.30) to verify (1.31). (b) Use (1.31) to verify (1.24).

The James–Stein estimator (1.23) shrinks each observed value z_i toward 0. We don't have to take 0 as the preferred shrinking point. A more general version of (1.8), (1.9) begins with

$$\mu_i \overset{\text{ind}}{\sim} \mathcal{N}(M, A) \quad \text{and} \quad z_i|\mu_i \overset{\text{ind}}{\sim} \mathcal{N}(\mu_i, \sigma_0^2) \tag{1.32}$$

for $i = 1, 2, \ldots, N$, where M and A are the mean and variance of the prior distribution. Then (1.10) and (1.20) become

$$z_i \overset{\text{ind}}{\sim} \mathcal{N}\left(M, A + \sigma_0^2\right) \quad \text{and} \quad \mu_i|z_i \overset{\text{ind}}{\sim} \mathcal{N}\left(M + B(z_i - M), B\sigma_0^2\right) \tag{1.33}$$

for $i = 1, 2, \ldots, N$, where

$$B = \frac{A}{A + \sigma_0^2}. \tag{1.34}$$

Now Bayes rule $\hat{\mu}_i^{(\text{Bayes})} = M + B(z_i - M)$ has the James–Stein empirical Bayes estimator

$$\hat{\mu}_i^{(\text{JS})} = \bar{z} + \left(1 - \frac{(N-3)\sigma_0^2}{S}\right)(z_i - \bar{z}), \tag{1.35}$$

with $\bar{z} = \sum z_i/N$ and $S = \sum(z_i - \bar{z})^2$. The theorem remains true as stated, except that we now require $N \geq 4$.

If the difference in (1.26) were tiny then $\hat{\boldsymbol{\mu}}^{(\text{JS})}$ would be no more than an interesting theoretical tidbit. In practice though, the gains from using $\hat{\boldsymbol{\mu}}^{(\text{JS})}$ can be substantial, and even, in favorable circumstances, enormous.

Table 1.1 illustrates one such circumstance. The batting averages z_i (number of successful hits divided by the number of tries) are shown for 18 major league baseball players early in the 1970 season. The true values μ_i are taken to be their averages over the remainder of the season, comprising about 370 more "at bats" each. We can imagine trying to predict the true values from the early results, using either $\hat{\mu}_i^{(\text{MLE})} = z_i$ or the James–Stein estimates (1.35) (with σ_0^2 equal to the binomial estimate $\bar{z}(1-\bar{z})/45$, $\bar{z} = 0.265$ the grand average[2]). The ratio of prediction errors is

$$\sum_1^{18}\left(\hat{\mu}_i^{(\text{JS})} - \mu_i\right)^2 \bigg/ \sum_1^{18}\left(\hat{\mu}_i^{(\text{MLE})} - \mu_i\right)^2 = 0.28, \tag{1.36}$$

indicating a tremendous advantage for the empirical Bayes estimates.

The initial reaction to the Stein phenomena was a feeling of paradox: Clemente, at the top of the table, is performing independently of Munson, near the bottom. Why should Clemente's good performance increase our prediction for Munson? It does for $\hat{\boldsymbol{\mu}}^{(\text{JS})}$ (mainly by increasing $\bar{z}$ in (1.35)), but not for $\hat{\boldsymbol{\mu}}^{(\text{MLE})}$. There is an implication of indirect evidence lurking *among* the players, supplementing the direct evidence of each player's own average. Formal Bayesian theory supplies the extra evidence through a prior distribution. Things are more mysterious for empirical Bayes methods, where the prior may exist only as a motivational device.

1.3 Estimating the Individual Components

Why haven't James–Stein estimators displaced MLEs in common statistical practice? The simulation study of Table 1.2 offers one answer. Here $N = 10$, with the ten μ_i values shown in the first column; $\mu_{10} = 4$ is much

[2] The z_i are binomial here, not normal, violating the conditions of the theorem, but the James–Stein effect is quite insensitive to the exact probabilistic model.

Table 1.1 *Batting averages* $z_i = \hat{\mu}_i^{(\mathrm{MLE})}$ *for 18 major league players early in the 1970 season;* μ_i *values are averages over the remainder of the season. The James–Stein estimates* $\hat{\mu}_i^{(JS)}$ (1.35) *based on the* z_i *values provide much more accurate overall predictions for the* μ_i *values. (By coincidence,* $\hat{\mu}_i$ *and* μ_i *both average* 0.265*; the average of* $\hat{\mu}_i^{(JS)}$ *must equal that of* $\hat{\mu}_i^{(\mathrm{MLE})}$*.)*

Name	Hits/AB	$\hat{\mu}_i^{(\mathrm{MLE})}$	μ_i	$\hat{\mu}_i^{(\mathrm{JS})}$
Clemente	18/45	.400	**.346**	.294
F. Robinson	17/45	.378	**.298**	.289
F. Howard	16/45	.356	**.276**	.285
Johnstone	15/45	.333	**.222**	.280
Berry	14/45	.311	**.273**	.275
Spencer	14/45	.311	**.270**	.275
Kessinger	13/45	.289	**.263**	.270
L. Alvarado	12/45	.267	**.210**	.266
Santo	11/45	.244	**.269**	.261
Swoboda	11/45	.244	**.230**	.261
Unser	10/45	.222	**.264**	.256
Williams	10/45	.222	**.256**	.256
Scott	10/45	.222	**.303**	.256
Petrocelli	10/45	.222	**.264**	.256
E. Rodriguez	10/45	.222	**.226**	.256
Campaneris	9/45	.200	**.286**	.252
Munson	8/45	.178	**.316**	.247
Alvis	7/45	.156	**.200**	.242
Grand Average		.265	**.265**	.265

different than the others. One thousand simulations of $\boldsymbol{z} \sim \mathcal{N}_{10}(\boldsymbol{\mu}, I)$ each gave estimates $\hat{\boldsymbol{\mu}}^{(\mathrm{MLE})} = \boldsymbol{z}$ and $\hat{\boldsymbol{\mu}}^{(\mathrm{JS})}$ (1.23). Average squared errors for each μ_i are shown. For example, $(\hat{\mu}_1^{(\mathrm{MLE})} - \mu_1)^2$ averaged 0.95 over the 1000 simulations, compared to 0.61 for $(\hat{\mu}_1^{(\mathrm{JS})} - \mu_1)^2$.

We see that $\hat{\mu}_i^{(\mathrm{JS})}$ gave better estimates than $\hat{\mu}_i^{(\mathrm{MLE})}$ for the first nine cases, but was much *worse* for estimating the outlying case μ_{10}. Overall, the total mean squared errors favored $\boldsymbol{\mu}^{(\mathrm{JS})}$, as they must.

Exercise 1.5 If we assume that the μ_i values in Table 1.2 were obtained from $\mu_i \overset{\mathrm{ind}}{\sim} \mathcal{N}(0, A)$, is the total error 8.13 about right?

The James–Stein theorem concentrates attention on the total squared error loss function $\sum(\hat{\mu}_i - \mu_i)^2$, without concern for the effects on individual cases. Most of those effects are good, as seen in Table 1.2, but genuinely

Table 1.2 *Simulation experiment:* $\boldsymbol{z} \sim \mathcal{N}_{10}(\boldsymbol{\mu}, I)$ *with* $(\mu_1, \mu_s, \ldots, \mu_{10})$ *as shown in first column.* $\text{MSE}_i^{(\text{MLE})}$ *is the average squared error* $(\hat{\mu}_i^{(\text{MLE})} - \mu_i)^2$*, likewise* $\text{MSE}_i^{(JS)}$*. Nine of the cases are better estimated by James–Stein, but for the outlying case 10,* $\hat{\mu}_{10}^{(JS)}$ *has nearly twice the error of* $\hat{\mu}_{10}^{(\text{MLE})}$.

	μ_i	$\text{MSE}_i^{(\text{MLE})}$	$\text{MSE}_i^{(\text{JS})}$
1	−.81	.95	.61
2	−.39	1.04	.62
3	−.39	1.03	.62
4	−.08	.99	.58
5	.69	1.06	.67
6	.71	.98	.63
7	1.28	.95	.71
8	1.32	1.04	.77
9	1.89	1.00	.88
10	4.00	1.08	2.04!!
Total Sqerr		10.12	8.13

unusual cases, like μ_{10}, can suffer. Baseball fans know that Clemente was in fact an extraordinarily good hitter, and shouldn't have been shrunk so drastically toward the mean of his less-talented cohort. Current statistical practice is quite conservative in protecting individual inferences from the tyranny of the majority, accounting for the continued popularity of stand-alone methods like $\hat{\boldsymbol{\mu}}^{(\text{MLE})}$. On the other hand, large-scale simultaneous inference, our general theme here, focuses on favorable group inferences.

Compromise methods are available that capture most of the group savings while protecting unusual individual cases. In the baseball example, for instance, we might decide to follow the James–Stein estimate (1.35) subject to the restriction of not deviating more than $D\,\sigma_0$ units away from $\hat{\mu}_i^{(\text{MLE})} = z_i$ (the so-called "limited translation estimator" $\hat{\mu}_i^{(D)}$):

$$\hat{\mu}_i^{(D)} = \begin{cases} \max\left(\hat{\mu}_i^{(\text{JS})}, \hat{\mu}_i^{(\text{MLE})} - D\sigma_0\right) & \text{for } z_i > \bar{z} \\ \min\left(\hat{\mu}_i^{(\text{JS})}, \hat{\mu}_i^{(\text{MLE})} + D\sigma_0\right) & \text{for } z_i \le \bar{z}. \end{cases} \tag{1.37}$$

Exercise 1.6 Graph $\hat{\mu}_i^{(D)}$ as a function of z_i for the baseball data.

Taking $D = 1$ says that $\hat{\mu}_i^{(D)}$ will never deviate more than $\sigma_0 = 0.066$ from z_i, so Clemente's prediction would be $\hat{\mu}_1^{(D)} = 0.334$ rather than $\hat{\mu}_1^{(\text{JS})} =$

0.294. This sacrifices some of the $\hat{\boldsymbol{\mu}}^{(\mathrm{JS})}$ savings relative to $\hat{\boldsymbol{\mu}}^{(\mathrm{MLE})}$, but not a great deal: it can be shown to lose only about 10% of the overall James–Stein advantage in the baseball example.

1.4 Learning from the Experience of Others

Bayes and empirical Bayes techniques involve learning from the experience of others, e.g., each baseball player learning from the other 17. This always raises the question, "Which others?" Chapter 10 returns to this question in the context of hypothesis testing. There we will have thousands of other cases, rather than 17, vastly increasing the amount of "other" experience.

Figure 1.1 diagrams James–Stein estimation, with case 1 learning from the $N-1$ others. We imagine that the others have been observed first, giving estimates $(\hat{M}, \hat{A})$ for the unknown Bayes parameters in (1.32) (taking $\sigma_0^2 = 1$). The estimated prior distribution $\mathcal{N}(\hat{M}, \hat{A})$ is then used to supplement the direct evidence $z_1 \sim \mathcal{N}(\mu_1, 1)$ for the estimation of μ_1. (Actually $\hat{\mu}_i^{(\mathrm{JS})}$ includes z_i as well as the others in estimating $(\hat{M}, \hat{A})$ for use on μ_1: it can be shown that this improves the accuracy of $\hat{\mu}_1^{(\mathrm{JS})}$.) Versions of this same diagram apply to the more intricate empirical Bayes procedures that follow.

Learning from the experience of others is not the sole property of the Bayes world. Figure 1.2 illustrates a common statistical situation. A total of $N = 157$ healthy volunteers have had their kidney function evaluated by a somewhat arduous medical procedure. The scores are plotted versus age, higher scores indicating better function, and it is obvious that function tends to decrease with age. (At one time, kidney donation was forbidden for donors exceeding 60, though increasing demand has relaxed this rule.) The heavy line indicates the least squares fit of function to age.

A potential new donor, age 55, has appeared, but it is not practical to evaluate his kidney function by the arduous medical procedure. Figure 1.2 shows two possible predictions: the starred point is the function score (−0.01) for the only 55-year-old person among the 157 volunteers, while the squared point reads off the value of the least square line (−1.46) at age = 55. Most statisticians, frequentist or Bayesian, would prefer the least squares prediction.

Tukey's evocative term "borrowing strength" neatly captures the regression idea. This is certainly "learning from the experience of others," but in a more rigid framework than Figure 1.1. Here there is a simple covari-

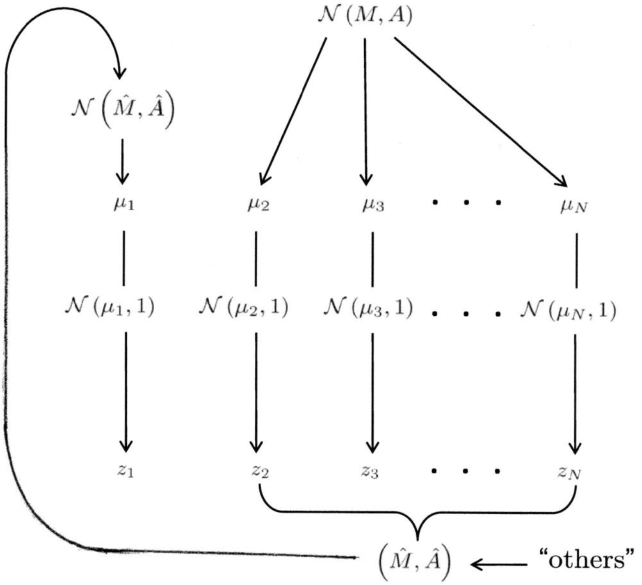

Figure 1.1 Schematic diagram of James–Stein estimation, showing case 1 learning from the experience of the other $N - 1$ cases.

ate, age, convincingly linking the volunteers with the potential donor. The linkage is more subtle in the baseball example.

Often the two methods can be combined. We might extend model (1.32) to

$$\mu_i \overset{\text{ind}}{\sim} \mathcal{N}(M_0 + M_1 \cdot \text{age}_i, A) \quad \text{and} \quad z_i \sim \mathcal{N}\left(\mu_i, \sigma_0^2\right). \tag{1.38}$$

The James–Stein estimate (1.35) takes the form

$$\hat{\mu}_i^{(\text{JS})} = \hat{\mu}_i^{(\text{reg})} + \left(1 - \frac{(N-4)\sigma_0^2}{S}\right)\left(z_i - \hat{\mu}_i^{(\text{reg})}\right), \tag{1.39}$$

where $\hat{\mu}_i^{(\text{reg})}$ is the linear regression estimate $(\hat{M}_0 + \hat{M}_1 \cdot \text{age}_i)$ and $S = \sum(z_i - \hat{\mu}_i^{(\text{reg})})^2$. Now $\hat{\mu}_i^{(\text{JS})}$ is shrunk toward the linear regression line instead of toward $\bar{z}$.

Exercise 1.7 For the kidney data, $S = 503$. Assuming $\sigma_0^2 = 1$, what is the James–Stein estimate for the starred point in Figure 1.2 (i.e., for the healthy volunteer, age 55)?

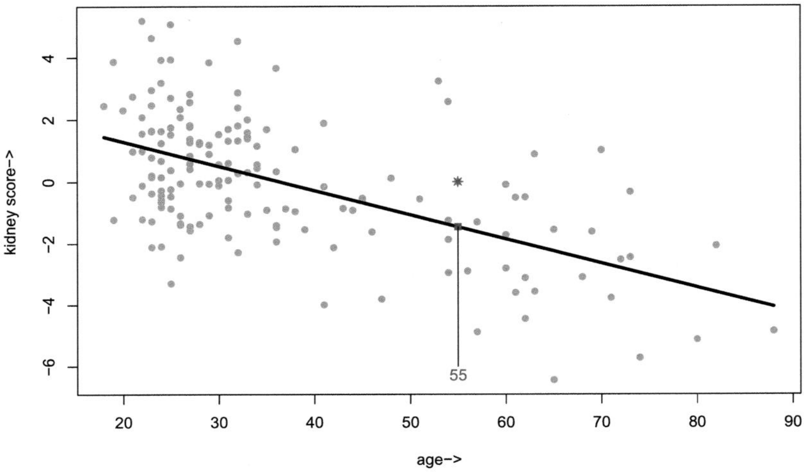

Figure 1.2 Kidney scores plotted versus age for 157 healthy volunteers. The least squares line shows the decrease of function with age. How should we predict the score of a potential donor, age 55?

1.5 Empirical Bayes Confidence Intervals

Returning to the situation in Section 1.1, suppose we have $N + 1$ independent normal observations z_i, with

$$\mu_i \overset{\text{ind}}{\sim} \mathcal{N}(0, A) \quad \text{and} \quad z_i|\mu_i \overset{\text{ind}}{\sim} \mathcal{N}(\mu_i, 1) \tag{1.40}$$

for $i = 0, 1, 2, \ldots, N$, and we want to assign a "confidence interval" to the parameter μ_0. The quotes are necessary here because we wish to take advantage of empirical Bayes information as in Figure 1.1, now with the "others" being $\boldsymbol{z} = (z_1, z_2, \ldots, z_N)$ and with (μ_0, z_0) playing the role of (μ_1, z_1) — taking us beyond classical confidence interval methodology.

If A were known we could calculate the Bayes posterior distribution for μ_0 according to (1.10),

$$\mu_0|z_0 \sim \mathcal{N}(Bz_0, B) \qquad [B = A/(A+1)], \tag{1.41}$$

yielding

$$\mu_0 \in Bz_0 \pm 1.96\sqrt{B} \tag{1.42}$$

as the obvious 95% posterior interval. A reasonable first try in the empirical

Bayes situation of Section 1.2 is to substitute the unbiased estimate

$$\hat{B} = 1 - \frac{N-2}{S} \qquad \left[S = \|\boldsymbol{z}\|^2\right] \tag{1.43}$$

into (1.41), giving the approximation

$$\mu_0 | z_0, \boldsymbol{z} \dot{\sim} \mathcal{N}\left(\hat{B}z_0, \hat{B}\right) \tag{1.44}$$

and similarly $\hat{B}z_0 \pm 1.96\sqrt{\hat{B}}$ for (1.42). In doing so, however, we have ignored the variability of $\hat{B}$ as an estimate of B, which can be substantial when N is small.

Here is a more accurate version of (1.44):

$$\mu_0 | z_0, \boldsymbol{z} \dot{\sim} \mathcal{N}\left(\hat{B}z_0, \hat{B} + \frac{2}{N-2}\left[z_0\left(1-\hat{B}\right)\right]^2\right) \tag{1.45}$$

and its corresponding posterior interval

$$\mu_0 \in \hat{B}z_0 \pm 1.96\left\{\hat{B} + \frac{2}{N-2}\left[z_0\left(1-\hat{B}\right)\right]^2\right\}^{\frac{1}{2}}. \tag{1.46}$$

Exercise 1.8 (a) Show that the relative length of (1.46) compared to the interval based on (1.44) is

$$\left\{1 + \frac{2}{N-2}\frac{z_0^2\left(1-\hat{B}\right)^2}{\hat{B}}\right\}^{\frac{1}{2}}. \tag{1.47}$$

(b) For $N = 17$ and $\hat{B} = 0.21$ (appropriate values for the baseball example), graph (1.47) for z_0 between 0 and 3.

Formula (1.45) can be justified by carefully following through a simplified version of Figure 1.1 in which $M = 0$, using familiar maximum likelihood calculations to assess the variability of $\hat{A}$ and its effect on the empirical Bayes estimation of μ_0 (called μ_1 in the figure).

Hierarchical Bayes methods offer another justification. Here the model (1.40) would be preceded by some Bayesian prior assumption on the *hyperparameter* A, perhaps A uniformly distributed over $(0, 10^6)$, chosen not to add much information beyond that in $\boldsymbol{z}$ to A's estimation. The term *objective Bayes* is used to describe such arguments, which are often insightful and useful. Defining $V = A + 1$ in model (1.40) and formally applying Bayes rule to the (impossible) prior that takes V to be uniformly distributed over $(0, \infty)$ yields exactly the posterior mean and variance in (1.45).

Notes

Herbert Robbins, paralleling early work by R. A. Fisher, I. J. Good, and Alan Turing (of Turing machine fame) developed a powerful theory of empirical Bayes statistical inference, some references being Robbins (1956) and Efron (2003). Robbins reserved the name "empirical Bayes" for situations where a genuine prior distribution like (1.8) was being estimated, using "compound Bayes" for more general parallel estimation and testing situations, but Efron and Morris (1973) hijacked "empirical Bayes" for James–Stein-type estimators.

Stein (1956) and James and Stein (1961) were written entirely from a frequentist point of view, which has much to do with their bombshell effect on the overwhelmingly frequentist statistical literature of that time. Stein (1981) gives the neat identity (1.28) and the concise proof of the theorem.

Limited translation estimates (1.37) were developed in Efron and Morris (1972), amid a more general theory of *relevance functions*, modifications of the James–Stein estimator that allowed individual cases to partially opt out of the overall shrinkage depending on how relevant the other cases appeared to be. Relevance functions for hypothesis testing will be taken up here in Chapter 10. Efron (1996) gives a more general version of Figure 1.1.

The kidney data in Figure 1.2 is from the Stanford nephrology lab of Dr. B. Myers; see Lemley et al. (2008). Morris (1983) gives a careful derivation of empirical Bayes confidence intervals such as (1.46), along with an informative discussion of what one should expect from such intervals.

2

Large-Scale Hypothesis Testing

Progress in statistics is usually at the mercy of our scientific colleagues, whose data is the "nature" from which we work. Agricultural experimentation in the early 20th century led Fisher to the development of analysis of variance. Something similar is happening at the beginning of the 21st century. A new class of "high throughput" biomedical devices, typified by the microarray, routinely produce hypothesis-testing data for thousands of cases at once. This is not at all the situation envisioned in the classical frequentist testing theory of Neyman, Pearson, and Fisher. This chapter begins the discussion of a theory of large-scale simultaneous hypothesis testing now under development in the statistics literature.

2.1 A Microarray Example

Figure 2.1 concerns a microarray example, the *prostate data*. Genetic expression levels for $N = 6033$ genes were obtained for $n = 102$ men, $n_1 = 50$ normal control subjects and $n_2 = 52$ prostate cancer patients. Without going into biological details, the principal goal of the study was to discover a small number of "interesting" genes, that is, genes whose expression levels differ between the prostate and normal subjects. Such genes, once identified, might be further investigated for a causal link to prostate cancer development.

The prostate data is a 6033×102 matrix X having entries[1]

$$x_{ij} = \text{expression level for gene } i \text{ on patient } j, \tag{2.1}$$

$i = 1, 2, \ldots, N$ and $j = 1, 2, \ldots, n$; with $j = 1, 2, \ldots, 50$ for the normal controls and $j = 51, 52, \ldots, 102$ for the cancer patients. Let $\bar{x}_i(1)$ and $\bar{x}_i(2)$ be the averages of x_{ij} for the normal controls and for the cancer patients.

[1] Obtained from oligonucleotide arrays.

The two-sample t-statistic for testing gene i is

$$t_i = \frac{\bar{x}_i(2) - \bar{x}_i(1)}{s_i}, \tag{2.2}$$

where s_i is an estimate of the standard error of the numerator,

$$s_i^2 = \frac{\sum_1^{50}\left(x_{ij} - \bar{x}(1)\right)^2 + \sum_{51}^{102}\left(x_{ij} - \bar{x}(2)\right)^2}{100} \cdot \left(\frac{1}{50} + \frac{1}{52}\right). \tag{2.3}$$

If we had only data from gene i to consider, we could use t_i in the usual way to test the null hypothesis

$$H_{0i} : \text{gene } i \text{ is "null"}, \tag{2.4}$$

i.e., that x_{ij} has the same distribution for the normal and cancer patients, rejecting H_{0i} if t_i looked too big in absolute value. The usual 5% rejection criterion, based on normal theory assumptions, would reject H_{0i} if $|t_i|$ exceeded 1.98, the two-tailed 5% point for a Student-t random variable with 100 degrees of freedom.

It will be convenient for our discussions here to use "z-values" instead of "t-values"; that is, we transform t_i to

$$z_i = \Phi^{-1}\left(F_{100}(t_i)\right), \tag{2.5}$$

where Φ and F_{100} are the cumulative distribution functions (abbreviated "cdf") for standard normal and t_{100} distributions. Under the usual assumptions of normal sampling, z_i will have a standard normal distribution if H_{0i} is true,

$$H_{0i} : z_i \sim \mathcal{N}(0, 1) \tag{2.6}$$

(called the *theoretical null* in what follows). The usual two-sided 5% test rejects H_{0i} for $|z_i| > 1.96$, the two-tailed 5% point for a $\mathcal{N}(0, 1)$ distribution.

Exercise 2.1 Plot $z = \Phi^{-1}(F_\nu(t))$ versus t for $-4 \leq t \leq 4$, for degrees of freedom $\nu = 25$, 50, and 100.

But of course we don't have just a single gene to test, we have $N = 6033$ of them. Figure 2.1 shows a histogram of the N z_i values, comparing it to a standard $\mathcal{N}(0, 1)$ density curve $c \cdot \exp\{-z^2/2\}/\sqrt{2\pi}$, with the multiplier c chosen to make the curve integrate to the same area as the histogram. If H_{0i} were true for every i, that is, if all of the genes were null, the histogram would match the curve. Fortunately for the investigators, it doesn't: it is too low near the center and too high in the tails. This suggests the presence

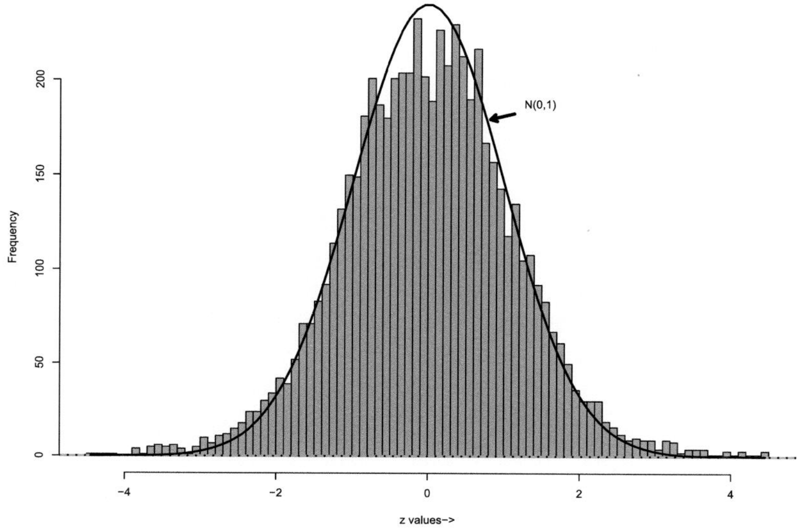

Figure 2.1 Prostate data; z-values testing 6033 genes for possible involvement with prostate cancer; curve is $\mathcal{N}(0, 1)$ theoretical null.

of some interesting non-null genes. How to identify the non-null cases dependably, without being misled by the effects of multiple inference, is the subject of intense current research.

A traditional approach to multiple inference uses the Bonferroni bound: we change the rejection level for each test from 0.05 to 0.05/6033. This amounts to rejecting H_{0i} only if $|z_i|$ exceeds 4.31, rather than 1.96. Now the total probability of mistakenly rejecting even one of the 6033 null hypotheses is less than 5%, but looking at Figure 2.1, 4.31 seems overly cautious. (Only six of the genes are declared non-null.) Empirical Bayes methods offer a less conservative approach to multiple testing.

2.2 Bayesian Approach

The *two-groups model* provides a simple Bayesian framework for multiple testing. We suppose that the N cases (the genes for the prostate study) are each either null or non-null with prior probability π_0 or $\pi_1 = 1 - \pi_0$, and with z-values having density either $f_0(z)$ or $f_1(z)$,

$$\begin{aligned} \pi_0 &= \Pr\{\text{null}\} & f_0(z) &= \text{density if null} \\ \pi_1 &= \Pr\{\text{non-null}\} & f_1(z) &= \text{density if non-null}. \end{aligned} \tag{2.7}$$

Ordinarily π_0 will be much bigger than π_1, say

$$\pi_0 \geq 0.90, \tag{2.8}$$

reflecting the usual purpose of large-scale testing: to reduce a vast collection of possibilities to a much smaller set of scientifically interesting prospects. If the assumptions underlying (2.6) are valid, then $f_0(z)$ is the standard normal density,

$$f_0(z) = \varphi(z) = e^{-\frac{1}{2}z^2} \Big/ \sqrt{2\pi} \tag{2.9}$$

while $f_1(z)$ might be some alternative density yielding z-values further away from 0.

Let F_0 and F_1 denote the probability distributions corresponding to f_0 and f_1 so that, for any subset $\mathcal{Z}$ of the real line,

$$F_0(\mathcal{Z}) = \int_{\mathcal{Z}} f_0(z)\,dz \quad \text{and} \quad F_1(\mathcal{Z}) = \int_{\mathcal{Z}} f_1(z)\,dz. \tag{2.10}$$

The *mixture density*

$$f(z) = \pi_0 f_0(z) + \pi_1 f_1(z) \tag{2.11}$$

has the mixture probability distribution

$$F(\mathcal{Z}) = \pi_0 F_0(\mathcal{Z}) + \pi_1 F_1(\mathcal{Z}). \tag{2.12}$$

(The usual cdf is $F((-\infty, z))$ in this notation, but later we will return to the less clumsy $F(z)$.) Under model (2.7), z has marginal density f and distribution F.

Suppose we observe $z \in \mathcal{Z}$ and wonder if it corresponds to the null or non-null arm of (2.7). A direct application of Bayes rule yields

$$\phi(\mathcal{Z}) \equiv \Pr\{\text{null}|z \in \mathcal{Z}\} = \pi_0 F_0(\mathcal{Z})/F(\mathcal{Z}) \tag{2.13}$$

as the posterior probability of nullity given $z \in \mathcal{Z}$. Following Benjamini and Hochberg's evocative terminology,[2] we call $\phi(\mathcal{Z})$ the *Bayes false discovery rate* for $\mathcal{Z}$: if we report $z \in \mathcal{Z}$ as non-null, $\phi(\mathcal{Z})$ is the chance that we've made a false discovery. We will also[3] write $\mathrm{Fdr}(\mathcal{Z})$ for $\phi(\mathcal{Z})$.

[2] Section 4.2 presents Benjamini and Hochberg's original frequentist development of false discovery rates, a name hijacked here for our Bayes/empirical Bayes discussion.

[3] A brief glossary of terms relating to false discovery rates appears at the end of the Notes for this chapter.

If $\mathcal{Z}$ is a single point z_0,

$$\phi(z_0) \equiv \Pr\{\text{null}|z = z_0\} = \pi_0 f_0(z_0)/f(z_0) \tag{2.14}$$

is the *local Bayes false discovery rate*, also written as fdr(z).

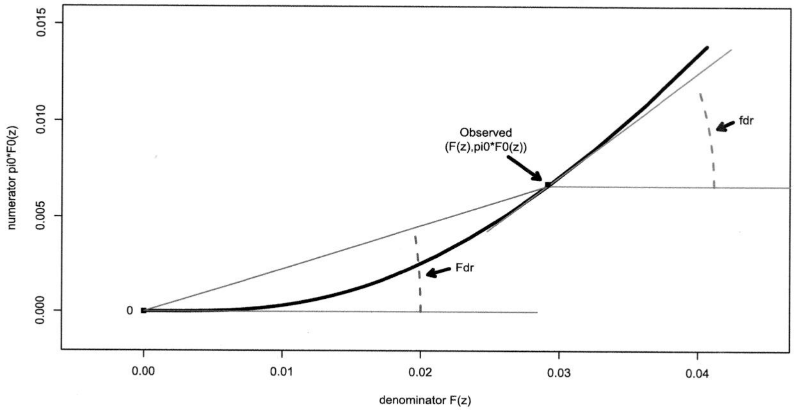

Figure 2.2 Relationship between Fdr(z) and fdr(z). Fdr(z) is the slope of the secant line connecting the origin with the point $(F(z), p_0 \cdot F_0(z_0))$, the denominator and numerator of (2.16); fdr(z) is the slope of the tangent line to the curve at that point.

Exercise 2.2 Show that

$$E_f\{\phi(z)|z \in \mathcal{Z}\} = \phi(\mathcal{Z}), \tag{2.15}$$

where E_f indicates conditional expectation with respect to the marginal density $f(z)$. In other words, Fdr($\mathcal{Z}$) is the conditional expectation of fdr(z) given $z \in \mathcal{Z}$.

In applications, $\mathcal{Z}$ is usually a tail interval $(-\infty, z)$ or (z, ∞). Writing $F(z)$ in place of $F((-\infty, z))$ for the usual cdf,

$$\phi((-\infty, z)) \equiv \text{Fdr}(z) = \pi_0 F_0(z)/F(z). \tag{2.16}$$

Plotting the numerator $\pi_0 F_0(z)$ versus the denominator $F(z)$ shows that Fdr(z) and fdr(z) are, respectively, secant and tangent, as illustrated in Figure 2.2, usually implying fdr(z) > Fdr(z) when both are small.

Exercise 2.3 Suppose

$$F_1(z) = F_0(z)^\gamma \qquad [\gamma < 1] \tag{2.17}$$

(often called *Lehmann alternatives*). Show that

$$\log\left\{\frac{\mathrm{fdr}(z)}{1-\mathrm{fdr}(z)}\right\} = \log\left\{\frac{\mathrm{Fdr}(z)}{1-\mathrm{Fdr}(z)}\right\} + \log\left(\frac{1}{\gamma}\right), \tag{2.18}$$

and that

$$\mathrm{fdr}(z) \doteq \mathrm{Fdr}(z)/\gamma \tag{2.19}$$

for small values of Fdr(z).

Exercise 2.4 We would usually expect $f_1(z)$, the non-null density in (2.7), to have heavier tails than $f_0(z)$. Why does this suggest, at least qualitatively, the shape of the curve shown in Figure 2.2?

2.3 Empirical Bayes Estimates

The Bayesian two-groups model (2.7) involves three quantities: the prior null probability π_0, the null density $f_0(z)$, and the non-null density $f_1(z)$. Of these, f_0 is known, at least if we believe in the theoretical null $\mathcal{N}(0,1)$ distribution (2.6), and π_0 is "almost known," usually being close enough to 1 as to have little effect on the false discovery rate (2.13). (In applications, π_0 is often taken equal to 1; Chapter 6 discusses the estimation of both π_0 and $f_0(z)$ in situations where the theoretical null is questionable.) This leaves $f_1(z)$, which is unlikely to be known to the statistician a priori.

There is, however, an obvious empirical Bayes approach to false discovery rate estimation. Let $\bar{F}(\mathcal{Z})$ denote the *empirical distribution* of the N z-values,

$$\bar{F}(\mathcal{Z}) = \#\{z_i \in \mathcal{Z}\}/N, \tag{2.20}$$

i.e., the proportion of the z_i values observed to be in the set $\mathcal{Z}$. Substituting into (2.13) gives an estimated false discovery rate,

$$\overline{\mathrm{Fdr}}(\mathcal{Z}) \equiv \bar{\phi}(\mathcal{Z}) = \pi_0 F_0(\mathcal{Z})\big/\bar{F}(\mathcal{Z}). \tag{2.21}$$

When N is large we expect $\bar{F}(\mathcal{Z})$ to be close to $F(\mathcal{Z})$, and $\overline{\mathrm{Fdr}}(\mathcal{Z})$ to be a good approximation for $\mathrm{Fdr}(\mathcal{Z})$.

Just how good is the question considered next. Figure 2.3 illustrates model (2.7): the N values $z_1, z_2, \dots, z_N$ are distributed to the null and non-null arms of the study in proportions π_0 and π_1. Let $N_0(\mathcal{Z})$ be the number of null z_i falling into $\mathcal{Z}$, and likewise $N_1(\mathcal{Z})$ for the non-null z_i in $\mathcal{Z}$. We can't observe $N_0(\mathcal{Z})$ or $N_1(\mathcal{Z})$, but we do get to see the total number $N_+(\mathcal{Z})$ in $\mathcal{Z}$,

$$N_+(\mathcal{Z}) = N_0(\mathcal{Z}) + N_1(\mathcal{Z}). \tag{2.22}$$

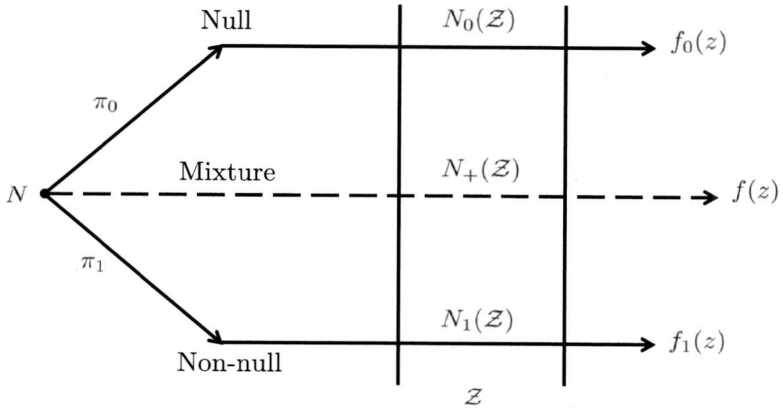

Figure 2.3 Diagram of two-groups model (2.7); N z-values are distributed to the two arms in proportions π_0 and π_1; $N_0(\mathcal{Z})$ and $N_1(\mathcal{Z})$ are numbers of nulls and non-nulls in $\mathcal{Z}$; the statistician observes $z_1, z_2, \ldots, z_N$ from mixture density $f(z)$ (2.11), and gets to see $N_+(\mathcal{Z}) = N_0(\mathcal{Z}) + N_1(\mathcal{Z})$.

Although $N_0(\mathcal{Z})$ is unobservable, we know its expectation

$$e_0(\mathcal{Z}) \equiv N\pi_0 F_0(\mathcal{Z}). \tag{2.23}$$

In this notation we can express $\overline{\text{Fdr}}(\mathcal{Z})$ as

$$\overline{\text{Fdr}}(\mathcal{Z}) = e_0(\mathcal{Z})/N_+(\mathcal{Z}). \tag{2.24}$$

As an example, the prostate data has $N_+(\mathcal{Z}) = 49$ z_i values in $\mathcal{Z} = (3, \infty)$, that is, exceeding 3; $e_0(\mathcal{Z})$ equals

$$6033 \cdot \pi_0 \cdot (1 - \Phi(3)) \tag{2.25}$$

under the theoretical null (2.6). The upper bound[4] $\pi_0 = 1$ gives $e_0(\mathcal{Z}) = 8.14$ and

$$\overline{\text{Fdr}}(\mathcal{Z}) = 8.14/49 = 0.166. \tag{2.26}$$

The implication is that about 1/6 of the 49 are false discoveries: if we report the list of 49 to the prostate study scientists as likely prospects, most of their subsequent work will not be in vain. Chapter 4 examines the logic behind this line of thought. Here we will only consider $\overline{\text{Fdr}}(\mathcal{Z})$ as an empirical Bayes point estimator for $\text{Fdr}(\mathcal{Z})$.

[4] Bounding π_0 by 1 in (2.25) does not imply that $\pi_1 = 0$ in (2.12) since the denominator $\bar{F}(\mathcal{Z})$ in (2.21) is estimated directly from the data regardless of π_0.

2.4 $\overline{\text{Fdr}}(\mathcal{Z})$ as a Point Estimate

We can express (2.13) as

$$\phi(\mathcal{Z}) = \text{Fdr}(\mathcal{Z}) = e_0(\mathcal{Z})/e_+(\mathcal{Z}), \tag{2.27}$$

where $e_+(\mathcal{Z}) = N \cdot F(\mathcal{Z})$ is the expected total count of z_i values in $\mathcal{Z}$. The quantity we would like to know, but can't observe, is the *false discovery proportion*

$$\text{Fdp}(\mathcal{Z}) = N_0(\mathcal{Z})/N_+(\mathcal{Z}), \tag{2.28}$$

the actual proportion of false discoveries in $\mathcal{Z}$. This gives us three quantities to consider:

$$\overline{\text{Fdr}}(\mathcal{Z}) = \frac{e_0(\mathcal{Z})}{N_+(\mathcal{Z})}, \quad \phi(\mathcal{Z}) = \frac{e_0(\mathcal{Z})}{e_+(\mathcal{Z})}, \quad \text{and} \quad \text{Fdp}(\mathcal{Z}) = \frac{N_0(\mathcal{Z})}{N_+(\mathcal{Z})}. \tag{2.29}$$

The next four lemmas discuss their relationship.

Lemma 2.1 *Suppose $e_0(\mathcal{Z})$ as defined in (2.23) is the same as the conditional expectation of $N_0(\mathcal{Z})$ given $N_1(\mathcal{Z})$. Then the conditional expectations of $\overline{\text{Fdr}}(\mathcal{Z})$ and $\text{Fdp}(\mathcal{Z})$ given $N_1(\mathcal{Z})$ satisfy*

$$E\left\{\overline{\text{Fdr}}(\mathcal{Z})|N_1(\mathcal{Z})\right\} \geq \phi_1(\mathcal{Z}) \geq E\left\{\text{Fdp}(\mathcal{Z})|N_1(\mathcal{Z})\right\}, \tag{2.30}$$

where

$$\phi_1(\mathcal{Z}) = \frac{e_0(\mathcal{Z})}{e_0(\mathcal{Z}) + N_1(\mathcal{Z})}. \tag{2.31}$$

Proof Writing $\overline{\text{Fdr}}(\mathcal{Z}) = e_0(\mathcal{Z})/(N_0(\mathcal{Z}) + N_1(\mathcal{Z}))$, Jensen's inequality gives $E\{\overline{\text{Fdr}}(\mathcal{Z})|N_1(\mathcal{Z})\} \geq \phi_1(\mathcal{Z})$. The condition on $e_0(\mathcal{Z})$ is satisfied if the number and distribution of the null case z-values does not depend on $N_1(\mathcal{Z})$. □

Exercise 2.5 Apply Jensen's inequality again to complete the proof.

Note The relationship in (2.30) makes the conventional assumption that $\text{Fdp}(\mathcal{Z}) = 0$ if $N_+(\mathcal{Z}) = 0$.

Lemma 2.1 says that for every value of $N_1(\mathcal{Z})$, the conditional expectation of $\overline{\text{Fdr}}(\mathcal{Z})$ exceeds that of $\text{Fdp}(\mathcal{Z})$, so that in this sense the empirical Bayes false discovery rate is a conservatively biased estimate of the actual false discovery proportion. Taking expectations over $N_1(\mathcal{Z})$, and reapplying Jensen's inequality, shows that[5]

$$\phi(\mathcal{Z}) \geq E\left\{\mathrm{Fdp}(\mathcal{Z})\right\}, \tag{2.32}$$

so that the Bayes Fdr $\phi(\mathcal{Z})$ is an upper bound on the expected Fdp. We also obtain $E\{\overline{\mathrm{Fdr}}(\mathcal{Z})\} \geq E\{\mathrm{Fdp}(\mathcal{Z})\}$, but this is uninformative since $E\{\overline{\mathrm{Fdr}}(\mathcal{Z})\} = \infty$ whenever $\Pr\{N_+(\mathcal{Z}) = 0\}$ is positive. In practice we would use $\overline{\mathrm{Fdr}}^{(\min)}(\mathcal{Z}) = \min(\overline{\mathrm{Fdr}}(\mathcal{Z}), 1)$ to estimate $\mathrm{Fdr}(\mathcal{Z})$, but it is not true in general that $E\{\overline{\mathrm{Fdr}}^{(\min)}(\mathcal{Z})\}$ exceeds $\phi(\mathcal{Z})$ or $E\{\mathrm{Fdp}(\mathcal{Z})\}$.

Exercise 2.6 Show that $E\{\min(\overline{\mathrm{Fdr}}(\mathcal{Z}), 2)\} \geq \phi(\mathcal{Z})$. *Hint*: Draw the tangent line to the curve $(N_+(\mathcal{Z}), \overline{\mathrm{Fdr}}(\mathcal{Z}))$ passing through the point $(e_+(\mathcal{Z}), \phi(\mathcal{Z}))$.

Standard delta-method calculations yield useful approximations for the mean and variance of $\overline{\mathrm{Fdr}}(\mathcal{Z})$, without requiring the condition on $e_0(\mathcal{Z})$ in Lemma 2.1.

Lemma 2.2 *Let $\gamma(\mathcal{Z})$ indicate the squared coefficient of variation of $N_+(\mathcal{Z})$,*

$$\gamma(\mathcal{Z}) = \mathrm{var}\left\{N_+(\mathcal{Z})\right\}\big/ e_+(\mathcal{Z})^2. \tag{2.33}$$

Then $\overline{\mathrm{Fdr}}(\mathcal{Z})/\phi(\mathcal{Z})$ has approximate mean and variance

$$\frac{\overline{\mathrm{Fdr}}(\mathcal{Z})}{\phi(\mathcal{Z})} \mathrel{\dot\sim} \left(1 + \gamma(\mathcal{Z}), \gamma(\mathcal{Z})\right). \tag{2.34}$$

Proof Suppressing $\mathcal{Z}$ from the notation,

$$\begin{aligned} \overline{\mathrm{Fdr}} = \frac{e_0}{N_+} &= \frac{e_0}{e_+}\, \frac{1}{1 + (N_+ - e_+)/e_+} \\ &\doteq \phi\left[1 - \frac{N_+ - e_+}{e_+} + \left(\frac{N_+ - e_+}{e_+}\right)^2\right], \end{aligned} \tag{2.35}$$

and $(N_+ - e_+)/e_+$ has mean and variance $(0, \gamma(\mathcal{Z}))$. □

Lemma 2.2 quantifies the obvious: the accuracy of $\overline{\mathrm{Fdr}}(\mathcal{Z})$ as an estimate of the Bayes false discovery rate $\phi(\mathcal{Z})$ depends on the variability of the

[5] Benjamini and Hochberg originally used *false discovery rate* as the name of $E\{\mathrm{Fdp}\}$, denoted by FDR; this runs the risk of confusing a rate with its expectation.

denominator $N_+(\mathcal{Z})$ in (2.24).[6] More specific results can be obtained if we supplement the two-groups model of Figure 2.3 with the assumption of independence,

$$\begin{array}{c}\textbf{Independence Assumption:}\\ \text{Each } z_i \text{ follows model (2.7) independently.}\end{array} \tag{2.36}$$

Then $N_+(\mathcal{Z})$ has binomial distribution

$$N_+(\mathcal{Z}) \sim \text{Bi}\,(N, F(\mathcal{Z})) \tag{2.37}$$

with squared coefficient of variation

$$\gamma(\mathcal{Z}) = \frac{1 - F(\mathcal{Z})}{NF(\mathcal{Z})} = \frac{1 - F(\mathcal{Z})}{e_+(\mathcal{Z})}. \tag{2.38}$$

We will usually be interested in regions $\mathcal{Z}$ where $F(\mathcal{Z})$ is small, giving $\gamma(\mathcal{Z}) \doteq 1/e_+(\mathcal{Z})$ and, from Lemma 2.2,

$$\overline{\text{Fdr}}(\mathcal{Z})\Big/\phi(\mathcal{Z}) \mathrel{\dot\sim} (1 + 1/e_+(\mathcal{Z}), 1/e_+(\mathcal{Z}))\,, \tag{2.39}$$

the crucial quantity for the accuracy of $\overline{\text{Fdr}}(\mathcal{Z})$ being $e_+(\mathcal{Z})$, the expected number of the z_i falling into $\mathcal{Z}$. For the prostate data with $\mathcal{Z} = (3, \infty)$ we can estimate $e_+(\mathcal{Z})$ by $N_+(\mathcal{Z}) = 49$, giving $\overline{\text{Fdr}}(\mathcal{Z})/\phi(\mathcal{Z})$ approximate mean 1.02 and standard deviation 0.14. In this case, $\overline{\text{Fdr}}(\mathcal{Z}) = 0.166$ (2.26), is nearly unbiased for the Bayes false discovery rate $\phi(\mathcal{Z})$, and has coefficient of variation about 0.14. A rough 95% confidence interval for $\phi(\mathcal{Z})$ is $0.166 \cdot (1 \pm 2 \cdot 0.14) = (0.12, 0.21)$. All of this depends on the independence assumption (2.36), which we will see in Chapter 8 is only a moderately risky assumption for the prostate data.

Neater versions of our previous results are possible if we add to the independence requirement the relatively innocuous assumption that the number of cases N is itself a Poisson variate, say

$$N \sim \text{Poi}(\eta). \tag{2.40}$$

Lemma 2.3 *Under the Poisson-independence assumptions* (2.36) *and* (2.40),

$$E\,\{\text{Fdp}(\mathcal{Z})\} = \phi(\mathcal{Z}) \cdot [1 - \exp\,(-e_+(\mathcal{Z}))]\,, \tag{2.41}$$

where now $e_+(\mathcal{Z}) = E\{N_+(\mathcal{Z})\} = \eta \cdot F(\mathcal{Z})$.

[6] Under mild conditions, both $\overline{\text{Fdr}}(\mathcal{Z})$ and $\text{Fdp}(\mathcal{Z})$ converge to $\phi(\mathcal{Z})$ as $N \to \infty$. However, this is less helpful than it seems, as the correlation considerations of Chapter 7 show. Asymptotics play only a minor role in what follows.

Proof With $N \sim \text{Poi}(\eta)$ in Figure 2.3, well-known properties of the Poisson distribution show that

$$N_0(\mathcal{Z}) \sim \text{Poi}\left(\eta\pi_0 F_0(\mathcal{Z})\right) \text{ independently of } N_1(\mathcal{Z}) \sim \text{Poi}\left(\eta\pi_1 F_1(\mathcal{Z})\right), \quad (2.42)$$

and

$$N_0(\mathcal{Z})\big|N_+(\mathcal{Z}) \sim \text{Bi}\left(N_+(\mathcal{Z}), \pi_0 F_0(\mathcal{Z})/F(\mathcal{Z})\right) \quad (2.43)$$

if $N_+(\mathcal{Z}) > 0$. But $\pi_0 F_0(\mathcal{Z})/F(\mathcal{Z}) = \phi(\mathcal{Z})$, and

$$\Pr\left\{N_+(\mathcal{Z}) = 0\right\} = \exp\left(-e_+(\mathcal{Z})\right),$$

giving

$$E\left\{\text{Fdp}(\mathcal{Z})\big|N_+(\mathcal{Z})\right\} = E\left\{\frac{N_0(\mathcal{Z})}{N_+(\mathcal{Z})}\bigg|N_+(\mathcal{Z})\right\} = \phi(\mathcal{Z}) \quad (2.44)$$

with probability $1 - \exp(-e_+(\mathcal{Z}))$ and, by definition, $E\{\text{Fdp}(\mathcal{Z})|N_+(\mathcal{Z}) = 0\} = 0$ with probability $\exp(-e_+(\mathcal{Z}))$. □

Applications of large-scale testing often have π_1, the proportion of non-null cases, very near 0, in which case a region of interest $\mathcal{Z}$ may have $e_+(\mathcal{Z})$ quite small. As (2.39) indicates, $\overline{\text{Fdr}}(\mathcal{Z})$ is then badly biased upward. A simple modification of $\overline{\text{Fdr}}(\mathcal{Z}) = e_0(\mathcal{Z})/N_+(\mathcal{Z})$ cures the bias. Define instead

$$\widetilde{\text{Fdr}}(\mathcal{Z}) = e_0(\mathcal{Z})/\left(N_+(\mathcal{Z}) + 1\right). \quad (2.45)$$

Lemma 2.4 *Under the Poisson-independence assumptions* (2.36) *and* (2.40),

$$E\left\{\widetilde{\text{Fdr}}(\mathcal{Z})\right\} = E\left\{\text{Fdp}(\mathcal{Z})\right\} = \phi(\mathcal{Z}) \cdot \left[1 - \exp\left(-e_+(\mathcal{Z})\right)\right]. \quad (2.46)$$

Exercise 2.7 Verify (2.46).

The Poisson assumption is more of a convenience than a necessity here. Under independence, $\widetilde{\text{Fdr}}(\mathcal{Z})$ is nearly unbiased for $E\{\text{Fdp}(\mathcal{Z})\}$, and both approach the Bayes false discovery rate $\phi(\mathcal{Z})$ exponentially fast as $e_+(\mathcal{Z})$ increases. In general, $\overline{\text{Fdr}}(\mathcal{Z})$ is a reasonably accurate estimator of $\phi(\mathcal{Z})$ when $e_+(\mathcal{Z})$ is large, say $e_+(\mathcal{Z}) \geq 10$, but can be both badly biased and highly variable for smaller values of $e_+(\mathcal{Z})$.

2.5 Independence versus Correlation

The independence assumption has played an important role in the literature of large-scale testing, particularly for false discovery rate theory, as discussed in Chapter 4. It is a dangerous assumption in practice!

Figure 2.4 illustrates a portion of the *DTI data*, a diffusion tensor imaging study comparing brain activity of six dyslexic children versus six normal controls. (DTI machines measure fluid flows in the brain and can be thought of as generating versions of magnetic resonance images.) Two-sample tests produced z-values at $N = 15\,443$ voxels (three-dimensional brain locations), with each $z_i \sim \mathcal{N}(0,1)$ under the null hypothesis of no difference between the dyslexic and normal children, as in (2.6).

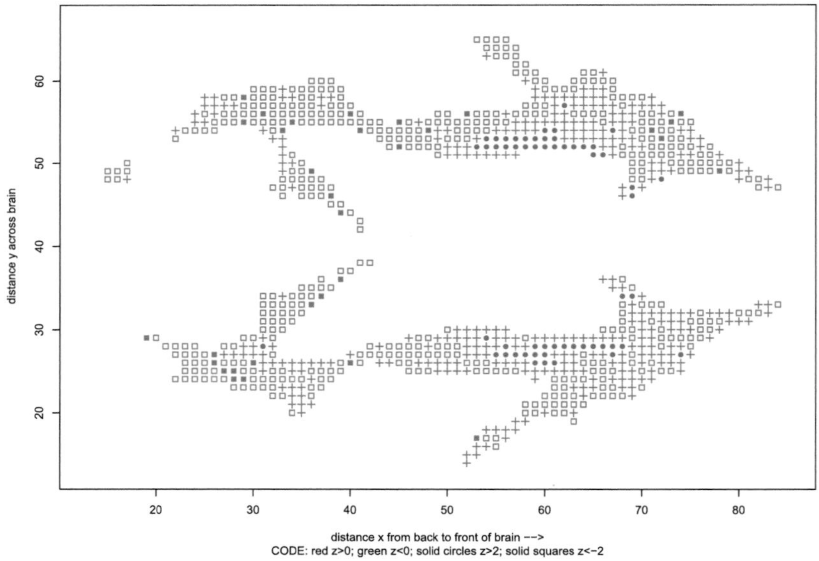

Figure 2.4 Diffusion tensor imaging study comparing brain activity in 6 normal and 6 dyslexic children; z-values for slice of $N = 15\,443$ voxels.

The figure shows only a single horizontal brain section containing 848 of the 15 443 voxels. Red indicates $z_i > 0$, green $z_i < 0$, with solid circles $z_i > 2$ and solid squares $z_i < -2$. Spatial correlation among the z-values is obvious: reds tend to be near other reds, and greens near greens. The symmetric patch of solid red circles near $x = 60$ is particularly striking.

Independence is an obviously bad assumption for the DTI data. We will see in Chapter 8 that it is an even worse assumption for many microarray studies, though not perhaps for the prostate data. There is no easy equivalent of brain geometry for microarrays. We can't draw evocative pictures like Figure 2.4, but correlation may still be a real problem. Chapter 7 takes up the question of how correlation affects accuracy calculations such as (2.39).

2.6 Learning from the Experience of Others II

Consider the Bayesian *hierarchical model*

$$\mu \sim g(\cdot) \quad \text{and} \quad z|\mu \sim \mathcal{N}(\mu, 1), \tag{2.47}$$

g indicating some prior density for μ, where we will take "density" to include the possibility of g including discrete components. The James–Stein estimator of Figure 1.1 assumes

$$g(\mu) = \mathcal{N}(M, A). \tag{2.48}$$

Model (2.47) also applies to simultaneous hypothesis testing: now take

$$g(\mu) = \pi_0 \Delta_0(\mu) + (1 - \pi_0) g_1(\mu) \tag{2.49}$$

where Δ_0 denotes a delta function at $\mu = 0$, and g_1 is a prior density for the non-null μ_i values. (This leads to a version of the two-groups model having $f_1(z) = \int \varphi(z - \mu) g_1(\mu) d\mu$.) For gene$_1$ of the prostate data, the "others" in Figure 1.1 are all of the other genes, represented by $z_2, z_3, \ldots, z_{6033}$. These must estimate π_0 and g_1 in prior distribution (2.49). Finally, the prior is combined with $z_1 \sim \mathcal{N}(\mu, 1)$ via Bayes theorem to estimate quantities such as the probability that gene$_1$ is null.

Clever constructions such as $\overline{\text{Fdr}}(\mathcal{Z})$ in (2.21) can finesse the actual estimation of $g(\mu)$, as further discussed in Chapter 11. The main point being made here is that gene$_1$ is learning from the other genes. "Which others?" is a crucial question, taken up in Chapter 10.

The fact that $g(\mu)$ is much smoother in (2.48) than (2.49) hints at estimation difficulties in the hypothesis testing context. The James–Stein estimator can be quite efficient even for N as small as 10, as in (1.25). Results like (2.39) suggest that we need N in the hundreds or thousands for accurate empirical Bayes hypothesis testing. These kinds of efficiency calculations are pursued in Chapter 7.

Notes

Benjamini and Hochberg's landmark 1995 paper introduced false discovery rates in the context of a now-dominant simultaneous hypothesis testing algorithm that is the main subject of Chapter 4. Efron et al. (2001) recast the fdr algorithm in an empirical Bayes framework, introducing the local false discovery rate. Storey (2002, 2003) defined the "positive false discovery rate,"

$$\text{pFdr}(\mathcal{Z}) = E\left\{N_0(\mathcal{Z})/N_+(\mathcal{Z})\middle|N_+(\mathcal{Z}) > 0\right\} \tag{2.50}$$

in the notation of Figure 2.3, and showed that if the z_i were i.i.d. (independent and identically distributed),

$$\text{pFdr}(\mathcal{Z}) = \phi(\mathcal{Z}),$$

(2.27). Various combinations of Bayesian and empirical Bayesian microarray techniques have been proposed, Newton et al. (2004) for example employing more formal Bayes hierarchical modeling. A version of the curve in Figure 2.2 appears in Genovese and Wasserman (2004), where it is used to develop asymptotic properties of the Benjamini–Hochberg procedure. Johnstone and Silverman (2004) consider situations where π_0, the proportion of null cases, might be much smaller than 1, unlike our applications here.

The two-groups model (2.7) is too basic to have an identifiable author, but it was named and extensively explored in Efron (2008a). It will reappear in several subsequent chapters. The more specialized structural model (2.47) will also reappear, playing a major role in the prediction theory of Chapter 11. It was the starting point for deep studies of multivariate normal mean vector estimation in Brown (1971) and Stein (1981).

The prostate cancer study was carried out by Singh et al. (2002). Figure 2.4, the DTI data, is based on the work of Schwartzman et al. (2005).

The t-test (or its cousin, the Wilcoxon test) is a favorite candidate for two-sample comparisons, but other test statistics have been proposed. Tomlins et al. (2005), Tibshirani and Hastie (2007), and Wu (2007) investigate analysis methods that emphasize occasional very large responses, the idea being to identify genes in which a *subset* of the subjects are prone to outlying effects.

The various names for false discovery-related concepts are more or less standard, but can be easily confused. Here is a brief glossary of terms.

Term(s)	**Definition**
$f_0(z)$ and $f_1(z)$	the null and alternative densities for z-values in the two-groups model (2.7)
$F_0(\mathcal{Z})$ and $F_1(\mathcal{Z})$	the corresponding probability distributions (2.10)
$f(z)$ and $F(\mathcal{Z})$	the mixture density and distribution (2.11)–(2.12)
$\text{Fdr}(\mathcal{Z})$ and $\phi(\mathcal{Z})$	two names for the Bayesian false discovery rate $\Pr\{\text{null}\|z \in \mathcal{Z}\}$ (2.13)
$\text{fdr}(z)$	the *local* Bayesian false discovery rate (2.14), also denoted $\phi(z)$
$\text{Fdp}(\mathcal{Z})$	the false discovery proportion (2.28), i.e., the proportion of z_i values in $\mathcal{Z}$ that are from the null distribution
$\bar{F}(\mathcal{Z})$	the empirical probability distribution $\#\{z_i \in \mathcal{Z}\}/N$ (2.20)
$\overline{\text{Fdr}}(\mathcal{Z})$	the empirical Bayes estimate of $\text{Fdr}(\mathcal{Z})$ obtained by substituting $\bar{F}(\mathcal{Z})$ for the unknown $F(\mathcal{Z})$ (2.21)
$\text{FDR}(\mathcal{Z})$	the expected value of the false discovery proportion $\text{Fdp}(\mathcal{Z})$
$N_0(\mathcal{Z})$, $N_1(\mathcal{Z})$, and $N_+(\mathcal{Z})$	the number of null, non-null, and overall z-values in $\mathcal{Z}$, as in Figure 2.3
$e_0(\mathcal{Z})$, $e_1(\mathcal{Z})$, and $e_+(\mathcal{Z})$	their expectations, as in (2.23)

3

Significance Testing Algorithms

Simultaneous hypothesis testing was a lively topic in the early 1960s, my graduate student years, and had been so since the end of World War II. Rupert Miller's book *Simultaneous Statistical Inference* appeared in 1966, providing a beautifully lucid summary of the contemporary methodology. A second edition in 1981 recorded only modest gains during the intervening years. This was a respite, not an end: a new burst of innovation in the late 1980s generated important techniques that we will be revisiting in this chapter.

Miller's book, which gives a balanced picture of the theory of that time, has three notable features:

1. It is overwhelmingly frequentist.
2. It is focused on control of α, the overall Type I error rate of a procedure.[1]
3. It is aimed at multiple testing situations with individual cases N between 2 and, say, 10.

We have now entered a scientific age in which $N = 10\,000$ is no cause for raised eyebrows. It is impressive (or worrisome) that the theory of the 1980s continues to play a central role in microarray-era statistical inference. Features 1 and 2 are still the norm in much of the multiple testing literature, despite the obsolescence of Feature 3. This chapter reviews part of that theory, particularly the ingenious algorithms that have been devised to control the overall Type I error rate (also known as FWER, the family-wise error rate). False discovery rate control, an approach which doesn't follow either Feature 1 or 2 and is better suited to the $N = 10\,000$ era, is taken up in Chapter 4. The material in this chapter is a digression from our chosen theme of empirical Bayes methods, and may be read lightly by those eager to get on with the main story.

[1] And, by extension, the construction of simultaneous confidence regions that have a guaranteed probability of containing *all* the relevant parameters.

3.1 p-Values and z-Values

First consider the classical "single-test" situation: we wish to test a single null hypothesis H_0 on the basis of observed data $\boldsymbol{x}$. For any value of α between 0 and 1 we construct a rejection region $\mathcal{R}_\alpha$ in the sample space of $\boldsymbol{x}$ such that

$$\Pr_0\{\boldsymbol{x} \in \mathcal{R}_\alpha\} = \alpha \qquad [\alpha \in (0,1)], \tag{3.1}$$

where $\Pr_0$ refers to the probability distribution of $\boldsymbol{x}$ under H_0. The regions $\mathcal{R}_\alpha$ decrease with α,

$$\mathcal{R}_\alpha \supseteq \mathcal{R}_{\alpha'} \qquad \text{for } \alpha > \alpha'. \tag{3.2}$$

The *p-value* $p(\boldsymbol{x})$ corresponding to $\boldsymbol{x}$ is defined as the smallest value of α such that $\boldsymbol{x} \in \mathcal{R}_\alpha$,

$$p(\boldsymbol{x}) = \inf_\alpha\{\boldsymbol{x} \in \mathcal{R}_\alpha\}. \tag{3.3}$$

Intuitively, the smaller is $p(\boldsymbol{x})$, the more decisively is H_0 rejected. There is a more-or-less agreed-upon scale for interpreting p-values, originally due to Fisher, summarized in Table 3.1. Using p-values instead of a fixed rule like "reject at the $\alpha = 0.05$ level" is a more informative mode of data summary.

Table 3.1 *Fisher's scale of evidence for interpreting p-values; for instance,* $p(\boldsymbol{x}) = 0.035$ *provides moderate to substantial grounds for rejecting* H_0.

α	.10	.05	.025	.01	.001
Evidence against H_0:	borderline	moderate	substantial	strong	overwhelming

For any value of u in (0, 1), the event $p(\boldsymbol{x}) \le u$ is equivalent to $\boldsymbol{x} \in \mathcal{R}_u$, implying

$$\Pr_0\{p(\boldsymbol{x}) \le u\} = \Pr_0\{\boldsymbol{x} \in \mathcal{R}_u\} = u. \tag{3.4}$$

In other words, under H_0 the random variable $P = p(\boldsymbol{x})$ has a uniform distribution[2] over the interval (0, 1),

$$H_0 : P = p(\boldsymbol{x}) \sim \mathcal{U}(0,1). \tag{3.5}$$

[2] Here we are ignoring discrete null distributions, like binomial $(n, \frac{1}{2})$, where there are minor difficulties with definition (3.3).

P-values serve as a universal language for hypothesis testing. They allow for general rules of interpretation, such as Fisher's scale, applying to all hypothesis-testing situations. Fisher's famous $\alpha = 0.05$ dictum for "significance" has been overused and abused, but has served a crucial purpose nevertheless in bringing order to scientific reporting.

In subsequent chapters we will be working mainly with z-values rather than p-values,

$$z(\boldsymbol{x}) = \Phi^{-1}\left(p(\boldsymbol{x})\right), \tag{3.6}$$

where Φ^{-1} is the inverse function of the standard normal cdf, as in (2.5). Z-values also enjoy a universal null hypothesis distribution,

$$H_0 : z(\boldsymbol{x}) \sim \mathcal{N}(0, 1), \tag{3.7}$$

called the *theoretical null* at (2.6).

Figure 3.1 shows p-values and z-values for the DTI data, as partially reported in Figure 2.4. Here there are $N = 15\,443$ p-values p_i and likewise 15 443 z-values z_i, obtained from voxel-wise two-sample t-tests. The t-tests each have 10 degrees of freedom, so

$$p_i = F_{10}(t_i) \quad \text{and} \quad z_i = \Phi^{-1}\left(F_{10}(t_i)\right) \tag{3.8}$$

in the notation of (2.4)–(2.5). There are some interesting comparisons between the two displays:

- Both histograms show sharp discrepancies from their theoretical null distributions (3.5) or (3.7): the dramatic right spike of the p_i's, and the corresponding heavy right tail of the z_i's.
- The spike is more striking to the eye, but pays the price of collapsing all of the detail evident in the z-value tail.
- There are also *central* discrepancies between the theoretical null distributions and the histograms: this is clearer in the bottom panel, where the histogram center is shifted a little to the left of the $\mathcal{N}(0, 1)$ curve.

Exercise 3.1 How does the shift effect appear in the p-value histogram?

- The z_i's, as monotone functions of the t_i's in (3.8), automatically maintain the signs of the t-tests, with positive effects mapped to the right and negative effects to the left; definition (3.8) does the same for the p-values by mapping large positive effects toward $p = 1$ and negative effects toward $p = 0$. (Of course we would transform $p_i = 0.98$ to 0.02, etc., for interpretation in Fisher's scale.)

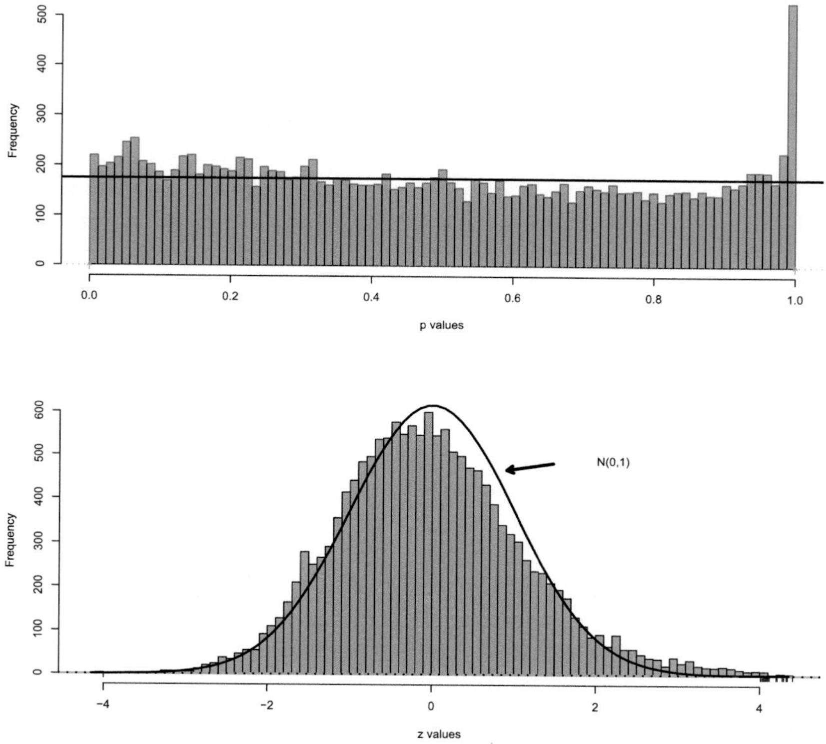

Figure 3.1 $N = 15\,443$ p-values (top) and z-values (bottom) for the DTI study partially displayed in Figure 2.4. Horizontal line in top panel indicates $\mathcal{U}(0,1)$ distribution; curve in bottom panel indicates theoretical $\mathcal{N}(0,1)$ null distribution. Small dashes at bottom right show the 18 z-values exceeding 4.

The literature tends to favor *two-sided* p-values, in our case

$$p_i^{(\text{two})} = \Pr\{|T_{10}| > |t_i|\}, \tag{3.9}$$

where T_{10} indicates a standard Student-t variate with 10 degrees of freedom. Besides losing the sign information in t_i, this is a potentially dangerous definition if the test statistic, unlike T_{10}, has an asymmetric null distribution. A better option, if two-sided p-values are essential, is to transform our first definition to

$$p_i^{(\text{two})} = 2 \cdot \min(p_i, 1 - p_i), \tag{3.10}$$

agreeing with (3.9) in the symmetric but not the asymmetric case. Two-sided testing makes some sense, sometimes, in classical single-test applications. It is less defensible in large-scale testing where, as in Figure 3.1, we can see the two sides behaving differently.

The author finds z-value histograms more informative than p-value histograms, but that is not the reason for the predominance of z-values in succeeding chapters: z-values allow us to bring the full power of normal theory to bear on large-scale inference problems. The `locfdr` algorithm of Chapter 5, for example, uses normal theory to estimate that the best-fit curve to the central peak in Figure 3.1 is $\mathcal{N}(-0.12, 1.06^2)$ rather than $\mathcal{N}(0, 1)$.

The original data matrix X for the DTI data is $15\,443 \times 12$, from which we hope to identify brain regions involved in dyslexia. In what follows we will be using a two-step analysis strategy: first we reduce each row of X to a single number p_i (or z_i); we then employ a testing algorithm to determine which of the N p_i values indicates non-null activity. Other methods that assume ANOVA-like structures for X are at least theoretically more informative, but I prefer the two-step approach. In my experience these huge data matrices show little respect for traditional analysis of variance structures. Using two-sample t-statistics t_i (or, if necessary, their Wilcoxon counterparts) as a starting point puts less strain on the statistical modeling. Nevertheless, modeling difficulties remain, as discussed in Chapter 10.

The use of p-values or z-values is not limited to two-sample situations. Suppose for example that y_j is some response variable, like a survival time, measured on subject j, $j = 1, 2, \ldots, n$. For each row i of X we can run a linear regression of y_j on x_{ij}, calculate the slope coefficient $\hat{\beta}_i$, and take p_i to be its attained significance level, computed in the usual normal-theory way. (This requires modeling across individual rows of X, but not over the entire matrix.) Some of the examples to come involve more elaborate versions of this tactic.

3.2 Adjusted p-Values and the FWER

The family-wise error rate FWER is defined as the probability of making at least one false rejection in a family of hypothesis-testing problems. Let $p_1, p_2, \ldots, p_N$ be the p-values obtained in tests of the corresponding family of null hypotheses $H_{01}, H_{02}, \ldots, H_{0N}$. For the DTI data, N equals $15\,443$, with H_{0i} being the hypothesis of no response distribution difference between dyslexics and controls at voxel i, and p_i the observed p-value for the two-sample procedure testing for that difference. Some, perhaps most, of the null hypotheses H_{0i} will be true.

The family-wise error rate is

$$\text{FWER} = \Pr\{\text{Reject } \textit{any} \text{ true } H_{0i}\}; \tag{3.11}$$

a FWER control procedure is an algorithm that inputs a family of p-values $(p_1, p_2, \ldots, p_N)$ and outputs the list of accepted and rejected null hypotheses, subject to the constraint

$$\text{FWER} \leq \alpha \tag{3.12}$$

for any preselected value of α.

Bonferroni's bound provides the classic FWER control method: we reject those null hypotheses for which

$$p_i \leq \alpha/N. \tag{3.13}$$

Let I_0 index the true null hypotheses, having N_0 members. Then

$$\begin{aligned} \text{FWER} &= \Pr\left\{\bigcup_{I_0}\left(p_i \leq \frac{\alpha}{N}\right)\right\} \leq \sum_{I_0} \Pr\left\{p_i \leq \frac{\alpha}{N}\right\} \\ &= N_0 \frac{\alpha}{N} \leq \alpha, \end{aligned} \tag{3.14}$$

verifying the FWER control property.[3] The crucial step in the top line follows from Boole's inequality $\Pr\{\bigcup A_i\} \leq \sum \Pr\{A_i\}$.

One way to think about the Bonferroni bound is that the individual p-value p_i for testing H_{0i} translates into the family-wise *adjusted p-value*

$$\tilde{p}_i = \{Np_i\}_1 \tag{3.15}$$

(where $\{x\}_1$ is short for $\min(x, 1)$); we reject H_{0i} in the family-wise context if $\tilde{p}_i$, rather than p_i, is less than α. Though not of much use here, the language of adjusted p-values is handy for some of the more complicated algorithms of the next section.

Let $\boldsymbol{x}$ indicate all the data available for testing the family of hypotheses $H_{01}, H_{02}, \ldots, H_{0N}$, and let $\text{FWER}_\alpha(\boldsymbol{x})$ be a FWER level-α test procedure based on $\boldsymbol{x}$. The general definition of adjusted p-value for case i is an analog of (3.3),

$$\tilde{p}_i(\boldsymbol{x}) = \inf_{\alpha} \left\{H_{0i} \text{ rejected by } \text{FWER}_\alpha(\boldsymbol{x})\right\}. \tag{3.16}$$

[3] Formula (3.14) demonstrates "strong control": FWER is bounded by α no matter what the pattern of true and false null hypotheses might be; "weak control" refers to methods that control FWER only if *all* the null hypotheses are true.

As an example, the *Šidák procedure* improves on the Bonferroni bound (3.13) by rejecting those hypotheses H_{0i} for which

$$p_i \leq 1 - (1 - \alpha)^{1/N}. \tag{3.17}$$

The corresponding adjusted p-value is

$$\tilde{p}_i = 1 - (1 - p_i)^N. \tag{3.18}$$

Exercise 3.2 (a) Verify (3.18). (b) Show that (3.17) improves on (3.13) in the sense of making it easier to reject every H_{0i} at any given level α. (c) Show that the Šidák procedure is FWER$_\alpha$ if the p-values $p_1, p_2, \dots, p_N$ are statistically independent.

Bonferroni's bound does not depend on independence, so Šidák's procedure cannot be considered a general improvement.[4] *Holm's procedure* is an example of a more elaborate testing strategy that *is* in fact a general improvement: let the ordered p-values be denoted by

$$p_{(1)} \leq p_{(2)} \leq p_{(3)} \leq \cdots \leq p_{(N)}, \tag{3.19}$$

and reject $H_{0(i)}$, the hypothesis corresponding to $p_{(i)}$, if

$$p_{(j)} \leq \frac{\alpha}{N - j + 1} \qquad \text{for } j = 1, 2, \dots, i. \tag{3.20}$$

The next section shows that (3.19)–(3.20) is FWER$_\alpha$. It is more powerful than Bonferroni's procedure since the rejection regions (3.20) are larger than the Bonferroni regions (3.13).

Exercise 3.3 Show that the adjusted p-value for Holm's procedure is

$$\tilde{p}_{(i)} = \max_{j \leq i} \left\{(N - j + 1)p_{(j)}\right\}_1 \tag{3.21}$$

where $\{x\}_1 \equiv \min(x, 1)$ as before.

Adjusted p-values share the p-value virtue of not requiring a rigid predefinition of the rejection level α. We can compute the $\tilde{p}_i$ values directly, as in (3.18) or (3.21), and see how "significant" each case is. The quotes are necessary here because $\tilde{p}_i$ usually does *not* follow a $\mathcal{U}(0, 1)$ distribution, or any other fixed distribution, when H_{0i} is true, so there is no universal interpretive scale such as Fisher's scale in Table 3.1.

[4] The validity of Šidák's procedure can be extended beyond independence, including to multivariate normal testing situations.

Exercise 3.4 Suppose that all N hypotheses H_{0i} are true and that the N p-values are mutually independent. Calculate the distribution of $\tilde{p}_{(1)} = \{Np_{(1)}\}_1$. What is the limiting distribution as N goes to infinity?

3.3 Stepwise Algorithms

FWER control is a hard-line frequentist concept, far removed from the more relaxed empirical Bayes methods that are our main interest here. Nevertheless, it continues to play an important role in the literature of large-scale significance testing. Stepwise algorithms, developed mainly in the late 1980s, represent the most successful, and ingenious, attack on FWER control. This section and the next review some of the methodology, with no attempt at completeness and only hints of the theoretical ideas involved. We return to the empirical Bayes world in Chapter 4, where the results here will be viewed in contrast to false discovery rate control methods.

A *step-down* procedure begins with the ordered p-values $p_{(1)} \leq p_{(2)} \leq \cdots \leq p_{(N)}$ as in (3.19), and from them defines a testing algorithm such that $H_{0(i)}$ can be rejected only if first $H_{0(j)}$ is rejected for $j = 1, 2, \ldots, i-1$. In other words, if $p_{(i)}$ is small enough to cause rejection, then so must be $p_{(1)}, p_{(2)}, \ldots, p_{(i-1)}$. Another way to say this is that the adjusted p-values are non-decreasing,

$$\tilde{p}_{(1)} \leq \tilde{p}_{(2)} \leq \cdots \leq \tilde{p}_{(N)}, \tag{3.22}$$

so that $\tilde{p}_{(i)} \leq \alpha$ implies $\tilde{p}_{(j)} \leq \alpha$ for $j = 1, 2, \ldots, i-1$. Step-down methods allow improvements on single-step procedures like (3.17). It can be shown, for example, that the step-down version of Šidák's procedure replaces (3.18) with

$$\tilde{p}_{(i)} = \max_{j \leq i} \left\{ 1 - \left(1 - p_{(j)}\right)^{N-j+1} \right\}. \tag{3.23}$$

Holm's method (3.20)–(3.21) was one of the first examples of a step-down procedure. Here is the proof that it satisfies the FWER control property.

Proof Let I_0 be the set of indices corresponding to true null hypotheses H_{0i}, $N_0 = \#I_0$ the number of members of I_0, and $i_0 = N - N_0 + 1$. Also let $\hat{i}$ be the stopping index for Holm's procedure, i.e., the maximum index satisfying (3.20). The event

$$A = \left\{p_{(i)} > \frac{\alpha}{N_0} \text{ for all } i \in I_0\right\} \Longrightarrow \left\{p_{(i_0)} > \frac{\alpha}{N_0} = \frac{\alpha}{N+1-i_0}\right\}$$
$$\Longrightarrow \left\{\hat{\imath} < i_0\right\} \Longrightarrow \left\{p_{(\hat{\imath})} < \frac{\alpha}{N_0}\right\} \equiv B. \quad (3.24)$$

However, the Bonferroni bound shows that $\Pr\{A\} \geq 1 - \alpha$, and B implies that none of the true null hypotheses have been rejected. □

Exercise 3.5 Explicate in detail the three steps in (3.24) and the conclusion that follows.

Holm's procedure illustrates a *closure principle* which is worth stating separately. Let I be a subset of the indices $\{1, 2, \ldots, N\}$, and $\mathcal{I}$ the statement that *all* of the null hypotheses in I are true,

$$\mathcal{I} = \bigcap_I H_{0(i)}. \quad (3.25)$$

If I' is a larger subset, $I' \supseteq I$, then logically $\mathcal{I}' \Rightarrow \mathcal{I}$. Suppose that for every subset I we have a level-α non-randomized test function $\phi_I(\boldsymbol{x}) : \phi_I(\boldsymbol{x})$ equals 1 or 0, with 1 indicating rejection of $\mathcal{I}$, satisfying

$$\Pr_{\mathcal{I}}\{\phi_I(\boldsymbol{x}) = 1\} \leq \alpha.$$

Now consider the simultaneous test function

$$\Phi_I(\boldsymbol{x}) = \min_{I' \supseteq I} \{\phi_{I'}(\boldsymbol{x})\}; \quad (3.26)$$

$\Phi_I(\boldsymbol{x})$ defines a rule that rejects $\mathcal{I}$ *if and only if* $\mathcal{I}'$ *is rejected at level* α *for every* I' *containing* I. But if $\mathcal{I}$ is true then $I \subseteq I_0$, the set of all true $H_{0(i)}$, and

$$\Pr_{\mathcal{I}_0}\{\phi_{\mathcal{I}_0}(\boldsymbol{x}) = 1\} \leq \alpha \quad \text{implies} \quad \Pr_{\mathcal{I}}\{\Phi_I(\boldsymbol{x}) = 1\} \leq \alpha. \quad (3.27)$$

In other words, the test Φ_I simultaneously controls the probability of rejecting *any* true subset $\mathcal{I}$ at level α.

The closure principle can be used to extend Bonferroni's bound to Holm's procedure. Let $I_i = \{i, i+1, \ldots, N\}$. In terms of the ordered p-values (3.19), Bonferroni's rule rejects $\mathcal{I}_i$ at level α if $p_{(i)} \leq \alpha/(N+1-i)$. Note that $I_j \supseteq I_i$ for $j \leq i$.

Exercise 3.6 Complete the proof that Holm's procedure is FWER_α.

All of our calculations have so far begun with the simple Bonferroni bound (3.13). If we are willing to assume *independence* among the original p-values $p_1, p_2, \ldots, p_N$, then a better bound, known as *Simes' inequality*, is available: when *all* the null hypotheses are true, then

$$\Pr\left\{p_{(i)} \geq \frac{\alpha i}{N} \text{ for } i = 1, 2, \ldots, N\right\} \geq 1 - \alpha, \tag{3.28}$$

with equality if the test statistics are continuous.

The proof of (3.28) begins by noting that in the independent continuous case, with all $H_{0(i)}$ true, $p_{(1)}, p_{(2)}, \ldots, p_{(N)}$ are the order statistics of N independent $\mathcal{U}(0, 1)$ variates, as in (3.5). A standard order statistic result then shows that given $p_{(N)}$, the largest p-value, the ratios

$$\left(p_{(1)}/p_{(N)}, p_{(2)}/p_{(N)}, \ldots, p_{(N-1)}/p_{(N)}\right) \tag{3.29}$$

are distributed as the order statistics from $(N - 1)$ independent $\mathcal{U}(0, 1)$ variates, while $p_{(N)}$ has cdf $p_{(N)}^N$.

Exercise 3.7 Use induction to verify Simes' inequality.

Starting from Simes' inequality, Hochberg used the closure principle to improve (i.e., raise) Holm's adjusted p-values (3.21) to

$$\tilde{p}_{(i)} = \min_{j \geq i} \left\{(N - j + 1)p_{(j)}\right\}_1, \tag{3.30}$$

with $\{x\}_1 \equiv \min(x, 1)$. This is not a general improvement though, since Simes' inequality depends on independence of the test statistics. (Some progress has been made in broadening its validity.) Algorithms such as (3.30), whose definitions depend on the *upper* tail of the sequence $p_{(1)} \leq p_{(2)} \leq \cdots \leq p_{(N)}$, are called "step-up procedures."

Exercise 3.8 Holm's step-down procedure (3.20) starts with $i = 1$ and keeps *rejecting* $H_{0(i)}$ until the first time $p_{(i)} > \alpha/(N - i + 1)$. Show that Hochberg's step-up procedure starts with $i = N$ and keeps *accepting* $H_{0(i)}$ until the first time $p_{(i)} \leq \alpha/(N-i+1)$. This shows that Hochberg's procedure is more powerful than Holm's, i.e., rejects more often at the same α level. The step-up/step-down nomenclature is unfortunate here.

3.4 Permutation Algorithms

The prize property of the Bonferroni bound, that it holds true regardless of the dependence structure of the data, puts it at a disadvantage if we happen to know the structure. Westfall and Young proposed step-down procedures that use permutation calculations to estimate dependence relationships, and then employ the estimated structure to improve on Holm's procedure.

Starting with the ordered p-values $p_{(1)} \leq p_{(2)} \leq \cdots \leq p_{(N)}$, as in (3.19), let $r_1, r_2, \ldots, r_N$ indicate the corresponding original indices,

$$p_{(j)} = p_{r_j}, \qquad j = 1, 2, \ldots, N. \tag{3.31}$$

Define

$$R_j = \{r_j, r_{j+1}, \ldots, r_N\} \tag{3.32}$$

and

$$\pi(j) = \text{Pr}_0 \left\{ \min_{k \in R_j}(P_k) \leq p_{(j)} \right\}. \tag{3.33}$$

Here $(P_1, P_2, \ldots, P_N)$ indicates a hypothetical realization of the unordered p-values $(p_1, p_2, \ldots, p_N)$ obtained under the *complete null hypothesis* [5] $\boldsymbol{H}_0$ that *all* of the H_{0i} are true; (3.33) is computed with $p_{(j)}$ fixed at its observed value. The *Westfall–Young step-down min-p* adjusted p-values are then defined by

$$\tilde{p}_{(i)} = \max_{j \leq i} \{\pi(j)\}. \tag{3.34}$$

To see the connection with Holm's procedure, notice that Boole's inequality implies

$$\pi(j) \leq \sum_{k \in R_j} \text{Pr}_0 \left\{ P_k \leq p_{(j)} \right\} = (N - j + 1) p_{(j)}. \tag{3.35}$$

Comparing (3.34) with (3.21) shows that the Westfall–Young adjusted p-values are smaller than Holm's values. The proof that (3.34) satisfies the FWER control property, that is, that

$$\text{Pr}\left\{\tilde{p}_i > \alpha \text{ for } all\ i \in I_0\right\} \geq 1 - \alpha, \tag{3.36}$$

for I_0 indexing the set of true hypotheses, is similar to the closure argument for Holm's procedure preceding Exercise 3.6. It does, however, involve an extra assumption, called "subset pivotality": that the vector $(P_i, i \in I_0)$ always follows the distribution it has under the complete null $\boldsymbol{H}_0$. In other words, the fact that some of the cases are non-null does not affect the joint distribution of the null p-values.

The min-p procedure can be difficult to implement. Westfall and Young also proposed a simpler variant, called "max-T." Let

$$t_{(1)} \geq t_{(2)} \geq \cdots \geq t_{(N)} \tag{3.37}$$

[5] Pr_0 now indicates probabilities computed under $\boldsymbol{H}_0$.

indicate ordered values of the original test statistics that gave the p-values (the two-sample t-statistics (2.2) for the prostate and DTI studies), with ordered values $t_{(j)} = t_{r_j}$. Also let $(T_1, T_2, \dots, T_N)$ represent a hypothetical unordered realization obtained under the complete null hypothesis $\boldsymbol{H}_0$. Now define[6]

$$\pi(j) = \Pr\left\{\max_{k \in R_j}(T_k) \geq t_{(j)}\right\}, \tag{3.38}$$

yielding adjusted p-values $\tilde{p}_{(i)}$ as in (3.34). If all T_i have the same cdf $F(T)$ then $p_i = 1 - F(t_i)$ and (3.38) is the same as (3.33), but otherwise the two procedures differ.

How can we evaluate $\pi(j)$ in (3.33) or (3.38)? In some situations, permutation methods provide straightforward and plausible answers. Consider the prostate data (2.1): the data matrix X is 6033×102, with its first 50 columns representing the healthy controls and the last 52 columns the cancer patients. X gave the 6033-vector $\boldsymbol{t}$ of two-sample t-statistics via row-wise application of (2.2)–(2.3).

Now let X^* be a version of X in which the columns have been randomly permuted: formally, if $\boldsymbol{J}^* = (j_1^*, j_2^*, \dots, j_n^*)$ is a randomly selected permutation of $(1, 2, \dots, n)$ then X^* has entries

$$x_{ij}^* = x_{iJ^*(j)} \qquad \text{for } j = 1, 2, \dots, n \text{ and } i = 1, 2, \dots, N. \tag{3.39}$$

Applying calculations (2.2)–(2.3) to X^* (and still considering the first 50 columns as controls and the last 52 as cancer patients) yields a 6033-vector of "permutation t-values"

$$\boldsymbol{T}^* = (T_1^*, T_2^*, \dots, T_N^*)'. \tag{3.40}$$

Independently repeating the permutation process some large number B times allows us to estimate (3.38) by simply counting the proportion of times $\max(T_k^*, k \in R_j)$ exceeds $t_{(j)}$,

$$\hat{\pi}(j) = \#\left\{\max_{k \in R_j}(T_k^*) > t_{(j)}\right\} \Big/ B. \tag{3.41}$$

Here R_j and $t_{(j)}$ retain their original values, only the T_k^* vary. Finally, the adjusted p-values $\tilde{p}_{(i)}$ are estimated from the $\hat{\pi}(j)$ as in (3.34). (The min-p calculation is more difficult to implement, explaining the greater popularity of max-T.)

The key idea is that the permutation distribution of $\boldsymbol{T}^*$ is a reasonable

[6] We could just as well define $\pi(j)$ with respect to two-sided or left-sided versions of (3.37)–(3.38).

stand-in for the hypothetical distribution of $\boldsymbol{T}$ we would obtain under the complete null hypothesis $\boldsymbol{H}_0$. This is plausible since

- permuting the columns of X destroys any true differences between controls and cancer patients, thereby enforcing $\boldsymbol{H}_0$;
- and since columns are permuted intact, the correlation structure between the rows (i.e., the genes) is maintained.

Permutation methods have played a major role in the large-scale testing literature. We will be discussing their virtues and limitations in several upcoming chapters.

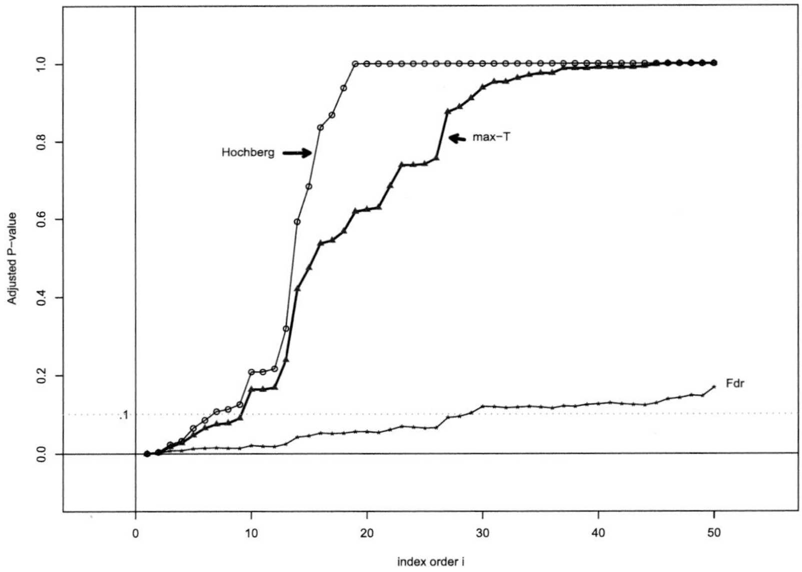

Figure 3.2 Adjusted p-values $\tilde{p}_{(i)}$ for the prostate data (right-sided), $N = 6033$; circles are Hochberg step-up procedure (3.30), triangles are max-T procedure (3.34)–(3.38). Also shown are estimated false discovery rates (3.42).

Figure 3.2 shows the max-T adjusted p-values $\tilde{p}_{(i)}$ for the prostate data. Here they are seen to be a substantial improvement over the Hochberg step-up values (3.30) (which are the same as Holm's values (3.21) in this case). For instance, the first nine of the max-T $\tilde{p}_{(i)}$ values are less than 0.1, versus only six for Hochberg.

Also shown in Figure 3.2 are the estimated false discovery rates $\overline{\text{Fdr}}_{(i)}$,

$$\overline{\text{Fdr}}_{(i)} = N \cdot \left[1 - \Phi\left(z_{(i)}\right)\right] \big/ \#\left\{z_j \geq z_{(i)}\right\} \tag{3.42}$$

where $z_{(1)} \geq z_{(2)} \geq \cdots \geq z_{(N)}$ are the ordered z-values and Φ the standard normal cdf. These follow definitions (2.23)–(2.24) with $\mathcal{Z} = [z_{(i)}, \infty)$, and π_0 taken to equal its upper bound 1. The striking fact is how much smaller $\overline{\text{Fdr}}_{(i)}$ is than either version of $\tilde{p}_{(i)}$. For $i = 20$, $\overline{\text{Fdr}}_{(20)} = 0.056$, while $\tilde{p}_{(20)} = 0.62$ for max-T. These results are not contradictory: there is a good chance that at least one of the 20 genes is null, but the expected number of nulls (0.056×20) is not much bigger than 1. The Fdr criterion has become popular because its more liberal conclusions are a good fit to modern applications having N in the thousands.

With $N = 6033$ and $i = 20$, as in Figure 3.2, Holm's method (3.20) and Hochberg's procedure are almost the same as the Bonferroni bound (3.13). Of course there are still plenty of small-N multiple testing problems, where these more sophisticated procedures come into their own.

3.5 Other Control Criteria

FWER control dominates the traditional multiple comparison literature, but other criteria have been proposed. Two other examples of quantities to control are the *per comparison error rate*

$$\text{PCER} = E\{\text{Number true null hypotheses rejected}\}/N \tag{3.43}$$

and the *expected error rate*

$$\text{EER} = E\{\text{Number wrong decisions}\}/N, \tag{3.44}$$

a wrong decision being rejection of H_{0i} when it should be accepted or vice versa. Neither PCER nor EER have attracted the attention accorded FWER.

The microarray era, with case sizes zooming up from $N = 10$ to $N = 10\,000$, has brought dissatisfaction with FWER control. Less conservative methods that still command scientific respectability are being developed, an interesting example being Lehmann and Romano's *k-FWER* criteria, which aims to control the probability of rejecting k or more true null hypotheses; $k = 1$ is the usual FWER, but choosing a larger value of k gives more generous results.

A simple extension of the Bonferroni bound (3.13) provides k-FWER control.

Theorem 3.1 *The procedure that rejects only those null hypotheses* H_{0i} *for which*

$$p_i \le k\alpha/N \tag{3.45}$$

controls k-FWER *at level* α*,*

$$\Pr\{k \text{ or more true } H_{0i} \text{ rejected}\} \le \alpha. \tag{3.46}$$

Proof Let I_0 index the true null hypotheses H_{0i} as in (3.14), with $N_0 = \#I_0$, and let $N_0(\text{rej})$ be the number of falsely rejected H_{0i}. Then

$$\begin{aligned}\Pr\{N_0(\text{reg}) \ge k\} \le E\{N_0(\text{reg})\}/k &= \sum_{i\in I_0} \Pr\{p_i \le k\alpha/N\}/k \\ &= \sum_{i\in I_0} \frac{k\alpha/N}{k} \qquad (3.47) \\ &= \frac{N_0}{N}\alpha \le \alpha. \end{aligned}$$

□

Exercise 3.9 Verify the first inequality above (*Markov's inequality*).

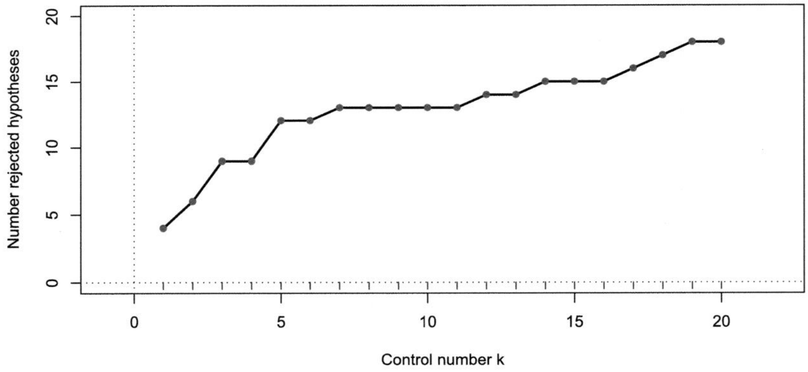

Figure 3.3 Number of rejections by k-FWER control criterion (3.45), $\alpha = 0.05$, for $k = 1, 2, \ldots, 20$; prostate data (right side).

Figure 3.3 applies k-FWER control, $\alpha = 0.05$, to the right side (that is, the positive z-values) of the prostate data. The graph traces the number of genes satisfying rejection criterion (3.45) for different choices of k, ranging from $k = 1$ to $k = 20$. At $k = 1$, the usual FWER criterion, four genes are rejected, climbing to 18 rejections at $k = 20$.

Once again the comparison with estimated false discovery rates (3.42) is startling: $\overline{\text{Fdr}}_{(18)} = 0.052$, so the estimated number of true null hypotheses among the first 18 rejected is less than one, implying k-FWER hasn't taken much advantage of its allowance of $k = 20$ errors; probably it has made only one or two. One of the main goals of Chapter 4 is to understand why FDR control permits such liberal conclusions.

Notes

Westfall and Young's (1993) book was a notable successor to Miller (1981), showing how modern computation could be used to good effect on multiple testing problems. Dudoit and van der Laan's (2008) book takes another large step in the direction of computation-intensive testing algorithms.

A series of ingenious papers produced the step-down and step-up algorithms of Section 3.3 and Section 3.4: Holm (1979), Simes (1986), Hommel (1988), and Hochberg (1988). The closure principle is nicely stated in Marcus et al. (1976), though its origins go back further. Dudoit et al. (2003) provide an excellent review of the whole theory.

An influential example of applying ANOVA methods to microarray data matrices appears in Kerr et al. (2000). Storey's *optimal discovery procedure* (2007) can be considered as an implementation of EER control (3.44). Efron and Gous (2001) provide a discussion of Fisher's interpretive scale for p-values, Table 3.1, and its Bayes-factor competitor, Jeffrey's scale. Lehmann and Romano (2005a) and Romano et al. (2008) discuss other control criteria besides k-FWER.

4

False Discovery Rate Control

Applied statistics is an inherently conservative enterprise, and appropriately so since the scientific world depends heavily on the consistent evaluation of evidence. Conservative consistency is raised to its highest level in classical significance testing, where the control of Type I error is enforced with an almost religious intensity. A p-value of 0.06 rather than 0.04 has decided the fate of entire pharmaceutical companies. Fisher's scale of evidence, Table 3.1, particularly the $\alpha = 0.05$ threshold, has been used in literally millions of serious scientific studies, and stakes a good claim to being the 20th century's most influential piece of applied mathematics.

All of this makes it more than a little surprising that a powerful rival to Type I error control has emerged in the large-scale testing literature. Since its debut in Benjamini and Hochberg's seminal 1995 paper, false discovery rate control has claimed an increasing portion of statistical research, both applied and theoretical, and seems to have achieved "accepted methodology" status in scientific subject-matter journals.

False discovery rate control moves us away from the significance-testing algorithms of Chapter 3, back toward the empirical Bayes context of Chapter 2. The language of classical testing is often used to describe FDR methods (perhaps in this way assisting their stealthy infiltration of multiple testing practice), but, as the discussion here is intended to show, both their rationale and results are quite different.

4.1 True and False Discoveries

We wish to test N null hypotheses

$$H_{01}, H_{02}, \ldots, H_{0N} \tag{4.1}$$

on the basis of a data set X, and have in mind some decision rule $\mathcal{D}$ that will produce a decision of "null" or "non-null" for each of the N cases.

Equivalently,[1] $\mathcal{D}$ *accepts* or *rejects* each H_{0i}, $i = 1, 2, \ldots, N$, on the basis of X. X is the 6033×102 matrix of expression values in the prostate data example of Section 2.1, giving the N z-values (2.5), while $\mathcal{D}$ might be the rule that rejects H_{0i} if $|z_i| \geq 3$ and accepts H_{0i} otherwise.

Decision

		Null	*Non-Null*	
Actual	*Null*	$N_0 - a$	a	N_0
	Non-Null	$N_1 - b$	b	N_1
		$N - R$	R	N

Figure 4.1 A decision rule $\mathcal{D}$ has rejected R out of N null hypotheses (4.1); a of these decisions were incorrect, i.e., they were "false discoveries," while b of them were "true discoveries." The false discovery proportion Fdp equals a/R.

Figure 4.1 presents a hypothetical tabulation of $\mathcal{D}$'s performance from the point of view of an omniscient oracle: N_0 of the N cases were actually null, of which $\mathcal{D}$ called a non-null (incorrectly) and $N_0 - a$ null (correctly); likewise, N_1 were actually non-null, with $\mathcal{D}$ deciding b of them correctly and $N_1 - b$ incorrectly. Of the $R = a + b$ total rejections, a were "false discoveries" and b "true discoveries," in the current terminology. The family-wise error rate of Section 2.2, FWER, equals $\Pr\{a > 0\}$ in terms of the figure.

N equals 1 in the classical single-case testing situation, so either N_0 or N_1 equals 1, with the other 0. Then

$$\Pr\{a = 1 | N_0 = 1\} = \alpha, \tag{4.2}$$

the Type I error rate, or size, of the decision rule, and

$$\Pr\{b = 1 | N_1 = 1\} = \beta, \tag{4.3}$$

[1] I am trying to avoid the term "significant" for the rejected cases as dubious terminology even in single-case testing, and worse in the false discovery rate context, preferring instead "interesting."

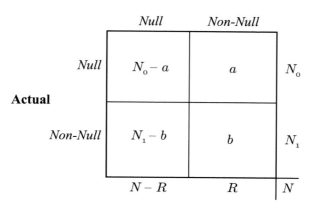

the rule's power.

Exercise 4.1 In a multiple testing situation with both N_0 and N_1 positive, show that

$$E\left\{\frac{a}{N_0}\right\} = \bar{\alpha} \quad \text{and} \quad E\left\{\frac{b}{N_1}\right\} = \bar{\beta}, \tag{4.4}$$

$\bar{\alpha}$ and $\bar{\beta}$ being the average size and power of the null and non-null cases, respectively.

Classical Fisherian significance testing is immensely popular because it requires so little from the scientist: only the choice of a test statistic and specification of its probability distribution when the null hypothesis is true. Neyman–Pearson theory adds the specification of a non-null distribution, the reward being the calculation of power as well as size. Both of these are calculated *horizontally* in the figure, that is restricting attention to either the null or non-null row, which is to say that they are frequentist calculations.

Large-scale testing, with N perhaps in the hundreds or thousands, opens the possibility of calculating *vertically* in the figure, in the Bayesian direction, without requiring Bayesian priors. The ratio a/R is what we called the *false discovery proportion* (2.28), the proportion of rejected cases that are actually null. Benjamini and Hochberg's testing algorithm, the subject of the next section, aims to control the expected value of a/R rather than that of a/N_0.

Exercise 4.2 Suppose that $z_1, z_2, \ldots, z_N$ are independent and identically distributed observations from the two-groups model (2.7) and that the decision rule rejects H_{0i} for $z_i \in \mathcal{Z}$, as illustrated in Figure 2.3. Show that a/R has a scaled binomial distribution given R (with $R > 0$),

$$a/R \sim \text{Bi}\left(R, \phi(\mathcal{Z})\right)/R, \tag{4.5}$$

with $\phi(\mathcal{Z}) = \text{Fdr}(\mathcal{Z})$ as in (2.13).

4.2 Benjamini and Hochberg's FDR Control Algorithm

We assume that our decision rule $\mathcal{D}$ produces a p-value p_i for each case i, so that p_i has a uniform distribution if H_{0i} is correct,

$$H_{0i} : p_i \sim \mathcal{U}(0, 1). \tag{4.6}$$

Denote the ordered p-values by

$$p_{(1)} \leq p_{(2)} \leq \cdots \leq p_{(i)} \leq \cdots \leq p_{(N)} \tag{4.7}$$

as in (3.19). Following the notation in Figure 4.1, let $R_{\mathcal{D}}$ be the number of cases rejected, $a_{\mathcal{D}}$ the number of those that are actually null, and $\text{Fdp}_{\mathcal{D}}$ the false discovery proportion

$$\text{Fdp}_{\mathcal{D}} = a_{\mathcal{D}}/R_{\mathcal{D}} \qquad [= 0 \text{ if } R_{\mathcal{D}} = 0]. \tag{4.8}$$

The Benjamini–Hochberg (BH) algorithm uses this rule: for a fixed value of q in (0, 1), let $i_{\max}$ be the *largest* index for which

$$p_{(i)} \leq \frac{i}{N}q, \tag{4.9}$$

and reject $H_{0(i)}$, the null hypothesis corresponding to $p_{(i)}$, if

$$i \leq i_{\max}, \tag{4.10}$$

accepting $H_{0(i)}$ otherwise.

Theorem 4.1 *If the p-values corresponding to the correct null hypotheses are independent of each other, then the rule* BH(q) *based on the BH algorithm controls the expected false discovery proportion at* q,

$$E\left\{\text{Fdp}_{\text{BH}(q)}\right\} = \pi_0 q \leq q \qquad \textit{where } \pi_0 = N_0/N. \tag{4.11}$$

A proof of Theorem 4.1 appears at the end of this section. The proportion of null cases $\pi_0 = N_0/N$ is unknown in practice though we will see that it can be estimated, so q is usually quoted as the control rate of BH(q).

There is a practical reason for the impressive popularity of BH(q): it is *much more liberal* in identifying non-null cases than the FWER algorithms of Chapter 3. Figure 4.2 illustrates the point by comparison with Hochberg's step-up procedure (3.30). BH(q) can also be described in step-up form: decrease i starting from $i = N$ and keep accepting $H_{0(i)}$ until the first time $p_{(i)} \leq q\,i/N$, after which all $H_{0(i)}$ are rejected. Hochberg's procedure instead uses $p_{(i)} \leq \alpha/(N - i + 1)$; see Exercise 3.8.

If we set $q = \alpha$, the ratio of the two thresholds is

$$\left(\frac{i}{N}\right) \bigg/ \left(\frac{1}{N - i + 1}\right) = i \cdot \left(1 - \frac{i - 1}{N}\right). \tag{4.12}$$

Usually only small values of i/N will be interesting, in which case BH(q) is approximately i times as liberal as Hochberg's rule.

The left panel of Figure 4.2 makes the comparison for $\alpha = q = 0.1$ and $N = 100$. The two threshold curves are equal at $i = 1$ where both take the Bonferroni value α/N, and at $i = N$ where both equal α. In between, BH(q)

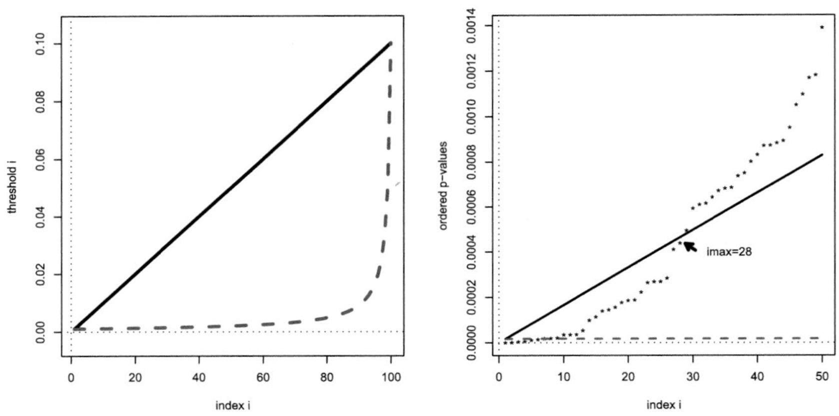

Figure 4.2 *Left panel*: Solid line is rejection boundary (4.9) for FDR control rule BH(q); dashed curve for Hochberg step-up FWER procedure, Exercise 3.8; $\alpha = q = 0.1,\ N = 100$. *Right panel*: Stars indicate p-values for the 50 largest z_i, prostate data (2.6); solid and dashed lines are rejection boundaries, BH(q) and Hochberg, $\alpha = q = 0.1,\ N = 6033$.

allows rejection at much larger values of $p_{(i)}$. The right panel shows the 50 smallest $p_{(i)}$ values for the prostate data, $p_{(i)} = F_{100}(-t_{(i)})$ in (2.5), and also the two rejection boundaries, again with $\alpha = q = 0.1$. Here $i_{\max} = 28$ genes are declared non-null by BH(q)(0.1) compared to 9 for the Hochberg 0.1 test.

Of course, rejecting more cases is only a good thing if they *should* be rejected. False discovery rate control is a more liberal rejecter than FWER: can we still trust its decisions? This is the question we will be trying to answer as we consider, in what follows, the pros and cons of the BH(q) rule and its underlying rationale. Here are a few preliminary comments:

- Theorem 4.1 depends on independence among the p-values of the null cases (the top row in Figure 4.1), usually an unrealistic assumption. This limitation can be removed if the rejection boundary (4.9) is lowered to

$$p_{(i)} \le \frac{i}{N}\frac{q}{l_i} \qquad \text{where } l_i = \sum_{j=1}^{i} \frac{1}{j}. \tag{4.13}$$

  However, (4.13) represents a severe penalty ($l_{28} = 3.93$ for instance) and is not really necessary. The independence condition in Theorem 4.1 can be weakened to *positive regression dependence* (PRD): roughly speaking, the assumption that the null-case z-values have non-negative correlations, though even PRD is unlikely to hold in practice. Fortunately, the empirical Bayes interpretation of BH(q) is *not* directly affected by dependence, as discussed in the next section.

- Theorem 4.1 depends on taking $a/R = 0$ when $R = 0$, that is, defining $0/0 = 0$ in Figure 4.1. Storey's "positive false discovery rate" criterion avoids this by only considering situations with $R > 0$, but doing so makes strict FDR control impossible: if $N_0 = N$, that is, if there are no non-null cases, then all rejections are false discoveries and $E\{\text{Fdp}\,|R > 0\} = 1$ for any rule $\mathcal{D}$ that rejects anything.
- Is it really satisfactory to control an error rate *expectation* rather than an error rate *probability* as in classical significance testing? The next two sections attempt to answer this question in empirical Bayes terms.
- How should q be chosen? The literature hasn't agreed upon a conventional choice, such as $\alpha = 0.05$ for single-case testing, though $q = 0.1$ seems to be popular. The empirical Bayes context of the next section helps clarify the meaning of q.
- The p-values in (4.9) are computed on the basis of an assumed null hypothesis distribution, for example $p_{(i)} = F_{100}(-t_{(i)})$ in the right panel of Figure 4.2, with F_{100} a Student-t cdf having 100 degrees of freedom. This is by necessity in classical single-case testing, where theory is the only possible source for a null distribution. Things are different in large-scale testing: empirical evidence may make it clear that the theoretical null is unrealistic. Chapter 6 discusses the proper choice of null hypotheses in multiple testing.

This last objection applies to all testing algorithms, not just to the Benjamini–Hochberg rule. The reason for raising it here relates to Theorem 4.1: its statement is so striking and appealing that it is easy to forget its limitations. Most of these turn out to be not too important in practice, except for the proper choice of a null hypothesis, which is crucial.

Proof of Theorem 4.1 For t in $(0, 1]$ define

$$R(t) = \#\{p_i \le t\}, \tag{4.14}$$

$a(t)$ the number of null cases with $p_i \le t$, false discovery proportion

$$\text{Fdp}(t) = a(t)/\max\{R(t), 1\}, \tag{4.15}$$

and

$$Q(t) = Nt/\max\{R(t), 1\}. \tag{4.16}$$

Also let

$$t_q = \sup_t \{Q(t) \le q\}. \tag{4.17}$$

Since $R(p_{(i)}) = i$ we have $Q(p_{(i)}) = Np_{(i)}/i$. This implies that the BH rule (4.9) can be re-expressed as

$$\text{Reject } H_{0(i)} \text{ for } p_{(i)} \le t_q. \tag{4.18}$$

Let $A(t) = a(t)/t$. It is easy to see that

$$E\{A(s)|A(t)\} = A(t) \qquad \text{for } s \le t, \tag{4.19}$$

and in fact $E\{A(s)|A(t') \text{ for } t' \ge t\} = A(t)$. In other words, $A(t)$ is a martingale as t decreases from 1 to 0. Then by the optional stopping theorem,

$$E\left\{A(t_q)\right\} = E\{A(1)\} = E\{a(1)/1\} = N_0, \tag{4.20}$$

the actual number of null cases.

Finally, note that (4.16) implies

$$\max\left\{R(t_q), 1\right\} = Nt_q/Q(t_q) = Nt_q/q, \tag{4.21}$$

so

$$\text{Fdp}(t_q) = \frac{q}{N}\frac{a(t_q)}{t_q}. \tag{4.22}$$

Then (4.20) gives

$$E\left\{\text{Fdp}(t_q)\right\} = \pi_0 q \qquad [\pi_0 = N_0/N] \tag{4.23}$$

which, together with (4.18), verifies Theorem 4.1. □

Exercise 4.3 Verify (4.19).

4.3 Empirical Bayes Interpretation

Benjamini and Hochberg's BH(q) procedure has an appealing empirical Bayes interpretation. Suppose that the p-values p_i correspond to real-valued test statistics z_i,

$$p_i = F_0(z_i) \qquad i = 1, 2, \dots, N, \tag{4.24}$$

where $F_0(z)$ is the cdf of a common null distribution applying to all N null hypotheses H_{0i}, for example, F_0 the standard normal cdf in (2.6). We can always take z_i to be p_i itself, in which case F_0 is the $\mathcal{U}(0,1)$ distribution.

Let $z_{(i)}$ denote the ith ordered value,

$$z_{(1)} \le z_{(2)} \le \cdots \le z_{(i)} \le \cdots \le z_{(N)}. \tag{4.25}$$

Then $p_{(i)} = F_0(z_{(i)})$ in (4.7), if we are interested in left-tailed p-values, or $p_{(i)} = 1 - F_0(z_{(i)})$ for right-tailed p-values.

Note that the empirical cdf of the z_i values,

$$\bar{F}(z) = \#\{z_i \le z\}/N \tag{4.26}$$

satisfies

$$\bar{F}(z_{(i)}) = i/N. \tag{4.27}$$

This means we can write the threshold condition for the BH rule (4.9) as

$$F_0(z_{(i)})\Big/\bar{F}(z_{(i)}) \le q \tag{4.28}$$

or

$$\pi_0 F_0(z_{(i)})\Big/\bar{F}(z_{(i)}) \le \pi_0 q. \tag{4.29}$$

However, $\pi_0 F_0(z)/\bar{F}(z)$ is the empirical Bayes false discovery rate estimate $\overline{\text{Fdr}}(z)$ (from (2.21) with $\mathcal{Z} = (-\infty, z)$).

We can now re-express Theorem 4.1 in empirical Bayes terms.

Corollary 4.2 *Let $i_{\max}$ be the largest index for which*

$$\overline{\text{Fdr}}(z_{(i)}) \le q \tag{4.30}$$

and reject $H_{0(i)}$ for all $i \le i_{\max}$, accepting $H_{0(i)}$ otherwise. Then, assuming that the z_i values are independent, the expected false discovery proportion of the rule equals q.

Exercise 4.4 Use Theorem 4.1 to verify Corollary 4.2.

Note With π_0 unknown it is usual to set it to its upper bound 1, giving $\overline{\text{Fdr}}(z_{(i)}) = F_0(z_{(i)})/\bar{F}(z_{(i)})$. This makes rule (4.30) conservative, with $E\{\text{Fdp}\} \le q$. But see Section 4.5.

Returning to the two-groups model (2.7), Bayes rule gives

$$\text{Fdr}(z) = \pi_0 F_0(z)/F(z) \tag{4.31}$$

as the posterior probability that case i is null given $z_i \le z$ (2.13). Section 2.4

shows $\overline{\mathrm{Fdr}}(z_i)$ to be a good estimate of $\mathrm{Fdr}(z_i)$ under reasonable conditions. A greedy empirical Bayesian might select

$$z_{\max} = \sup_z \left\{ \overline{\mathrm{Fdr}}(z) \le q \right\} \tag{4.32}$$

and report those cases having $z_i \le z_{\max}$ as "having estimated probability q of being null." Corollary 4.2 justifies the greedy algorithm in frequentist terms: if the z_i values are independent, then the expected null proportion of the reported cases will equal q. It is always a good sign when a statistical procedure enjoys both frequentist and Bayesian support, and the BH algorithm passes the test.

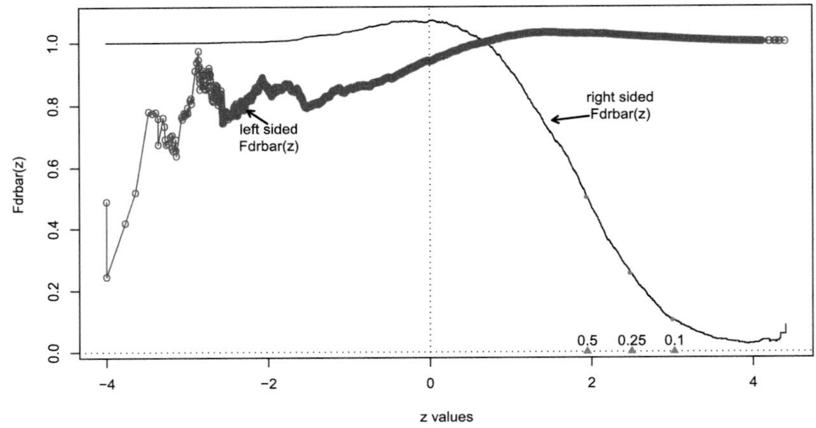

Figure 4.3 Left-sided and right-sided values of $\overline{\mathrm{Fdr}}(z)$ for the DTI data of Section 2.5; triangles indicate values of z having $\overline{\mathrm{Fdr}}^{(c)}(z)$ (4.33) equal 0.5, 0.25 and 0.1; 192 voxels have $z_{(i)}$ exceeding 3.02, the $q = 0.1$ threshold.

Figure 4.3 graphs $\overline{\mathrm{Fdr}}(z)$ and the analogous right-sided quantity

$$\overline{\mathrm{Fdr}}^{(c)}(z) = \pi_0 F_0^{(c)}(z) \big/ \bar{F}^{(c)}(z) \tag{4.33}$$

for the DTI data of Section 2.5 (setting π_0 to 1 in (4.29) and (4.33)), where $F^{(c)}(z)$ indicates the complementary cdf $1 - F(z)$. There is just the barest hint of anything interesting on the left, but on the right, $\mathrm{Fdr}^{(c)}(z)$ gets quite small. For example, 192 of the voxels reject their null hypotheses, those having $z_i \ge 3.02$ at the $q = 0.1$ threshold.

Exercise 4.5 I set $\pi_0 = 1$ in (4.33). How does that show up in Figure 4.3?

The empirical Bayes viewpoint clarifies some of the questions raised in the previous section.

- *Choice of q* Now q is an estimate of the Bayes probability that a rejected null hypothesis H_{0i} is actually correct. It is easy to explain to a research colleague that $q = 0.1$ means an estimated 90% of the rejected cases are true discoveries. The uncomfortable moment in single-case testing, where it has to be confessed that $\alpha = 0.05$ rejection does *not* imply a 95% chance that the effect is genuine, is happily avoided.

- *Independence assumption* $\overline{\text{Fdr}}(z) = \pi_0 F_0(z)/\bar{F}(z)$ is an accurate estimate of the Bayes false discovery rate $\text{Fdr}(z) = \pi_0 F_0(z)/F(z)$ (2.13) whenever $\bar{F}(z)$, the empirical cdf, is close to $F(z)$. This does *not* require independence of the z-values, as shown in Section 2.4. $\overline{\text{Fdr}}(z)$ is upwardly biased for estimating $\text{Fdr}(z)$, and also for estimating the expected false discovery proportion, and in this sense it is always conservative. Lemma 2.2 shows that the upward bias is small under reasonable conditions. Roughly speaking, $\overline{\text{Fdr}}(z)$ serves as an unbiased estimate of $\text{Fdr}(z)$, and of FDR$= E\{\text{Fdp}\}$, even if the z_i are correlated.

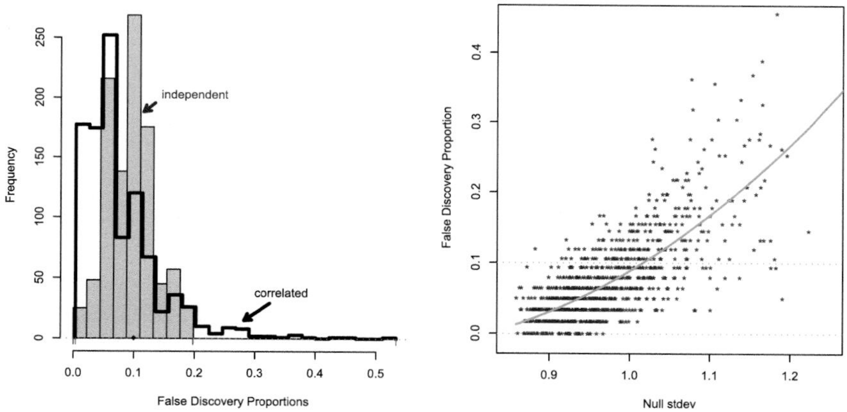

Figure 4.4 *Left panel*: Solid histogram Fdp for BH rule, $q = 0.1$, 1000 simulations of model (4.34) with z_i values independent; line histogram for z_i values correlated, root mean square correlation $= 0.1$. *Right panel*: Fdp values for correlated simulations plotted against $\hat{\sigma}_0$, the empirical standard deviation of the null z_i values; smooth curve is quadratic regression.

The price for correlation is paid in the *variability* of $\overline{\text{Fdr}}(z)$ as an estimate of $\text{Fdr}(z)$, as illustrated in Figure 4.4. Our simulation model involved $N =$ 3000 z-values, with

$$\begin{aligned} z_i &\sim \mathcal{N}(0,1), & i &= 1,2,\dots,2850 \\ \text{and} \quad z_i &\sim \mathcal{N}(2.5,1), & i &= 2851,\dots,3000 \end{aligned} \tag{4.34}$$

so $\pi_0 = 0.95$. Two runs of 1000 simulations each were made, the first with the z_i values independent and the second with substantial correlation: the root mean square value of all $3000 \cdot 2999/2$ pairs of correlations equaled 0.1.[2] For each simulation, the rule $\text{BH}(q)$, $q = 0.1$, was applied (right-sided) and the actual false discovery proportion Fdp observed.

The left panel of Figure 4.4 compares a histogram of the 1000 Fdp values under independence with that for the correlated simulations. The expected Fdp is controlled below $q = 0.1$ in both cases, averaging 0.095 under independence and 0.075 under correlation (see Table 4.1). Control is achieved in quite different ways, though: correlation produces a strongly asymmetric Fdp distribution, with more very small or very large values. The BH algorithm continues to control the expectation of Fdp under correlation, but $\overline{\text{Fdr}}$ becomes a less accurate estimator of the true Fdr.

Table 4.1 *Means and standard deviations (in parentheses) for the simulation experiments of Figure 4.4.*

	Fdp	$\hat{\sigma}_0$	# Rejected
Uncorrelated	**.095** (.038)	1.00 (.014)	64.7 (3.7)
Correlated	**.075** (.064)	.97 (.068)	63.1 (7.5)

In Figure 4.4's right panel, the Fdp values for the 1000 correlated simulations are plotted versus $\hat{\sigma}_0$, the empirical standard deviation of the 2850 null z-values. Correlation greatly increases the variability of $\hat{\sigma}_0$, as discussed in Chapter 7. Fdp tends to be greater or less than the nominal value 0.1 as $\hat{\sigma}_0$ is greater or less than 1.0, *varying by a factor of 10.*

In practice, $\hat{\sigma}_0$ isn't observable. However, it is "almost observable" in some situations, in which case the overall control level q can be misleading: if we know that $\hat{\sigma}_0$ is much greater than 1, then there is good reason to believe that Fdp is greater than q. This point is investigated in Chapter 6 in terms of the *empirical null distribution.*

[2] The correlation structure is described in Section 8.2.

• *FDR control as a decision criterion* The BH algorithm only controls the expectation of Fdp. Is this really sufficient for making trustworthy decisions? Part of the answer must depend upon the accuracy of $\overline{\text{Fdr}}$ as an estimate of Fdr (4.31) or of FDR $= E\{\text{Fdp}\}$. This same question arises in single-case testing where the concept of *power* is used to complement Type I error control. Chapters 5 and 7 discuss accuracy and power considerations for false discovery rate control methods.

• *Left-sided, right-sided, and two-sided inferences* For the DTI data of Figure 4.3, BH(q)(0.1) rejects zero voxels on the left and 192 voxels on the right. However, if we use two-sided p-values, $p_i = 2 \cdot \Phi(-|z_i|)$, only 110 voxels are rejected by BH(q)(0.1), all from among the 192. From a Bayesian point of view, two-sided testing only blurs the issue by making posterior inferences over larger, less precise rejection regions $\mathcal{Z}$. The *local* false discovery rate (2.14) provides the preferred Bayesian inference. Chapter 5 concerns estimation of the local fdr.

Exercise 4.6 For the prostate data, the left-tailed, right-tailed, and two-tailed BH(q) rules reject 32, 28, and 60 genes at the $q = 0.1$ level. The rejection regions are $z_i \leq -3.26$ on the left, $z_i \geq 3.36$ on the right, and $|z_i| \geq 3.29$ two-sided. Why is two-sided testing less wasteful here than in the DTI example?

• *False negative rates* Looking at Figure 4.1, it seems important to consider the false negative proportion

$$\text{Fnp} = (N_1 - b)/(N - R) \tag{4.35}$$

as well as Fdp. The expectation of Fnp is a measure of Type II error for $\mathcal{D}$, indicating the rule's power. It turns out that the Bayes/empirical Bayes interpretation of the false discovery rate applies to both Fdp and Fnp.

Suppose that rule $\mathcal{D}$ rejects H_{0i} for $z_i \in \mathcal{R}$, and accepts H_{0i} for z_i in the complementary region $\mathcal{A}$. Following notation (2.13),

$$1 - \phi(\mathcal{A}) = \Pr\{\text{non-null}|z \in \mathcal{A}\} \tag{4.36}$$

is the Bayes posterior probability of a Type II error. The calculations in Section 2.3 and Section 2.4 apply just as well to $\overline{\text{Fdr}}(\mathcal{A})$ as $\overline{\text{Fdr}}(\mathcal{R})$: under the conditions stated there, for instance in Lemma 2.2, $1 - \overline{\text{Fdr}}(\mathcal{A})$ will accurately estimate $1 - \phi(\mathcal{A})$, the Bayesian false negative rate. Chapter 5 uses this approach to estimate power in multiple testing situations.

4.4 Is FDR Control "Hypothesis Testing"?

The Benjamini–Hochberg BH(q) rule is usually presented as a multiple hypothesis testing procedure. This was our point of view in Section 4.2, but not in Section 4.3, where the *estimation* properties of $\overline{\text{Fdr}}$ were emphasized. It pays to ask in what sense false discovery rate control is actually hypothesis testing.

Here we will fix attention on a given subset $\mathcal{R}$ of the real line, e.g., $\mathcal{R} = [3, \infty)$. We compute

$$\overline{\text{Fdr}}(\mathcal{R}) = e_0(\mathcal{R})/R, \tag{4.37}$$

where $e_0(\mathcal{R})$ is the expected number of null cases falling in $\mathcal{R}$ and R is the observed number of z_i in $\mathcal{R}$. We might then follow the rule of rejecting all the null hypotheses H_{0i} corresponding to z_i in $\mathcal{R}$ if $\overline{\text{Fdr}}(\mathcal{R}) \leq q$, and accepting all of them otherwise. Equivalently, we reject all the H_{0i} for z_i in $\mathcal{R}$ if

$$R \geq e_0(\mathcal{R})/q. \tag{4.38}$$

It is clear that (4.38) cannot be a test of the FWER-type null hypothesis that at least one of the R hypotheses is true,

$$H_0(\text{union}) = \bigcup_{i:z_i \in \mathcal{R}} H_{0i}. \tag{4.39}$$

Rejecting H_0(union) implies we believe *all* the H_{0i} for z_i in $\mathcal{R}$ to be incorrect (that is, all should be rejected). But if, say, $R = 50$ then $\overline{\text{Fdr}}(\mathcal{R}) = 0.1$ suggests that about five of the R H_{0i} are correct.

Exercise 4.7 Calculate the probability that H_0(union) is correct if $R = 50$ and $\phi(\mathcal{R}) = 0.1$, under the assumptions of Lemma 2.2.

In other words, the $\overline{\text{Fdr}}$ rule (4.38) is too *liberal* to serve as a test of H_0(union). Conversely, it is too *conservative* as a test of

$$H_0(\text{intersection}) = \bigcap_{i:z_i \in \mathcal{R}} H_{0i} \tag{4.40}$$

which is the hypothesis that all of the R null hypotheses are correct. Rejecting H_0(intersection) says we believe at least one of the R cases is non-null.

Under the Poisson-independence assumptions of Lemma 2.2, H_0(intersection) implies

$$R \sim \text{Poi}\left(e_0(\mathcal{R})\right). \tag{4.41}$$

The obvious level-α test[3] rejects H_0(intersection) for

$$R \geq Q_\alpha, \tag{4.42}$$

the upper $1-\alpha$ quantile of a Poi($e_0(\mathcal{R})$) variate. A two-term Cornish–Fisher expansion gives the approximation

$$Q_\alpha = e_0(\mathcal{R}) + \sqrt{e_0(\mathcal{R})}z_\alpha + \left(z_\alpha^2 - 1\right)\big/6 \tag{4.43}$$

with z_α the standard normal quantile $\Phi^{-1}(1-\alpha)$. (Increasing (4.43) to the nearest integer makes test (4.42) conservative.) Table 4.2 compares the minimum rejection values of R from (4.38) and (4.42) for $q = \alpha = 0.1$. It is clear that (4.38) is far more conservative.

The inference of $\overline{\text{Fdr}}$ outcome (4.32) lies somewhere between "all R cases are true discoveries" and "at least one of the R is a true discovery." I prefer to think of $\overline{\text{Fdr}}$ as an estimate rather than a test statistic: a quantitative assessment of the proportion of false discoveries among the R candidates.

Table 4.2 *Rejection thresholds for R, Fdr test* (4.38), *and* H_0(intersection) *test* (4.42); $q = \alpha = 0.1$. *As a function of* $e_0(\mathcal{R})$, *the expected number of null cases in R. (Rounding* H_0(intersection) *upward gives conservative level-*α *tests.)*

$e_0(\mathcal{R})$	1	2	3	4	6	8
H_0(intersection)	2.39	3.92	5.33	6.67	9.25	11.73
Fdr	10	20	30	40	60	80

Exercise 4.8 For the DTI data of Figure 4.3, 26 of the 15 443 z-values are less than -3.0. How strong is the evidence that at least one of the 26 is non-null? (Assume independence and set π_0 to its upper bound 1.)

4.5 Variations on the Benjamini–Hochberg Algorithm

The BH algorithm has inspired a great deal of research and development in the statistics literature, including some useful variations on its original form. Here we will review just two of these.

[3] This test is a form of Tukey's "higher criticism."

• *Estimation of* π_0 The estimated false discovery rate $\overline{\text{Fdr}}(z) = \pi_0 F_0(z) / \bar{F}(z)$ appearing in Corollary 4.2 requires knowing π_0, the actual proportion of null cases. Rather than setting π_0 equal to its upper bound 1 as in the original BH procedure, we can attempt to estimate it from the collection of observed z-values.

Returning to the two-groups model (2.7), suppose we believe that $f_1(z)$ is zero for a certain subset $\mathcal{A}_0$ of the sample space, perhaps those points near zero,

$$f_1(z) = 0 \qquad \text{for } z \in \mathcal{A}_0; \tag{4.44}$$

that is, all the non-null cases must give z-values outside of $\mathcal{A}_0$ (sometimes called the *zero assumption*). Then the expected value of $N_+(\mathcal{A}_0)$, the observed number of z_i values in $\mathcal{A}_0$, is

$$E\{N_+(\mathcal{A}_0)\} = \pi_0 N \cdot F_0(\mathcal{A}_0), \tag{4.45}$$

suggesting the estimators

$$\hat{\pi}_0 = N_+(\mathcal{A}_0) / (N \cdot F_0(\mathcal{A}_0)) \tag{4.46}$$

and

$$\widehat{\text{Fdr}}(z) = \hat{\pi}_0 F_0(z) \Big/ \bar{F}(z). \tag{4.47}$$

Using $\widehat{\text{Fdr}}(z)$ in place of $\overline{\text{Fdr}}(z) = F_0(z)/\bar{F}(z)$ in Corollary 4.2 increases the number of discoveries (i.e., rejections). It can be shown that the resulting rule still satisfies $E\{\text{Fdp}\} \leq q$ under the independence assumption even if (4.44) isn't valid.

We might take $\mathcal{A}_0$ to be the central α_0 proportion of the f_0 distribution on the grounds that all the "interesting" non-null cases should produce z-values far from the central region of f_0. If f_0 is $\mathcal{N}(0, 1)$ in (2.7) then $\mathcal{A}_0$ is the interval

$$\mathcal{A}_0 = \left[\Phi^{-1}(0.5 - \alpha_0/2), \Phi^{-1}(0.5 + \alpha_0/2)\right] \tag{4.48}$$

with Φ the standard normal cdf. Figure 4.5 graphs $\hat{\pi}_0$ as a function of α_0 for the prostate and DTI data.

Nothing in Figure 4.5 suggests an easy way to select the appropriate $\mathcal{A}_0$ region, particularly not for the prostate data. Part of the problem is the assumption that $f_0(z)$ is the theoretical null $\mathcal{N}(0, 1)$ density. The central peak of the prostate data seen in Figure 2.1 is slightly wider, about $\mathcal{N}(0, 1.06^2)$, which affects calculation (4.46). Chapter 6 discusses methods that estimate π_0 in conjunction with estimation of the mean and variance of f_0.

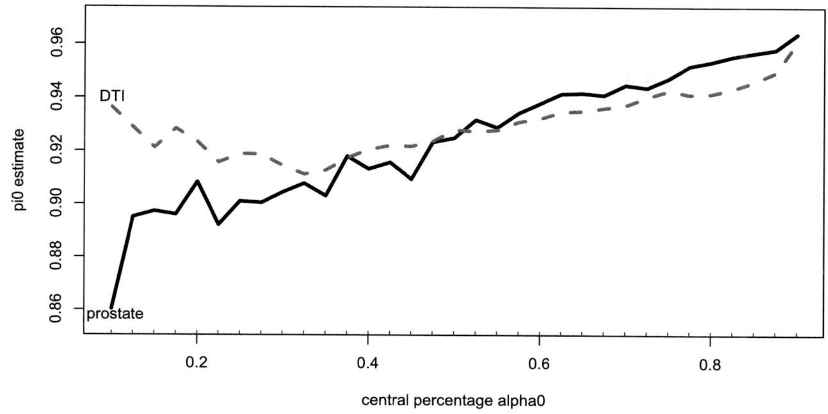

Figure 4.5 Estimated values of π_0 (4.46) for the prostate and DTI data sets; $\mathcal{A}_0$ as in (4.48); α_0 ranging from 0.1 to 0.9.

Exercise 4.9 How would the prostate data $\hat{\pi}_0$ values in Figure 4.5 change if we took f_0 to be $\mathcal{N}(0, 1.06^2)$?

The exact choice of $\hat{\pi}_0$ in (4.47) is not crucial: if we are interested in values of Fdr near $q = 0.1$ then the difference between $\hat{\pi}_0 = 0.9$ and $\hat{\pi}_0 = 1$ is quite small. A much more crucial and difficult issue is the appropriate choice of the null density f_0, the subject of Chapter 6.

• *Significance analysis of microarrays* SAM, the significance analysis of microarrays, is a popular Fdr-like program originally intended to identify interesting genes in microarray experiments. Microarray studies have nothing in particular to do with SAM's workings, but we will use them for illustration here. Suppose that X is an $N \times n$ matrix of expression levels as in Section 2.1 that we have used to produce an N-vector of z-values $\boldsymbol{z} = (z_1, z_2, \ldots, z_N)'$. For the sake of definiteness, assume that X represents a two-sample study, say healthy versus sick subjects, and that the z_i are normalized t-values as in (2.2)–(2.5). (SAM actually handles more general experimental layouts and summary statistics: in particular, there need be no theoretical null assumption (2.6).)

1 The algorithm begins by constructing some number B of $N \times n$ matrices X^*, each of which is a version of X in which the columns have been randomly permuted as in (3.40). Each X^* yields an N-vector $\boldsymbol{z}^*$ of z-values calculated in the same way as $\boldsymbol{z}$.

2 Let $\boldsymbol{Z}$ be the ordered version of $\boldsymbol{z}$, and likewise $\boldsymbol{Z}^{*b}$ the ordered version of $\boldsymbol{z}^{*b}$, the bth $\boldsymbol{z}^*$ vector, $b = 1, 2, \ldots, B$. Define

$$\bar{Z}_i = \sum_{b=1}^{B} Z_i^{*b} \Big/ B \tag{4.49}$$

so $\bar{Z}_i$ is the average of the ith largest values of $\boldsymbol{z}^{*b}$.

3 Plot Z_i versus $\bar{Z}_i$ for $i = 1, 2, \ldots, N$. The upper panel of Figure 4.6 shows the $(\bar{Z}_i, Z_i)$ plot for the prostate data of Figure 2.1. (This amounts to a *QQ*-plot of the actual z-values versus the permutation distribution.)

4 For a given choice of a positive constant Δ, define

$$\begin{aligned} c_{\text{up}}(\Delta) &= \min\{Z_i : Z_i - \bar{Z}_i \geq \Delta\} \\ \text{and} \quad c_{\text{lo}}(\Delta) &= \max\{Z_i : \bar{Z}_i - Z_i \geq \Delta\}. \end{aligned} \tag{4.50}$$

In words, $c_{\text{up}}(\Delta)$ is the first Z_i value at which the $(\bar{Z}_i, Z_i)$ curve exits the band $\bar{Z}_i + \Delta$, and similarly for $c_{\text{lo}}(\Delta)$. In the top panel of Figure 4.6, $\Delta = 0.7$, $c_{\text{up}}(\Delta) = 3.29$, and $c_{\text{lo}}(\Delta) = -3.34$.

5 Let $R(\Delta)$ be the number of z_i values outside of $[c_{\text{lo}}(\Delta), c_{\text{up}}(\Delta)]$,

$$R(\Delta) = \#\left\{z_i \geq c_{\text{up}}(\Delta)\right\} + \#\left\{z_i \leq c_{\text{lo}}(\Delta)\right\}, \tag{4.51}$$

and likewise

$$R^*(\Delta) = \#\left\{z_i^{*b} \geq c_{\text{up}}(\Delta)\right\} + \#\left\{z_i^{*b} \leq c_{\text{lo}}(\Delta)\right\}, \tag{4.52}$$

the sums in (4.52) being over all $N \cdot B$ permutation z-values.

6 Finally, define the false discovery rate corresponding to Δ as

$$\overline{\text{Fdr}}(\Delta) = \frac{R^*(\Delta)/NB}{R(\Delta)/N} = \frac{1}{B}\frac{R^*(\Delta)}{R(\Delta)}. \tag{4.53}$$

The SAM program calculates $\overline{\text{Fdr}}(\Delta)$ for a range of Δ values in a search for the Δ that produces a pre-chosen value $\overline{\text{Fdr}}(\Delta) = q$. For the prostate data, $\Delta = 0.7$ gave $\text{Fdr}(\Delta) = 0.1$; $R(\Delta) = 60$ genes were identified as significant, 28 on the right and 32 on the left.

Definition (4.53) of $\overline{\text{Fdr}}(\Delta)$ is equivalent to our previous usage at (4.28) or (2.21): the rejection region

$$\mathcal{R}(\Delta) = \left\{z \notin \left[c_{\text{lo}}(\Delta), c_{\text{up}}(\Delta)\right]\right\} \tag{4.54}$$

has empirical probability $\bar{F}(\Delta) = R(\Delta)/N$; similarly, $R^*(\Delta)/NB$ is the null estimate $\bar{F}_0(\Delta)$, the proportion of the $N \cdot B$ z_i^{*b} values in $\mathcal{R}(\Delta)$, so

$$\overline{\text{Fdr}}(\Delta) = \bar{F}_0(\Delta)\big/\bar{F}(\Delta), \tag{4.55}$$

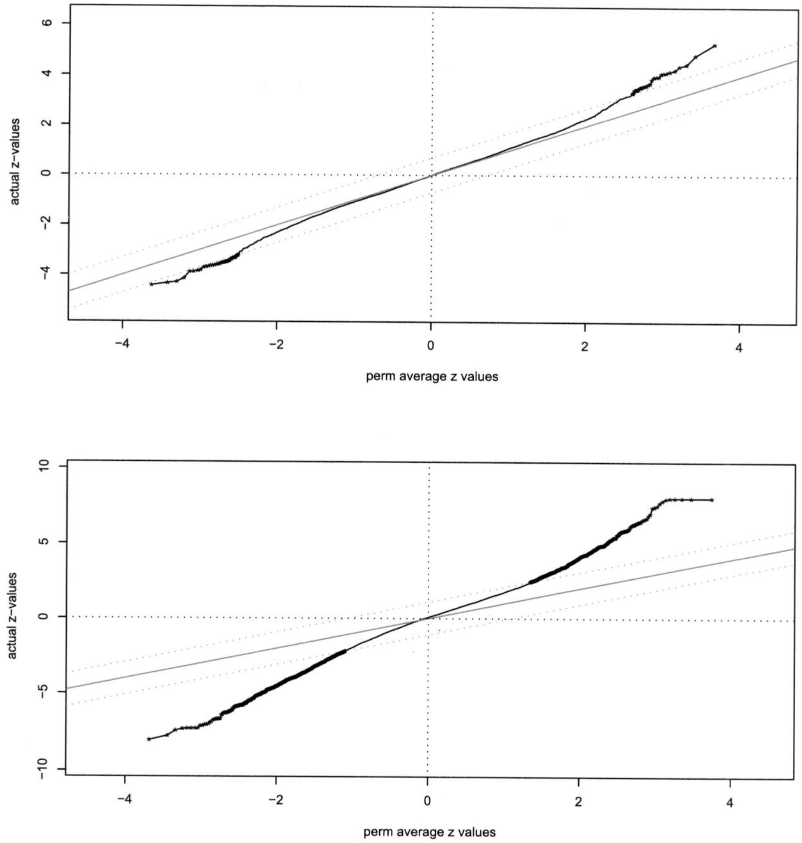

Figure 4.6 SAM plots for the prostate data (top) and the leukemia data (bottom). Starred points indicate “significant” genes at the $q = 0.1$ level: 60 in the top panel, 1660 in the bottom.

which is (4.28), setting $\pi_0 = 1$ and using the permutation z_i^{*b} values instead of a theoretical distribution to define the nulls.

Despite the disparaged “significance” terminology, the output of SAM is closer to empirical Bayes estimation than hypothesis testing; that is, the statistician gets more than a simple yes/no decision for each gene. The two-sided nature of the procedure is unfortunate from a Bayesian perspective, but this can be remedied by choosing Δ separately for positive and negative z-values.

The bottom panel of Figure 4.6 concerns the *leukemia data*, another microarray study featured in Chapter 6. Here there are $N = 7128$ genes whose expression levels are measured on $n = 72$ patients, 47 with a less severe and 25 with a more severe form of leukemia. Two-sample t-tests have led to z-values as in (2.1)–(2.5) (now with 70 degrees of freedom rather than 100). The SAM plot reveals a serious problem: unlike the prostate panel, the leukemia plot doesn't match the solid 45° line near $z = 0$, crossing it instead at a sharp angle.

We will see in Chapter 6 that the histogram of the 7128 leukemia z-values, unlike Figure 2.1, is much wider at the center than a $\mathcal{N}(0, 1)$ distribution. However, the permutation null distributions are almost perfectly $\mathcal{N}(0, 1)$ in both cases, a dependable phenomenon it turns out. This casts doubt on the appropriateness of $\bar{F}_0(\Delta)$ in the numerator of $\overline{\text{Fdr}}(\Delta)$ (4.55) and the identification of 1660 "significant" leukemia genes. The appropriate choice of a null distribution is the crucial question investigated in Chapter 6.

Exercise 4.10 Suppose the z-value histogram is approximately $\mathcal{N}(0, \sigma_0^2)$ near $z = 0$ while the permutation distribution is $\mathcal{N}(0, 1)$. What will be the angle of crossing of the SAM plot?

4.6 $\overline{\text{Fdr}}$ and Simultaneous Tests of Correlation

When dealing with t-statistics, as in the prostate study (2.2), the false discovery rate estimator $\overline{\text{Fdr}}$ (2.21) has a nice geometrical interpretation in terms of clustering on the hypersphere. This interpretation allows us to use the BH algorithm to answer a different kind of question: Given a case of interest, say gene 610 in the prostate study, which of the other $N - 1$ cases is unusually highly correlated with it? "Unusual" has the meaning here of being in the rejection set of a simultaneous testing procedure.

It is easier to describe the main idea in terms of one-sample rather than two-sample t-tests. Suppose that X is an $N \times n$ matrix with entries x_{ij}. For each row $\boldsymbol{x}_i$ of X we compute t_i, the one-sample t-statistic,

$$t_i = \frac{\bar{x}_i}{\hat{\sigma}_i / \sqrt{n}} \qquad \left[\bar{x}_i = \frac{\sum_1^n x_{ij}}{n}, \ \hat{\sigma}_i^2 = \frac{\sum_1^n \left(x_{ij} - \bar{x}_i\right)^2}{n - 1} \right]. \tag{4.56}$$

We wish to test which if any of the N t_i values are unusually large. (X might arise in a paired comparison microarray study where x_{ij} is the difference in expression levels, Treatment minus Placebo, for gene i in the jth pair of subjects.)

Let

$$\boldsymbol{u} = (1, 1, \dots, 1)' \big/ \sqrt{n} \tag{4.57}$$

be the unit vector lying along the direction of the main diagonal in n-dimensional space. The angle θ_i between $\boldsymbol{u}$ and

$$\boldsymbol{x}_i = (x_{i1}, x_{i2}, \dots, x_{in})' \tag{4.58}$$

has cosine

$$\cos(\theta_i) = \tilde{\boldsymbol{x}}_i' \boldsymbol{u} \tag{4.59}$$

where

$$\tilde{\boldsymbol{x}}_i = \boldsymbol{x}_i / \|\boldsymbol{x}_i\| = \boldsymbol{x}_i \Big/ \left(\textstyle\sum_{j=1}^n x_{ij}^2 \right)^{1/2} \tag{4.60}$$

is the scale multiple of $\boldsymbol{x}_i$ having unit length. A little bit of algebra shows that t_i is a monotonically decreasing function of θ_i,

$$t_i = \sqrt{n-1} \cos(\theta_i) \Big/ \left[1 - \cos(\theta_i)^2 \right]^{1/2}. \tag{4.61}$$

Exercise 4.11 Verify (4.61).

The unit sphere in n dimensions,

$$\mathcal{S}_n = \left\{ \boldsymbol{v} : \sum_{i=1}^{n} v_i^2 = 1 \right\} \tag{4.62}$$

can be shown to have *area* (i.e., $(n-1)$-dimensional Lebesgue measure)

$$A_n = 2\pi^{n/2} / \Gamma(n/2). \tag{4.63}$$

With $n = 3$ this gives the familiar result $A_3 = 4\pi$. Under the null hypothesis,

$$H_{0i} : x_{ij} \overset{\text{ind}}{\sim} \mathcal{N}(0, \sigma_0^2) \qquad j = 1, 2, \dots, n, \tag{4.64}$$

the vector $\boldsymbol{x}_i$ is known to have spherical symmetry around the origin $\mathbf{0}$; that is, $\tilde{\boldsymbol{x}}_i$ is randomly distributed over $\mathcal{S}_n$, with its probability of falling into any subset $\mathcal{R}$ on $\mathcal{S}_n$ being proportional to the $(n-1)$-dimensional area $A(\mathcal{R})$ of $\mathcal{R}$.

Putting this together, we can calculate[4] p_i, the one-sided p-value of the one-sample t-test for H_{0i}, in terms of θ_i:

$$p_i = A(\mathcal{R}_{\theta_i}) / A_N \equiv \tilde{A}(\theta_i). \tag{4.65}$$

Here $\mathcal{R}_\theta$ indicates a spherical cap of angle θ on $\mathcal{S}_n$ centered at $\boldsymbol{u}$, while $\tilde{A}(\theta_i)$ is the cap's area relative to the whole sphere. (Taking $\boldsymbol{u}$ as the north

[4] We are following Fisher's original derivation of the Student t-distribution.

pole on a globe of the Earth, $\mathcal{R}_\theta$ with $\theta = 23.5°$ is the region north of the Arctic circle.)

Small values of θ_i correspond to small p-values p_i. If $\theta(n,\alpha)$ defines a cap having relative area α, perhaps $\alpha = 0.05$, then the usual α-level t-test rejects H_{0i} for $\theta_i \leq \theta(n,\alpha)$. Intuitively, under the alternative hypothesis $x_{ij} \overset{\text{ind}}{\sim} \mathcal{N}(\mu_i, \sigma_0^2)$ for $j = 1, 2, \dots, n$, $\tilde{\boldsymbol{x}}_i$ will tend to fall nearer $\boldsymbol{u}$ if $\mu_i > 0$, rejecting H_{0i} with probability greater than α.

Exercise 4.12 Calculate $\theta(n, 0.05)$ for $n = 5, 10, 20$, and 40. *Hint*: Work backwards from (4.61), using a table of critical values for the t-test.

Getting back to the simultaneous inference problem, we observe N points $\tilde{\boldsymbol{x}}_1, \tilde{\boldsymbol{x}}_2, \dots, \tilde{\boldsymbol{x}}_N$ on $\mathcal{S}_n$ and wonder which of them, if any, lie unusually close to $\boldsymbol{u}$. We can rephrase the Benjamini–Hochberg procedure BH(q) to provide an answer. Define $\overline{\text{Fdr}}(\theta)$ to be $\overline{\text{Fdr}}(\mathcal{Z})$ in (2.21) with $\mathcal{Z} = \mathcal{R}_\theta$ and let $N_+(\theta)$ denote the number of points $\tilde{\boldsymbol{x}}_i$ in $\mathcal{R}_\theta$. Then

$$\overline{\text{Fdr}}(\theta) = N\pi_0\tilde{A}(\theta)\big/N_+(\theta) \tag{4.66}$$

as in (2.24).

Corollary 4.2 now takes this form: ordering the θ_i values from smallest to largest,

$$\theta_{(1)} \leq \theta_{(2)} \leq \cdots \leq \theta_{(i)} \leq \cdots \leq \theta_{(N)}, \tag{4.67}$$

let i_{max} be the largest index for which

$$\frac{\tilde{A}\left(\theta_{(i)}\right)}{N_+\left(\theta_{(i)}\right)/N} \leq q \tag{4.68}$$

and reject $H_{0(i)}$ for $i \leq i_{\text{max}}$. Assuming independence of the points, the expected proportion of null cases among the i_{max} rejectees will be less than q (actually equaling $\pi_0 q$). The empirical Bayes considerations of Section 4.3 suggest the same bound, even under dependence.

Let

$$\hat{\theta}(q) = \theta_{(i_{\text{max}})}, \quad R(q) = N_+\left(\hat{\theta}(q)\right), \quad \text{and} \quad \mathcal{R}(q) = \mathcal{R}\left(\hat{\theta}(q)\right). \tag{4.69}$$

The BH algorithm BH(q) rejects the $R(q)$ cases having $\theta_i \leq \hat{\theta}(q)$, that is, those having $\tilde{\boldsymbol{x}}_i$ within the spherical cap $\mathcal{R}(q)$, as illustrated in Figure 4.7.

Exercise 4.13 Show that $R(q)/N$, the observed proportion of points in $\mathcal{R}(q)$, is at least $1/q$ times the relative area $\tilde{A}(\hat{\theta}(q))$. (So if $q = 0.1$ there are at least ten times as many points in $\mathcal{R}(q)$ as there would be if all the null hypotheses were correct.)

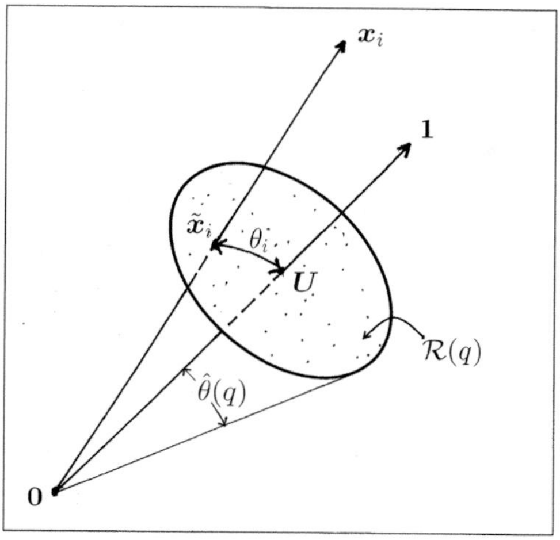

Figure 4.7 Spherical cap rejection region $\mathcal{R}(q)$ for BH procedure BH(q); H_{0i} is rejected since $\theta_i \leq \hat{\theta}(q)$. Dots indicate the other rejected cases. The number of points in $\mathcal{R}(q)$ is at least $1/q$ times larger than the expected number if all N null hypotheses were correct.

The same procedure can be used for the simultaneous testing of correlations. Suppose we are interested in a particular case i_0 and wonder which if any of the other $N - 1$ cases are unusually highly correlated with case i_0. Define $\boldsymbol{x}^\dagger_{ij}$ to be the centered version of $\boldsymbol{x}_i$,

$$\boldsymbol{x}^\dagger_{ij} = x_{ij} - \bar{x}_i \qquad j = 1, 2, \ldots, n, \tag{4.70}$$

and let $\boldsymbol{x}^\dagger_{i_0}$ play the role of $\mathbf{1}$ in Figure 4.7. Then

$$\begin{aligned} \cos(\theta_i) &= \boldsymbol{x}^{\dagger\prime}_i \boldsymbol{x}^\dagger_{i_0} \Big/ \left[\|\boldsymbol{x}^\dagger_i\| \cdot \|\boldsymbol{x}^\dagger_{i_0}\| \right] \\ &= \widehat{\text{cor}}(i, i_0), \end{aligned} \tag{4.71}$$

the Pearson sample correlation coefficient between $\boldsymbol{x}_i$ and $\boldsymbol{x}_{i_0}$.

Following through definitions (4.67)–(4.68) gives a BH(q) simultaneous test for the $N - 1$ null hypotheses

$$H_{0i} : \text{cor}(i, i_0) = 0, \qquad i \neq i_0. \tag{4.72}$$

Thinking of the vectors $\boldsymbol{x}_i^\dagger/\|\boldsymbol{x}_i^\dagger\|$ as points on the $(n-1)$-dimensional sphere

$$\mathcal{S}_{n-1}^\dagger = \left\{ \boldsymbol{v} : \sum_1^n v_i = 0, \ \sum_1^n v_i^2 = 1 \right\}, \tag{4.73}$$

the test amounts to checking for high-density clusters near $\boldsymbol{x}_{i_0}^\dagger/\|\boldsymbol{x}_{i_0}^\dagger\|$. Different choices of i_0 let us check for clusters all over $\mathcal{S}_{n-1}^\dagger$, i.e., for groups of correlated cases. (*Note*: The test can be carried out conveniently by first computing

$$t_i = \sqrt{\nu}\,\widehat{\text{cor}}(i, i_0) \Big/ \left[1 - \widehat{\text{cor}}(i, i_0)^2\right]^{1/2} \quad \text{and} \quad p_i = 1 - F_\nu(t_i) \tag{4.74}$$

with $\nu = n - 2$, the degrees of freedom, and F_ν the Student-t cdf, and then applying BH(q) to the p_i values. Using $p_i = F_\nu(t_i)$ checks for large *negative* correlations.)

Correlation testing was applied to the prostate data of Section 2.1. Definition (4.70) was now modified to subtract either the control or cancer patient mean for gene i, as appropriate, with $\nu = n - 3 = 99$ in (4.74). Gene 610 had the largest z-value (2.5) among the $N = 6033$ genes, $z_{610} = 5.29$, with estimated $\overline{\text{Fdr}}^{(c)}(z_{610}) = 0.0007$ (using (4.33) with $\pi_0 = 1$). Taking $i_0 = 610$ in tests (4.72) produced only gene 583 as highly correlated at level $q = 0.10$; taking $i_0 = 583$ gave genes 637 and 610, in that order, as highly correlated neighbors; $i_0 = 637$ gave 14 near neighbors, etc. Among the cases listed in Table 4.3 only genes 610 and 637 had $\overline{\text{Fdr}}^{(c)}(z_i) \le 0.50$. One might speculate that gene 637, which has low $\overline{\text{Fdr}}^{(c)}$ and a large number of highly correlated neighbors, is of special interest for prostate cancer involvement, even though its z-value 3.29 is not overwhelming, $\overline{\text{Fdr}}^{(c)}(z_{637}) = 0.105$.

Table 4.3 *Correlation clusters for prostate data using* BH(q) *with* $q = 0.10$. *Taking* $i_0 = 610$ *gave only gene 583 as highly correlated;* $i_0 = 583$ *gave genes 637 and 610, etc. Genes are listed in order of* $\widehat{\text{cor}}(i, i_0)$*, largest values first. Gene 637 with* $z_i = 3.29$ *and* $\overline{\text{Fdr}}^{(c)}(z_i) = 0.105$ *is the only listed gene besides 610 with* $\overline{\text{Fdr}}^{(c)} \le 0.50$.

610 $\longrightarrow$ **583** $\longrightarrow$ (**637***, 610*)

$\downarrow$

(583, **837**, 878, 5674, 1048, 1618, 1066, 610*, and 5 others)

$\downarrow$

(878, 637*, 1963, 376, 5674)

The two-sample t-test has almost the same "points on a sphere" description as the one-sample test: $\boldsymbol{x}_i$ is replaced by $\boldsymbol{x}_i^\dagger = (x_{ij} - \bar{x}_i)$ (4.70), $\mathcal{S}_n$ is replaced by $\mathcal{S}_{n-1}^\dagger$ (4.73), and $\mathbf{1}_n$, the vector of n 1's, is replaced by

$$\mathbf{1}^\dagger \equiv (-\mathbf{1}_{n_1}/n_1, \mathbf{1}_{n_2}/n_2). \tag{4.75}$$

Everything then proceeds as in (4.65) forward, as illustrated in Figure 4.7 (remembering that $\tilde{A}(\theta)$ now refers to the relative areas on an $(n-1)$-dimensional sphere). The same picture applies to more general regression z-values, as mentioned at the end of Section 3.1.

Notes

The true and false discovery terminology comes from Soric (1989) along with a suggestion of the evocative table in Figure 4.1. Benjamini and Hochberg credit Simes (1986) with an early version of the BH algorithm (4.9) and (3.29), but the landmark FDR control theorem (Theorem 4.1) is original to Benjamini and Hochberg (1995). The neat martingale proof of Theorem 4.1 comes from Storey et al. (2004), as does the result that $\widehat{\text{Fdr}}(z)$ (4.47) can be used to control FDR. Efron et al. (2001) presented an empirical Bayes interpretation of false discovery rates (emphasizing local fdr) while Storey (2002) developed a more explicitly Bayes approach. The positive regression dependence justification for the BH(q) algorithm appears in Benjamini and Yekutieli (2001). Lehmann and Romano (2005a) develop an algorithm that controls the *probability* that Fdp exceeds a certain threshold, rather than $E\{\text{Fdp}\}$. False negative rates are extensively investigated in Genovese and Wasserman (2002). Section 1 of Efron (1969) discusses the geometric interpretation of the one-sample t-test. Donoho and Jin (2009) apply Tukey's higher criticism to large-scale selection problems where genuine effects are expected to be very rare.

5

Local False Discovery Rates

Classic single-case hypothesis testing theory depended on the interpretation of tail-area statistics, that is, on p-values. The post-World War II boom in multiple testing continued to feature p-values, with their dominance extending into the large-scale testing era, as witnessed in Chapters 3 and 4. Even false discovery rate control, which strays far from significance testing and Type I error, is phrased in terms of p-values in its Benjamini–Hochberg formulation.

It is by necessity that tail areas are featured in single-case testing: one can sensibly interpret the probability that test statistic z exceeds 1.96, but $\text{Prob}\{z = 1.96\} = 0$ is uninformative. Large-scale testing, however, allows the possibility of *local* inference in which outcomes such as $z = 1.96$ are judged on their own terms and not with respect to the hypothetical possibility of more extreme results. This is the intent of local false discovery rates, a Bayesian idea implemented by empirical Bayes methods in large-scale testing situations.

5.1 Estimating the Local False Discovery Rate

As in Section 2.2, we begin with the Bayesian *two-groups* model, in which each of the N cases is either null or non-null, with prior probability π_0 or π_1,

$$\begin{aligned} \pi_0 &= \Pr\{\text{null}\} \qquad & f_0(z) &= \text{null density} \\ \pi_1 &= \Pr\{\text{non-null}\} \qquad & f_1(z) &= \text{non-null density}, \end{aligned} \tag{5.1}$$

as diagrammed in Figure 2.3. Assumption (2.9), that π_0 is near 1, is not necessary as long as $f_0(z)$ is known; (2.9) won't be our assumption here, but in Chapter 6, where f_0 must be estimated, we will need π_0 near 1.

The local false discovery rate (2.14) is

$$\text{fdr}(z) = \Pr\{\text{null}|z\} = \pi_0 f_0(z)/f(z) \tag{5.2}$$

where $f(z)$ is the *mixture density*

$$f(z) = \pi_0 f_0(z) + \pi_1 f_1(z). \tag{5.3}$$

Of the three quantities in (5.2), $f_0(z)$ is assumed known in this chapter while π_0 can be estimated as in Section 4.5 (or set equal to 1 with little harm if $\pi_0 \geq 0.90$). That leaves $f(z)$ to be estimated from the z-values $\boldsymbol{z} = (z_1, z_2, \ldots, z_N)$, all of which, by definition of the mixture density, come from $f(z)$.

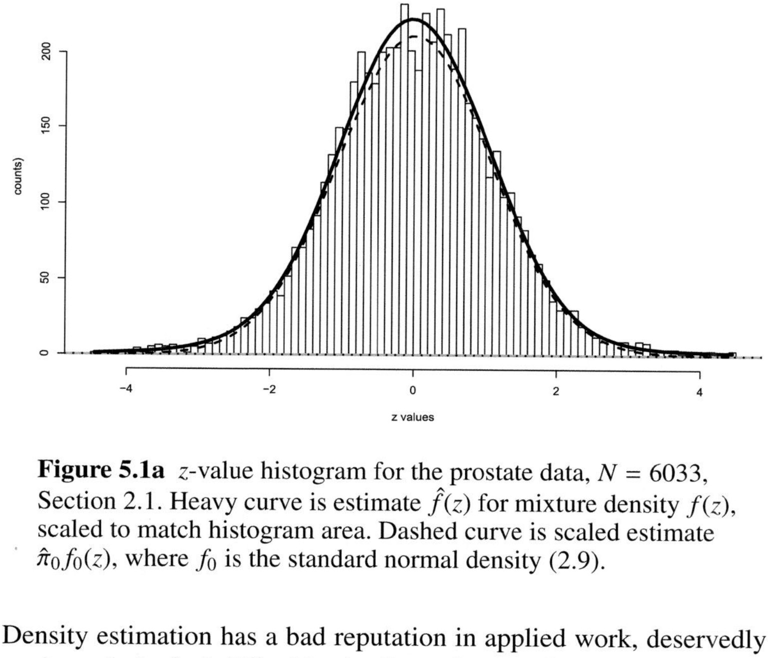

Figure 5.1a z-value histogram for the prostate data, $N = 6033$, Section 2.1. Heavy curve is estimate $\hat{f}(z)$ for mixture density $f(z)$, scaled to match histogram area. Dashed curve is scaled estimate $\hat{\pi}_0 f_0(z)$, where f_0 is the standard normal density (2.9).

Density estimation has a bad reputation in applied work, deservedly so given its pathological difficulties in discontinuous situations. Z-value densities, however, tend to be quite smooth, as discussed in Chapter 7. This can be seen in Figures 5.1a and 5.1b, which show z-value histograms for the prostate and DTI data sets. The heavy curves are estimates $\hat{f}(z)$ of $f(z)$ obtained as smooth fits to the heights of the histogram bars by a Poisson regression method described in Section 5.2.

The null probability π_0 in the numerator of (5.2) was estimated taking $\alpha_0 = 0.50$ in (4.48), that is, using the central 50% of the F_0-distribution;

$$\hat{\pi}_0 = 0.932 \text{ (prostate data)}, \qquad \hat{\pi}_0 = 0.935 \text{ (DTI)}. \tag{5.4}$$

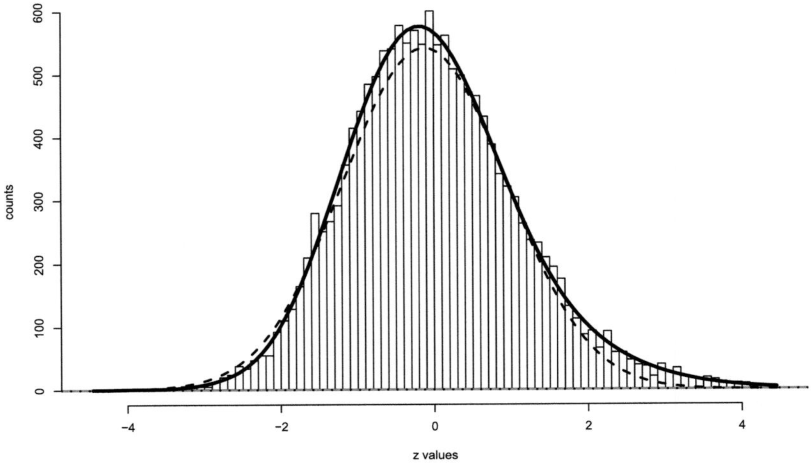

Figure 5.1b z-value histogram for the DTI data, $N = 15\,443$, Section 2.4; curves as in Figure 5.1a.

We can then estimate $\text{fdr}(z)$ by

$$\widehat{\text{fdr}}(z) = \hat{\pi}_0 f_0(z)/\hat{f}(z) \tag{5.5}$$

with $f_0(z)$ the $\mathcal{N}(0,1)$ density $\varphi(z)$ (2.9). The heavy curves in Figure 5.2 show $\widehat{\text{fdr}}(z)$. More precisely, they show $\min\{\widehat{\text{fdr}}(z), 1\}$. Equation (5.5) can exceed 1 because $\hat{\pi}_0$ isn't a perfect estimate of π_0 and, more seriously, because $\varphi(z)$ isn't a perfect estimate of f_0. The latter can be seen in Figure 5.1b where $f_0(z)$ is shifted noticeably to the right of the histogram high point. Chapter 6 deals with much worse mismatches.

The estimate $\hat{f}(z)$ can be integrated to give a smoothed cdf estimate $\hat{F}(z)$, and then a smoothed version of the tail area false discovery rate,

$$\widehat{\text{Fdr}}(z) = \hat{\pi}_0 F_0(z)/\hat{F}(z), \tag{5.6}$$

here with $F_0(z) = \Phi(z)$, the standard normal cdf, or likewise a right-sided estimate as in (4.33). These are also shown in Figure 5.2. $\widehat{\text{Fdr}}(z)$ tends to be a little less variable than the non-parametric version $\overline{\text{Fdr}}(z) = \hat{\pi}_0 F_0(z)/\bar{F}(z)$ (4.26), especially in the extreme tails.

A conventional threshold for reporting "interesting" cases is

$$\widehat{\text{fdr}}(z_i) \leq 0.20. \tag{5.7}$$

This was achieved for $z \leq -3.37$ and $z \geq 3.34$ for the prostate data and $z \geq 3.05$ for the DTI data. $\widehat{\text{Fdr}}(z)$ was about half of $\widehat{\text{fdr}}(z)$ at these points,

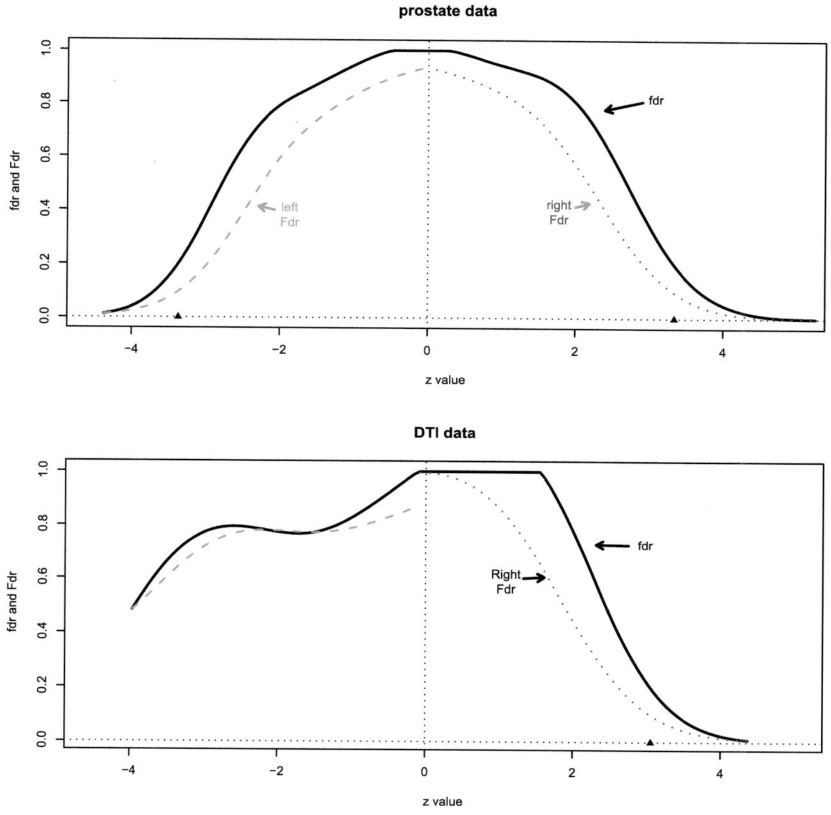

Figure 5.2 Heavy curves are estimated local false discovery rates $\widehat{\mathrm{fdr}}(z)$ (5.5); dashed and dotted curves show estimated tail-area false discovery rates $\widehat{\mathrm{Fdr}}(z)$ (5.6). The small triangles indicate values at which $\widehat{\mathrm{fdr}}(z) = 0.20$; 27 genes in each tail had $\widehat{\mathrm{fdr}}(z_i) \leq 0.20$ for the prostate data; 184 voxels, all in the right tail, had $\widehat{\mathrm{fdr}}(z_i) \leq 0.20$ for the DTI data.

roughly following (2.17)–(2.19) with $\gamma = 2$. Criterion (5.7) was effectively about the same as the $q = 0.1$ $\overline{\mathrm{Fdr}}$ threshold.

Exercise 5.1 Show that the criterion $\mathrm{fdr}(z) \leq 0.2$ is equivalent to

$$\frac{f_1(z)}{f_0(z)} \geq 4\,\frac{\pi_0}{\pi_1}. \tag{5.8}$$

Criterion (5.7) is based on the admittedly subjective grounds that reporting fdr values greater than 0.2 is dangerously prone to wasting investigators' resources. Like $\alpha = 0.05$ or $q = 0.1$, overuse of (5.7) is subject to Emerson's "foolish consistency" hobgoblin criticism, and perhaps no such criterion is necessary for interpreting a Bayesian posterior probability. My own experience nevertheless is that investigators often expect a definitive list of likely prospects, and (5.7) provides a reasonable way for the statistician to oblige.

If we assume $\pi_0 \geq 0.90$ as in (2.8), then (5.8) implies

$$\frac{f_1(z)}{f_0(z)} \geq 36. \tag{5.9}$$

The ratio $f_1(z)/f_0(z)$ is called the *Bayes factor* against the null hypothesis. Equation (5.9) requires a much stronger level of evidence against the null than in classical single-test practice: suppose we observe $z \sim \mathcal{N}(\mu, 1)$ and wish to test $H_0 : \mu = 0$ versus $\mu = 2.80$, a familiar scenario for power calculations since rejecting H_0 for $z \geq 1.96$ yields two-sided size 0.05 and power 0.80. Here the critical Bayes factor is only $f_{2.80}(1.96)/f_0(1.96) = 4.80$. We might rationalize (5.9) as necessary conservatism in guarding against multiple-testing fallacies. All of this relates to the question of power versus size, as discussed in Section 5.4.

5.2 Poisson Regression Estimates for $f(z)$

The smooth estimates $\hat{f}(z)$ for the mixture density $f(z)$, seen in Figure 5.1a, were obtained using maximum likelihood estimation (MLE) in flexible exponential family models.[1] As a first example, suppose f belongs to the J-parameter family

$$f(z) = \exp\left\{\sum_{j=0}^{J} \beta_j z^j\right\}. \tag{5.10}$$

The constant β_0 in (5.10) is determined from $\boldsymbol{\beta} = (\beta_1, \beta_2, \ldots, \beta_J)$ by the requirement that $f(z)$ integrate to 1 over the range of z. The choice $J = 2$ makes (5.10) into the normal family $\mathcal{N}(\mu, \sigma^2)$.

Exercise 5.2 Write the $\mathcal{N}(\mu, \sigma^2)$ family in the form (5.10) showing the expressions for β_0, β_1, and β_2.

[1] Appendix A gives a brief review of exponential family theory, including Lindsey's method as used in what follows.

If we wish to detect differences between $f(z)$ and the theoretical null density $f_0(z) = \varphi(z)$, we will certainly need J bigger than 2. As J goes to infinity, family (5.10) grows to approximate all densities on the line. This is the non-parametric ideal, but not an efficient choice when we expect $f(z)$ to be quite smooth. The program `locfdr` discussed below defaults to $J = 7$. Relying on family (5.10) with $J = 7$ puts us, very roughly, midway between traditional parametric and non-parametric estimation.

Maximum likelihood estimation in family (5.10) seems like it would require special software. This isn't the case. *Lindsey's method*, an algorithm based on discretizing the z_i values, obtains maximum likelihood estimates $\hat{\beta}$ using standard Poisson regression techniques. We partition the range $\mathcal{Z}$ of the z_i values into K bins of equal width d,

$$\mathcal{Z} = \bigcup_{k=1}^{K} \mathcal{Z}_k. \tag{5.11}$$

The histograms in Figure 5.1a used $K = 90$ bins of width $d = 0.1$ over range $\mathcal{Z} = [-4.5, 4.5]$. Define y_k as the count in the kth bin,

$$y_k = \#\{z_i \in \mathcal{Z}_k\} \tag{5.12}$$

and let

$$x_k = \text{ centerpoint of } \mathcal{Z}_k \tag{5.13}$$

so $x_1 = -4.45,\ x_2 = -4.35, \ldots, x_{90} = 4.45$ in Figure 5.1a. The expected value of y_k is approximately

$$\nu_k = Nd\, f(x_k) \tag{5.14}$$

where N is the total number of cases, respectively 6033 and 15443 in Figures 5.1a and 5.1b. Lindsey's method assumes that the y_k are independent Poisson counts

$$y_k \overset{\text{ind}}{\sim} \text{Poi}(\nu_k) \qquad k = 1, 2, \ldots, K, \tag{5.15}$$

and then fits $\hat{f}$ via a regression model for ν_k as a function of x_k. In model (5.10),

$$\log(\nu_k) = \sum_{j=0}^{J} \beta_j x_k^j; \tag{5.16}$$

(5.15)–(5.16) is a standard Poisson generalized linear model (GLM). Program `locfdr`[2] is written in the language `R` which permits a one-line call

[2] Appendix B describes `locfdr`.

for the MLE $\hat{\boldsymbol{\beta}} = (\hat{\beta}_1, \hat{\beta}_2, \ldots, \hat{\beta}_J)$. Except for the effects of discretization, which are usually negligible, $\hat{\boldsymbol{\beta}}$ is identical to the MLE we would obtain by direct maximization of $\prod_1^N f(z_i)$ in (5.10), assuming independence.

Lindsey's method has the nice effect of moving density estimation into the more familiar realm of regression: we are fitting a smooth function $f_{\boldsymbol{\beta}}(x_k)$ to the counts y_k. (Technically, this is done by minimizing the sum of Poisson deviances,

$$\sum_{k=1}^{K} D\left(y_k, N d f_{\boldsymbol{\beta}}(x_k)\right), \qquad D(y, \nu) = 2y\left[\log\left(\frac{y}{\nu}\right) - \left(1 - \frac{y}{\nu}\right)\right], \tag{5.17}$$

over the choice of $\boldsymbol{\beta}$.) Independence of the z_i or of the y_k in (5.15) is not required; $\hat{f}$ tends to be consistent and close to unbiased for f even under dependence, the penalty for dependence being increased variability, as described in Chapter 7. The fit can be examined by eye or more formally, and adjusted as necessary. Form (5.10) for $f(x)$ isn't required. By default, `locfdr` replaces the polynomial exponent with a natural spline basis. Doing so had very little effect on Figure 5.1a and Figure 5.2, nor did reducing J from 7 to 5.

Our interest in $\hat{f}(z)$ arises from its use in the denominator of $\widehat{\text{fdr}}(z) = \hat{\pi}_0 f_0(z)/\hat{f}(z)$. The examples in Figure 5.2 are typical: $\widehat{\text{fdr}}(z)$ is about 1 near $z = 0$ and either does or does not decline toward zero in the tails. The transition points from high to low $\widehat{\text{fdr}}$, denoted (following criterion (5.7)) by the small triangles in Figure 5.2, claim our attention. It is near such points where the accuracy of $\hat{f}(z)$ is most important.

It seems, intuitively, that fdr should be harder to estimate than Fdr, but that is not the case, at least not within the purview of models like (5.10). Figure 5.3 compares the standard deviation of $\log(\widehat{\text{fdr}}(z))$ with those of $\log(\widehat{\text{Fdr}}(z))$ and $\log(\overline{\text{Fdr}}(z))$. The situation involves $N = 6000$ independent observations from the two-groups model (2.7),

$$\begin{aligned} \pi_0 &= 0.95 \qquad & z &\sim \mathcal{N}(0, 1) \\ \pi_1 &= 0.05 \qquad & z &\sim \mathcal{N}(2.5, 1); \end{aligned} \tag{5.18}$$

$\widehat{\text{fdr}}(z)$, $\widehat{\text{Fdr}}(z)$, and $\overline{\text{Fdr}}(z)$ were estimated as in (5.5), (5.6), and Section 4.3, except that $\pi_0 = 0.95$ was assumed known in all three cases. Both $\widehat{\text{fdr}}$ and $\widehat{\text{Fdr}}$ used a natural spline basis with $J = 5$ degrees of freedom. Their standard deviations were computed from the theoretical formulas in Chapter 7, and checked with simulations.

In this situation, $\widehat{\text{fdr}}(z)$ is seen to be a little *less* variable than $\widehat{\text{Fdr}}(z)$, which in turn is less variable than the non-parametric version $\overline{\text{Fdr}}(z)$. All

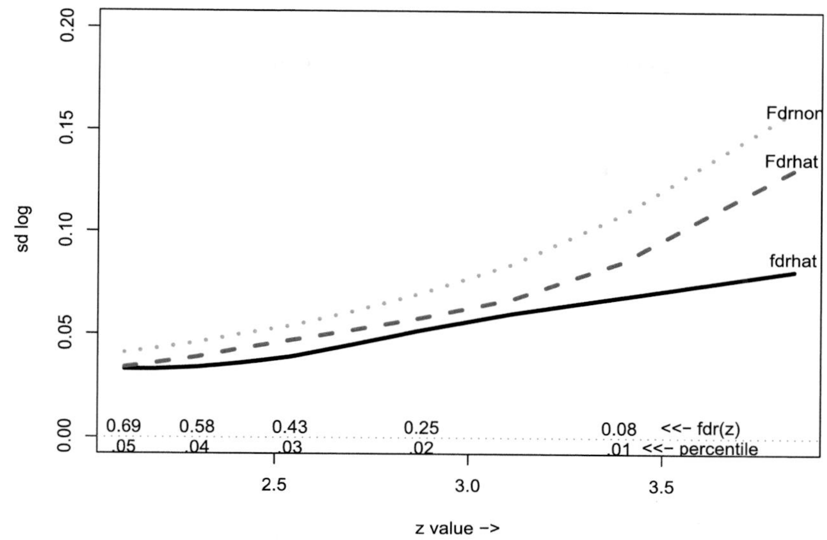

Figure 5.3 Comparison of standard deviations for $\log(\widehat{\mathrm{fdr}}(z))$, $\log(\widehat{\mathrm{Fdr}}(z))$, and $\log(\overline{\mathrm{Fdr}}(z))$ for $N = 6000$ independent observations from model (5.18). Small numbers show true values of fdr(z) at the indicated right percentiles of the mixture distribution. Standard deviations are based on the results in Chapter 7.

three estimates are satisfyingly accurate: at the 0.01 right percentile of the mixture distribution, $\mathrm{sd}\{\log \widehat{\mathrm{fdr}}(z)\} = 0.068$, implying that $\widehat{\mathrm{fdr}}(z)$ has coefficient of variation only about 7%. Correlation among the z_i values makes the standard deviations worse, as demonstrated in Chapter 7, but the overall comparison still shows the three estimates to be of roughly equal variability, with an edge to $\widehat{\mathrm{fdr}}(z)$ in the extreme tails.

Exercise 5.3 If X is a positive random variable with mean μ and variance σ^2, show that $\mathrm{CV} \equiv \sigma/\mu$, the *coefficient of variation* of X, approximately equals $\mathrm{sd}\{\log(X)\}$ when both are small.

5.3 Inference and Local False Discovery Rates

Switching attention from Fdr to fdr moves us still closer to Bayesian estimation theory: it is more appropriate from the Bayesian point of view to estimate $\mathrm{fdr}(z) = \Pr\{\mathrm{null}|z\}$ than the tail-area version $\Pr\{\mathrm{null}|Z \geq z\}$.

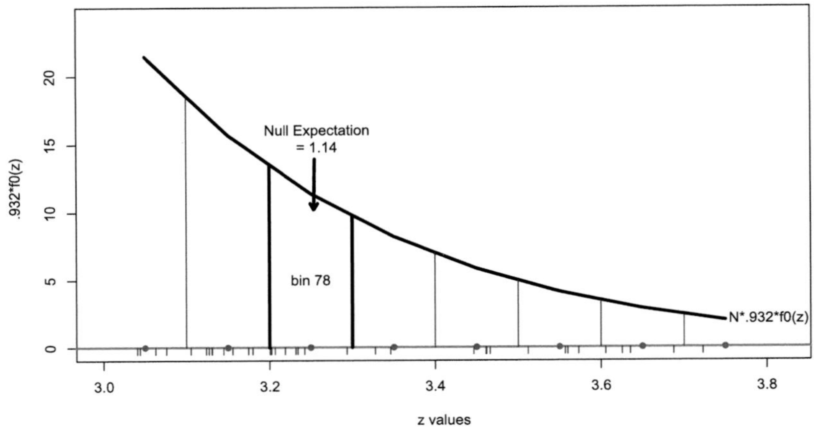

Figure 5.4 Right tail of the prostate data histogram, Figure 5.1a. Heavy curve is estimated $\widehat{\text{fdr}}$ numerator $N \cdot 0.932 \cdot f_0(z)$ (5.5). Hash marks indicate observed values z_i. Bin 78 contains 8 z_i values and has null expectation (area under the curve) of 1.14, giving raw estimate $\overline{\text{fdr}}_{78} = 1.14/8 = 0.143$. Smoothing the counts using $\hat{f}(z)$ gave $\widehat{\text{fdr}}_{78} = 0.254$.

Figure 5.4 uses the prostate data to illustrate the simple idea behind the estimate $\widehat{\text{fdr}}(z)$. Histogram bin $k = 78$, $z \in [3.2, 3.3)$, is observed to contain $y_k = 8$ z-values; the expected null count is 1.14, leading to the raw estimate

$$\overline{\text{fdr}}_{78} = 1.14/8 = 0.143. \tag{5.19}$$

However, the counts y_k are highly variable. Replacing them with smooth values $\hat{y}_k = N \cdot d \cdot \hat{f}(x_k)$, equaling 4.49 for $k = 78$, gave

$$\widehat{\text{fdr}}_{78} = 1.14/4.49 = 0.254. \tag{5.20}$$

Table 5.1 provides data for bins 76 through 83, emphasizing the importance of smoothing.

The use of local false discovery rates raises several issues concerning both technique and inference.

- *More general structure* The two-groups model (5.1) can be generalized to allow the cases $i = 1, 2, \ldots, N$ to behave differently:

$$\begin{aligned} \pi_{i0} &= \Pr\{\text{case } i \text{ null}\} & f_{i0}(z) &= \text{case } i \text{ null density} \\ \pi_{i1} &= \Pr\{\text{case } i \text{ non-null}\} & f_{i1}(z) &= \text{case } i \text{ non-null density.} \end{aligned} \tag{5.21}$$

Table 5.1 *Data for bins* $k = 76$ *through* 83, *prostate data, Figure 5.4;* e_{0k} *is null expected count (area under the curve);* y_k *is observed count,* $\hat{y}_k$ *smoothed version* $N \cdot d \cdot \hat{f}(x_k)$; $\overline{\text{fdr}}_k = e_{0k}/y_k$, $\widehat{\text{fdr}}_k = e_{0k}/\hat{y}_k$.

k	x_k	e_{0k}	y_k	$\overline{\text{fdr}}_k$	$\hat{y}_k$	$\widehat{\text{fdr}}_k$
76	3.05	2.14	4	.536	5.96	.359
77	3.15	1.57	9	.175	5.14	.306
78	**3.25**	**1.14**	**8**	**.143**	**4.49**	**.254**
79	3.35	.82	2	.410	3.97	.207
80	3.45	.58	4	.146	3.55	.165
81	3.55	.41	4	.103	3.20	.128
82	3.65	.29	4	.072	2.92	.098
83	3.75	.20	1	.198	2.69	.074

Define

$$\pi_0 = \frac{1}{N}\sum_{i=1}^{N} \pi_{i0} \qquad f_0(z) = \frac{1}{N}\sum_{i=1}^{N} \frac{\pi_{i0}}{\pi_0} f_{i0}(z)$$
$$\pi_1 = \frac{1}{N}\sum_{i=1}^{N} \pi_{i1} \qquad f_1(z) = \frac{1}{N}\sum_{i=1}^{N} \frac{\pi_{i1}}{\pi_1} f_{i1}(z). \tag{5.22}$$

Marginally (that is, averaging over $i = 1, 2, \ldots, N$) we are back in the two-groups situation, with $\pi_0, \pi_1, f_0(z), f_1(z)$, and $f(z) = \pi_0 f_0(z) + \pi_1 f_1(z)$ having the same meanings as in (5.1) and Figure 2.3, and with $\text{fdr}(z) = \pi_0 f_0(z)/f(z) = \Pr\{\text{null}|z\}$ (5.2).

Exercise 5.4 (i) Verify the statements above. (ii) Why is $\hat{f}(z)$, obtained as in Section 5.2, a valid estimate for $f(z)$ under model (5.21)? *Hint*: (5.14).

These results can be visualized in terms of Figure 5.4. Under model (5.21) the expected number of null cases falling into bin k is

$$e_0(k) \doteq d \sum_{i=1}^{N} \pi_{i0} f_{i0}(x_k) = N d\, \pi_0 f_0(x_k), \tag{5.23}$$

d = bin width, the same approximation we would use for model (5.1). The estimate $\widehat{\text{fdr}}(z) = \hat{\pi}_0 f_0(z)/\hat{f}(z)$ is obtained entirely from the empirical distribution of the N z_i values so model (5.21) gives the same results as (5.1), at least as long as we ignore distinctions among the cases. An assumption such as $f_0(z) \sim \mathcal{N}(0, 1)$ is still valid if it applies to the *average* null case, not necessarily to each of them. These same comments hold true for $\overline{\text{Fdr}}(z)$ and $\widehat{\text{Fdr}}(z)$.

• *Using prior knowledge* All of this assumes that model (5.21) is invisible to us and that we can only work marginally with the z_i values. If (5.21) were known, then

$$\text{fdr}_i(z_i) = \pi_{i0} f_{i0}(z_i)/f_i(z_i) = \Pr\{\text{case } i \text{ null}|z_i\} \tag{5.24}$$

would be more relevant than $\text{fdr}(z_i) = \pi_0 f_0(z_i)/f(z_i)$.

Exercise 5.5 Suppose that in model (5.21) $f_{i0}(z)$ and $f_{i1}(z)$ do not depend on i. Show that

$$\text{fdr}_i(z_i) = \text{fdr}(z_i)\frac{r_i}{1-(1-r_i)\,\text{fdr}(z_i)} \qquad \text{when } r_i = \frac{\pi_{i0}/(1-\pi_{i0})}{\pi_0/(1-\pi_0)}. \tag{5.25}$$

Gene 637 of the prostate data, Table 4.3, had $\widehat{\text{fdr}}(z_{637}) = 0.105$. If gene 637 was on a short list of "hot prospects" provided a priori by the investigator, we could assume $r_{637} < 1$ (i.e., that $\pi_{637,0}$ is less than the average π_{i0}), perhaps taking $r_{637} = 0.5$, in which case (5.25) would yield $\widehat{\text{fdr}}_{637}(z_{637}) = 0.055$. Chapter 10 concerns working with models like (5.21) in situations where covariate information is available to assist with the estimation of $\text{fdr}_i(z_i)$.

• *Exchangeability* The right-sided tail area Fdr for the prostate data has $\widehat{\text{Fdr}}(3.2) = 0.108$. We might report the 36 genes having $z_i \geq 3.2$ as interesting prospects for further prostate cancer investigation, since fewer than four of them are expected to be null. A tacit exchangeability assumption is at work here: each of the 36 is implied to have probability about 1/9 of being null. This ignores the fact that some of the 36 z_i's are much greater than others, ranging from 3.20 to 5.29.

The local fdr puts less strain on exchangeability. We can interpret $\widehat{\text{fdr}}_k$ in Table 5.1 as saying that the eight cases with z_i in $[3.2, 3.3)$ each have probability 0.25 of being null, the two cases in $[3.3, 3.4)$ each have probability 0.21, etc. Of course, exchangeability is lost if we have different prior knowledge about the cases, say, different values of r_i in (5.25). There we used Bayes theorem to appropriately adjust $\text{fdr}(z_i)$.

• *Scaling properties* Fdr and fdr methods scale in a nice way as the number N of cases changes. Suppose case 1 of the prostate data had $z_1 = 3.25$, giving $\widehat{\text{fdr}}(z_1) = 0.25$. What would $\widehat{\text{fdr}}(3.25)$ be if the investigator had measured twice as many genes? The answer is probably about the same, at least if we believe model (5.1): doubling N will give more accurate estimates of $f(z)$ and π_0 but major changes in $\hat{\pi}_0 f_0(z)/\hat{f}(z)$ are unlikely. Even letting N go to infinity shouldn't greatly change $\widehat{\text{fdr}}(z)$, as it converges to $\text{fdr}(z) = \pi_0 f_0(z)/f(z)$ in the usual asymptotic manner.

In contrast, inferences from the FWER methods of Chapter 3 depend crucially on the number of cases N. The right-sided p-value for $z_1 = 3.25$ is

$$p_1 = 1 - \Phi(3.25) = 0.00058 \tag{5.26}$$

under the theoretical null $\mathcal{N}(0, 1)$ distribution. Bonferroni's method (3.13) would declare case 1 non-null at significance level α if

$$p_1 \leq \alpha/N, \tag{5.27}$$

so doubling N lowers by half the threshold for rejection; p_1 would be *significant* if N were $< \alpha/p_1$ but not otherwise, the significance condition in (5.26) being $N \leq 86$ for $\alpha = 0.05$.

Exercise 5.6 Using the two-groups model, show that as $n \to \infty$, Holm's procedure (3.20) requires, asymptotically,

$$p_1 \leq \frac{\alpha}{N(1 - P_1)} \qquad \text{with } P_1 = 1 - F(z_1) \tag{5.28}$$

as a necessary but not sufficient condition for rejection. (For Hochberg's step-up procedure, Exercise 3.8, condition (5.28) is sufficient but not necessary.)

Let $z_{(1)}$ be the smallest (most negative) of the z_i values, with corresponding p-value $p_{(1)} = F_0(z_{(1)})$. Then $\overline{\text{Fdr}}(z_{(1)}) = \pi_0 F_0(z_{(1)})/\bar{F}(z_{(1)})$ equals $N\pi_0 p_{(1)}$. This will be less than the control level q if

$$p_{(1)} \leq \frac{1}{N}\frac{q}{\pi_0}, \tag{5.29}$$

which is *the same as the Bonferroni bound* with $\alpha = q/\pi_0$. In other words, the nice scaling properties of false discovery rate methods break down for extreme inferences. We will see this happening in the "snp" example of Chapter 11.

Another limitation concerns *relevance*. Model (5.1) taken literally implies an infinite reservoir of relevant cases. In practice though, if we doubled the number of genes in the prostate study, the new ones — not previously included on the microarray roster — might be less relevant to inferences about gene 1. Chapter 10 concerns the relevance question, which applies to all multiple inference procedures, not just false discovery rates.

• *Models with more structure* Exponential family models like (5.10) can be elaborated to handle situations with more structure than the two-groups model (5.11). Suppose the N cases are naturally partitioned into M classes, perhaps representing different regions of the brain in the DTI example. We could apply `locfdr` separately to each class, but this invites estimation problems in the smaller classes. A more efficient approach expands (5.10) to

$$\log\{f(z)\} = \sum_{j=0}^{J} \beta_j z^j + \gamma_{1m} z + \gamma_{2m} z^2, \tag{5.30}$$

m indicating the class, with $\sum_m \gamma_{1m} = \sum_m \gamma_{2m} = 0$, effectively allowing different means and variances for $\hat{f}$ in each group while retaining common tail behavior. Similar methods are discussed in Chapter 10.

• *Combining Fdr and fdr* It is not necessary to choose between $\widehat{\text{Fdr}}$ and $\widehat{\text{fdr}}$; they can be used in tandem. For example, $\widehat{\text{Fdr}}(3.20) = 0.108$ for the prostate data, right-sided, applies to all 36 z_i's exceeding 3.20 while $\widehat{\text{fdr}}(3.25) = 0.254$ applies to those in $[3.2, 3.3)$, etc., as suggested earlier, giving the investigator both a list of likely prospects and quantitative differentiation within the list. Program `locfdr` automatically puts $\widehat{\text{Fdr}}$ and $\widehat{\text{fdr}}$ in relationship (2.15),

$$\widehat{\text{Fdr}}(z) = \int_{-\infty}^{z} \widehat{\text{fdr}}(x)\hat{f}(x)\,dx \Big/ \int_{-\infty}^{z} \hat{f}(x)\,dx, \tag{5.31}$$

and similarly for the right-sided $\widehat{\text{Fdr}}$.

• *Bayes limitation* The empirical Bayes inference that $\widehat{\text{fdr}}(z_i)$ estimates $\Pr\{\text{case } i \text{ null}|z_i\}$ (5.2) comes with a caveat: it is not necessarily an estimate of

$$\Pr\{\text{case } i \text{ null}|z_1, z_2, \dots, z_N\}; \tag{5.32}$$

(5.32) can be different than (5.2) if the z-values are correlated. Chapter 9 on *enrichment* concerns methods for making inferences on groups of cases, rather than one at a time.

• *Expected false and true positives* Local false discovery rates are directly related to "EFP" and "ETP", the expected number of false positives and true positives of a decision procedure. Suppose that z_i follows model (5.21) for $i = 1, 2, \ldots, N$ and that we intend to reject the ith null hypothesis for $z_i \geq c_i$. The size and power of the ith test are

$$\alpha_i = \int_{c_i}^{\infty} f_{i0}(z_i)\, dz_i \quad \text{and} \quad \beta_i = \int_{c_i}^{\infty} f_{i1}(z_i)\, dz_i \tag{5.33}$$

giving

$$\text{EFP} = \sum_{i=1}^{N} w_i \pi_{i0} \alpha_i \quad \text{and} \quad \text{ETP} = \sum_{i=1}^{N} w_i \pi_{i1} \beta_i \tag{5.34}$$

where w_i is the prior probability of case i (which we can take to be $1/N$ without affecting the following calculations).

We wish to maximize ETP for a given value of EFP, by an optimum choice of the cutoff values $\boldsymbol{c} = (c_1, c_2, \ldots, c_N)$. Since

$$\frac{\partial \text{EFP}}{\partial c_i} = \sum_{i=1}^{N} w_i \pi_{i0} \frac{\partial \alpha_i}{\partial c_i} = -\sum_{i=1}^{N} w_i \pi_{i0} f_{i0}(c_i) \tag{5.35}$$

and similarly $\partial \text{ETP}/\partial c_i = -\sum w_i \pi_{i1} f_{i1}(c_i)$, a standard Lagrange multiplier argument shows that at the optimal value of $\boldsymbol{c}$,

$$\pi_{i1} f_{i1}(c_i) = \lambda \pi_{i0} f_{i0}(c_i) \qquad \text{for } i = 1, 2, \ldots, N \tag{5.36}$$

for some constant λ. In terms of (5.24) this is equivalent to

$$\text{fdr}_i(c_i) = 1/(1+\lambda). \tag{5.37}$$

That is, the decision rule that maximizes ETP for a given value of EFP *rejects each null hypothesis at the same threshold value of its local false discovery rate.* If all the cases follow the same two-groups model (5.1), then the optimum rule rejects for large values of $\text{fdr}(z_i)$ (5.2).

5.4 Power Diagnostics

The FWER techniques of Chapter 3 are aimed at the control of Type I errors, the rejection of null hypotheses that are actually correct. Discussion of false discovery rates, as the name suggests, similarly tends to focus on a form of Type I error control. However, Fdr theory also has something useful to say about *power*, the probability of rejecting null hypotheses that *should be* rejected. This section discusses some simple power diagnostics based on local false discovery rates.

Starting from the two-groups model (5.1), define the *local true discovery rate* $\text{tdr}(z)$ to be

$$\text{tdr}(z) = \Pr\{\text{non-null}|z\} = 1 - \text{fdr}(z) = \pi_1 f_1(z)/f(z), \tag{5.38}$$

the last equality following from (5.2)–(5.3). An fdr estimate $\widehat{\text{fdr}}(z) = \hat{\pi}_0 f_0(z) / \hat{f}(z)$ also gives a tdr estimate

$$\widehat{\text{tdr}}(z) = 1 - \widehat{\text{fdr}}(z) = \hat{\pi}_1 \hat{f}_1(z)/\hat{f}(z) \tag{5.39}$$

where

$$\hat{\pi}_1 = 1 - \hat{\pi}_0 \quad \text{and} \quad \hat{f}_1(z) = \frac{\hat{f}(z) - \hat{\pi}_0 f_0(z)}{1 - \hat{\pi}_0}. \tag{5.40}$$

Exercise 5.7 Verify (5.38) and (5.40).

Count y_k (5.12) includes all cases in bin k, null or non-null. We cannot, of course, separate y_k into nulls and non-nulls, but we can compute an approximation for the *expected* non-null count,

$$y_{1k} = \widehat{\text{tdr}}_k \cdot y_k \qquad \left[\widehat{\text{tdr}}_k = \widehat{\text{tdr}}(x_k)\right]. \tag{5.41}$$

The rationale for (5.41) is based on the fact that tdr_k is the proportion of non-null cases in bin k.

Exercise 5.8 Assuming the two-groups model (5.1) and independence among the z_i, what is the distribution of the number of nulls in bin k, *given* y_k? How does this support definition (5.41)?

The non-null counts y_{1k} can suffer from excess variability due to "histogram noise" in the observed counts y_k. An improved version is

$$\hat{y}_{1k} = Nd\,\widehat{\text{tdr}}_k \cdot \hat{f}_k \qquad \left[\hat{f}_k = \hat{f}(x_k)\right] \tag{5.42}$$

as in (5.14), called the *smoothed non-null counts* in what follows.

The vertical bars in Figure 5.5 are non-null counts y_{1k} for the prostate data using the histogram bins pictured in Figure 5.1a. The solid curve traces the smoothed non-null counts y_{1k}. There is one important difference here from the previous discussion. Rather than taking $f_0(z)$ to be the $\mathcal{N}(0,1)$ theoretical null (2.9), the estimate of $\text{fdr}(z) = \pi_0 f_0(z)/f(z)$ used

$$\hat{f}_0(z) \sim \mathcal{N}(0, 1.06^2); \tag{5.43}$$

(5.43) is the *empirical null* discussed in Chapter 6, the normal distribution that best fits the z-value histogram near its center. This change increased $\hat{\pi}_0$ from 0.932 in (5.4) to

$$\hat{\pi}_0 = 0.984. \tag{5.44}$$

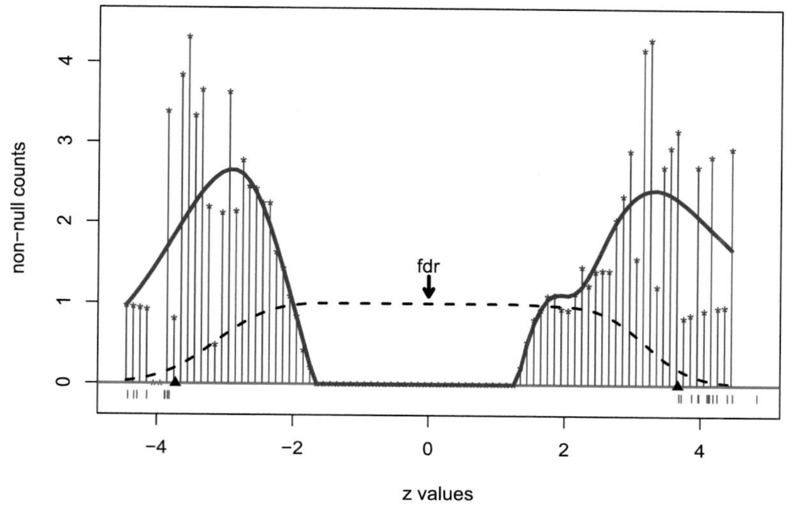

Figure 5.5 Vertical bars indicate non-null counts y_{1k} for the prostate data, taking $\hat{\pi}_0 = 0.984$ and $\hat{f}_0 \sim \mathcal{N}(0, 1.06^2)$; histogram bins as in (5.12)–(5.13). Heavy curve follows smoothed non-null counts (5.42); dashed curve is $\widehat{\text{fdr}}(z)$. Twenty-three of the z_i have $\widehat{\text{fdr}}(z_i) \leq 0.20$, indicated by small hash marks. Most of the non-null counts occur when $\widehat{\text{fdr}}(z)$ is high, indicating low power.

The difference between (5.43) and $f_0 \sim \mathcal{N}(0, 1)$ becomes crucial in power calculations, which in general are more delicate than questions relating to Type I error.

The estimate $\widehat{\text{fdr}}(z)$ (broken curve) is less optimistic than its counterpart in Figure 5.2. Now only 23 genes, those with z_i less than -3.73 or greater than 3.67, have $\widehat{\text{fdr}}(z_i) \leq 0.20$. The total smoothed non-null count $\sum \hat{y}_{1k}$ equals 105, which is a little greater than $(1 - \hat{\pi}_0) \cdot N = 97$ because some small negative values near $z = 0$ have been suppressed.

Exercise 5.9 Show that $\int_{-\infty}^{\infty} \text{tdr}(z) f(z)\, dz = \pi_1$. Why does this suggest $\sum_k y_{1k} \doteq (1 - \hat{\pi}_0) \cdot N$?

Of the 105 total smoothed non-null counts, only 26.8 occur in the regions where $\widehat{\text{fdr}}(z) \leq 0.20$, about 26%. The bulk occur where $\widehat{\text{fdr}}(z)$ is high, more than half with $\widehat{\text{fdr}}(z) > 0.50$. So if we try to report more than the list of 23 genes as interesting prospects, we soon expose the investigators to high probabilities of disappointment. In other words, the prostate study has low power.

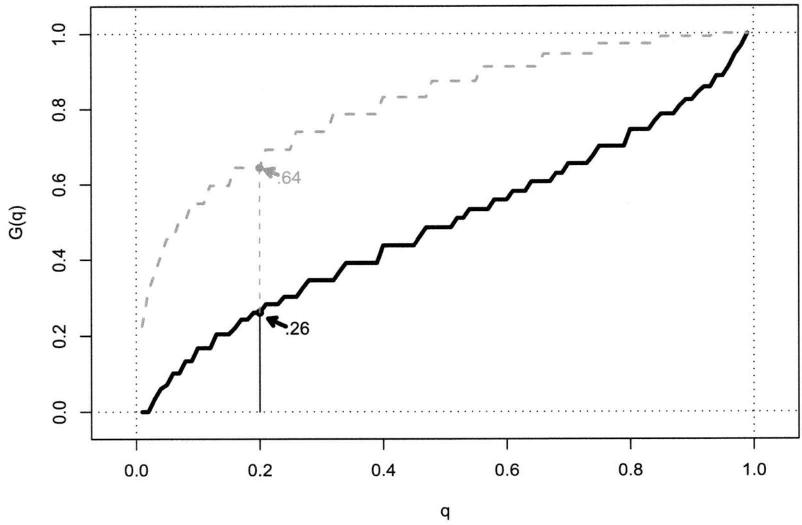

Figure 5.6 Heavy curve is smoothed non-null cdf $\hat{G}(q)$ for the estimated local false discovery rate; for example, $\hat{G}(0.2) = 0.26$ indicates that only 26% of the $\hat{y}_{1k}$ values occurred where $\widehat{\text{fdr}}_k \leq 0.20$. Dashed curve is $\hat{G}(q)$ for a simulated sample of $N = 6000$ independent z_i values, 95% from $\mathcal{N}(0, 1)$ and 5% from $\mathcal{N}(3, 1)$. Here 64% of the $\hat{y}_{1k}$ values had $\widehat{\text{fdr}}_k \leq 0.20$.

Figure 5.6 shows the entire non-null cdf of $\widehat{\text{fdr}}(z)$,

$$\Pr_1\left\{\widehat{\text{fdr}} \leq q\right\} = \sum_{k:\widehat{\text{fdr}}_k \leq q} \hat{y}_{1k} \Bigg/ \sum_k \hat{y}_{1k} \qquad [q \in (0, 1)]. \tag{5.45}$$

For example, $G(0.2) = 0.26$ and $G(0.5) = 0.49$. A simulated high-powered situation is shown for comparison, in which there are $N = 6000$ independent z_i's, $\pi_0 = 95\%$ having $z_i \sim \mathcal{N}(0, 1)$ and $\pi_1 = 5\%$ with $z_i \sim \mathcal{N}(3, 1)$. It had $\hat{G}(0.2) = 0.64$. In a typical realization of this situation, a majority of the non-null cases would be reported on the list of those with $\widehat{\text{fdr}}(z_i) \leq 0.20$, compared with about one-quarter for the prostate study.

A simple but useful diagnostic summary statistic[3] for power is $\widehat{E\,\text{fdr}}_1$,

[3] Program `locfdr` returns $\widehat{E\,\text{fdr}}_1$ and also separate versions of (5.46) applying to the left and right tails.

the expectation of $\widehat{\text{fdr}}_k$ under the $\hat{y}_{1k}$ distribution,

$$\widehat{E\,\text{fdr}}_1 = \sum_k \widehat{\text{fdr}}_k \hat{y}_{1k} \Big/ \sum_k \hat{y}_{1k}. \tag{5.46}$$

Low values of $\widehat{E\,\text{fdr}}_1$ indicate high power — that a typical non-null case will occur where the local false discovery rate is low — and high values indicate low power. For the prostate study, $\widehat{E\,\text{fdr}}_1 = 0.51$, indicating that a typical non-null case could expect to have $\widehat{\text{fdr}}$ about 0.5. For the simulated sample, $\widehat{E\,\text{fdr}}_1 = 0.28$.

Table 5.2 *Simulation study of* $\widehat{E\,\text{fdr}}_1$, *power diagnostic* (5.46). *Model* (5.1), $\pi_0 = 0.95$, $f_0 \sim \mathcal{N}(0,1)$, *and* $\pi_1 = 0.05$, $f_1 \sim \mathcal{N}(3,1)$; *number of cases* $N = 1500, 3000$, *or* 6000. *Boldface shows mean over 100 simulations, with standard deviations in parentheses. True* $E\,\text{fdr}_1 = 0.329$.

N	$\widehat{E\,\text{fdr}}_1$	$\hat{\pi}_0$
1500	**.266** (.046)	**.965** (.008)
3000	**.272** (.038)	**.963** (.007)
6000	**.281** (.028)	**.962** (.005)

Table 5.2 reports on a small simulation study of $\widehat{E\,\text{fdr}}_1$. As in Figure 5.6, each simulation had N independent z_i values, $\pi_0 = 0.95$ of them $\mathcal{N}(0,1)$ and $\pi_1 = 0.05$ from $\mathcal{N}(3,1)$, $N = 1500, 3000$, or 6000. The true value is $E\,\text{fdr}_1 = 0.329$. We see that $\widehat{E\,\text{fdr}}_1$ is quite stable but biased downward. Most of the bias can be traced to the upward bias of $\hat{\pi}_0$, which reduces $\widehat{\text{tdr}}(z)$ (5.39). Since

$$\widehat{E\,\text{fdr}}_1 = \sum_k \widehat{\text{tdr}}_k \cdot \widehat{\text{fdr}}_k \cdot \hat{f}_k \Big/ \sum_k \widehat{\text{fdr}}_k \cdot \hat{f}_k, \tag{5.47}$$

this leads to underestimates of $\widehat{E\,\text{fdr}}_1$. Tracing things back further, the upward bias of $\hat{\pi}_0$ arises from taking literally the zero assumption (4.44).

Exercise 5.10 Numerically verify the "true value" $E\,\text{fdr}_1 = 0.329$.

Investigators often object that their pre-experimental favorites have not appeared on the reported $\widehat{\text{fdr}} \le 0.20$ (or $\overline{\text{Fdr}} \le 0.10$) list. Low power is a likely culprit. If the prostate study were rerun from scratch, another list of 23 or so genes might be reported, barely overlapping with the original list.

Z-values, even ones from genuinely non-null cases, are quite variable (see Section 7.4), and in low-power situations a case has to be lucky as well as good to be reported. The Bayesian formula (5.25) may help salvage some of the investigators' hot prospects.

Finally, it should be noted that all of these power diagnostics are computed from the data at hand, the original set of z-values, and do not require prior knowledge of the situation. This is one of the advantages of large-scale studies.

Notes

Local false discovery rates were introduced as a natural extension of tail area Fdrs in Efron et al. (2001). Techniques both more and less parametric than Lindsey's method have been proposed for estimating $f(z)$ and $\text{fdr}(z)$, all of which seem to perform reasonably well: Allison et al. (2002), Pan et al. (2003), Pounds and Morris (2003), Heller and Qing (2003), Broberg (2004), Aubert et al. (2004) and Liao et al. (2004). Recent methods, which also encompass Chapter 6's problem of estimating the null density, include Jin and Cai (2007) and Muralidharan (2010), the former using Fourier series and the latter based on normal mixture models.

Estimation of the mixture density $f(z)$ was a problem of considerable interest in the period following Robbins' path-breaking empirical Bayes papers. That literature was more mathematics-oriented than our current computation-intensive efforts. A nice summary and example is found in Singh (1979), where the asymptotics of kernel estimates for $f(z)$ are carefully elaborated.

Power diagnostics are discussed in Section 3 of Efron (2007b). A familiar power-related question is also considered there: how much increased power could we obtain by increasing the number n of subjects? Answers to such a question are necessarily speculative, but simple techniques provide at least a rough guide. For the prostate data they suggest that tripling n would halve $E\,\text{fdr}_1$.

It isn't necessary that the individual summary statistics z_i be scalars. Tibshirani and Efron (2002) analyze an example in which each z_i is two-dimensional. The null density f_0 is more difficult to define and estimate in this case, but the fdr definition (5.2) still makes sense.

Storey (2007) suggests maximizing ETP as a function of EFP. His "optimum discovery procedure" proposes a different rule than (5.37), applicable to a restricted class of testing procedures.

6

Theoretical, Permutation, and Empirical Null Distributions

In classical significance testing, the null distribution plays the role of devil's advocate: a standard that the observed data must exceed in order to convince the scientific world that something interesting has occurred. We observe, say, $z = 2$, and note that in a hypothetical "long run" of observations from a $\mathcal{N}(0, 1)$ distribution less than 2.5% of the draws would exceed 2, thereby discrediting the uninteresting null distribution as an explanation.

Considerable effort has been expended trying to maintain the classical model in large-scale testing situations, as seen in Chapter 3, but there are important differences that affect the role of the null distribution when the number of cases N is large:

- With $N = 10\,000$ for example, the statistician has his or her own "long run" in hand. This diminishes the importance of theoretical null calculations based on mathematical models. In particular, it may become clear that the classical null distribution appropriate for a single-test application is in fact wrong for the current situation.
- Scientific applications of single-test theory most often suppose, or hope for, *rejection* of the null hypothesis, perhaps with power = 0.80. Large-scale studies are usually carried out with the expectation that most of the N cases will *accept* the null hypothesis, leaving only a small number of interesting prospects for more intensive investigation.
- Sharp null hypotheses, such as $H_0 : \mu = 0$ for $z \sim \mathcal{N}(\mu, 1)$, are less important in large-scale studies. It may become clear that most of the N cases have small, uninteresting but non-zero values of μ, leaving just a few genuinely interesting cases to identify. As we will discuss, this results in a broadening of classical null hypotheses.
- Large-scale studies allow empirical Bayes analyses, where the null distribution is put into a probabilistic context with its non-null competitors (as seen in Chapters 4 and 5).
- The line between estimation and testing is blurred in large-scale studies.

Large N isn't infinity and empirical Bayes isn't Bayes, so estimation efficiency of the sort illustrated in Figure 5.3 plays a major role in large-scale testing.

The theoretical null distribution provides a reasonably good fit for the prostate and DTI examples of Figure 5.1a and Figure 5.1b. This is a less-than-usual occurrence in my experience. A set of four large-scale studies are presented next in which the theoretical null has obvious troubles.[1] We will use them as trial cases for the more flexible methodology discussed in this chapter.

6.1 Four Examples

Figure 6.1 displays z-values for four large-scale testing studies, in each of which the theoretical null distribution is incompatible with the observed data. (A fifth, artificial, example ends the section.) Each panel displays the following information:

- the number of cases N and the histogram of the N z-values;
- the estimate $\hat{\pi}_{00}$ of the null proportion π_0 (2.7) obtained as in (4.46), (4.48) with $\alpha_0 = 0.5$, using the theoretical null density $f_0(z) \sim \mathcal{N}(0, 1)$;
- estimates $(\hat{\delta}_0, \hat{\sigma}_0, \hat{\pi}_0)$ for π_0 and the empirical null density $\hat{f}_0(z) \sim \mathcal{N}(\hat{\delta}_0, \hat{\pi}_0)$ obtained by the MLE method of Section 6.3, providing a normal curve fit to the central histogram;
- a heavy solid curve showing the empirical null density, scaled to have area $\hat{\pi}_0$ times that of the histogram (i.e., $Nd\hat{\pi}_0 \cdot \hat{f}_0(z)$, where d is the bin width);
- a light dotted curve proportional to $\hat{\pi}_{00} \cdot \varphi(z)$ ($\varphi(z) = \exp\{-\frac{1}{2}z^2\}/\sqrt{2\pi}$);
- small triangles on the x-axis indicating values at which the local false discovery rate $\widehat{\text{fdr}}(z) = \hat{\pi}_0 \hat{f}_0(z)/\hat{f}(z)$ based on the empirical null equals 0.2 (with $\hat{f}(z)$ as described in Section 5.2 using a natural spline basis with $J = 7$ degrees of freedom);
- small hash marks below the x-axis indicating z-values with $\widehat{\text{fdr}}(z_i) \le 0.2$;
- the number of cases for which $\widehat{\text{fdr}}(z_i) \le 0.2$.

What follows are descriptions of the four studies.

[1] Questions about the proper choice of a null distribution are not restricted to large-scale studies. They arise prominently in analysis of variance applications, for instance in whether to use the interaction or residual sums of squares for testing main effects in a two-way replicated layout.

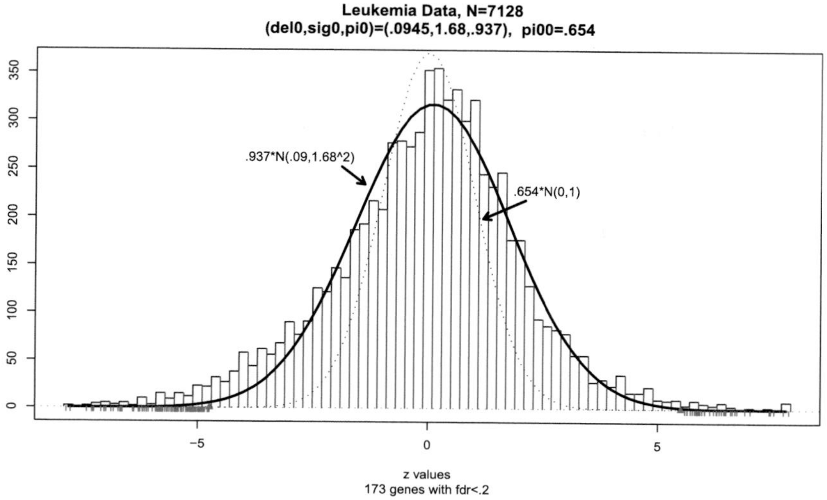

Figure 6.1a z-value histogram for leukemia study. Solid curve is empirical null, dotted curve theoretical $\mathcal{N}(0, 1)$ null. Hash marks indicate z-values having $\widehat{\text{fdr}} \le 0.2$. The theoretical null greatly underestimates the width of the central histogram.

A. Leukemia study

High-density oligonucleotide microarrays provided expression levels on $N = 7128$ genes for $n = 72$ patients, $n_1 = 45$ with ALL (acute lymphoblastic leukemia) and $n_2 = 27$ with AML (acute myeloid leukemia); the latter has the worse prognosis. The raw expression levels on each microarray, X_{ij} for gene i on array j, were transformed to a *normal scores* value

$$x_{ij} = \Phi^{-1}\left(\frac{\text{rank}(X_{ij}) - 0.5}{N}\right), \tag{6.1}$$

$\text{rank}(X_{ij})$ being the rank of X_{ij} among the N raw scores on array j, and Φ the standard normal cdf. This was necessary to eliminate response disparities among the n microarrays as well as some wild outlying values.[2] Z-values z_i were then obtained from two-sample t-tests comparing AML with ALL patients as in (2.1)–(2.5), now with 70 degrees of freedom.

We see that the z-value histogram is highly overdispersed compared to a $\mathcal{N}(0, 1)$ theoretical null. The empirical null is $\mathcal{N}(0.09, 1.68^2)$ with $\hat{\pi}_0 = 0.937$; 173 of the 7128 genes had $\widehat{\text{fdr}}(z) \le 0.20$. If we insist on using the

[2] Some form of standardization is almost always necessary in microarray studies.

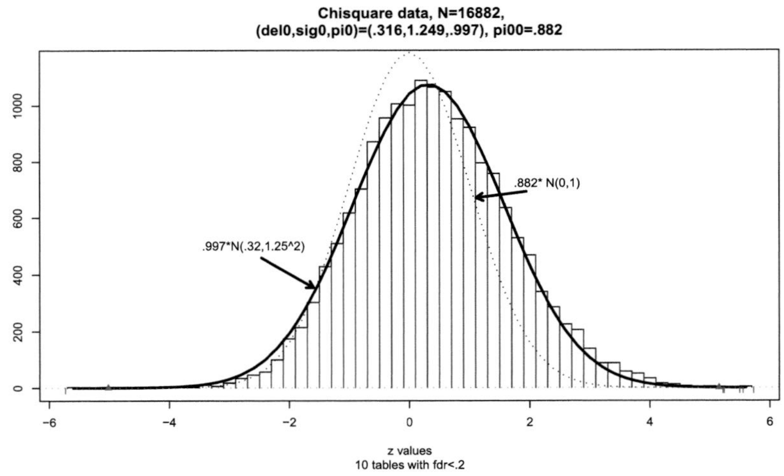

Figure 6.1b z-value histogram for chi-square study as in Figure 6.1a. Here the theoretical null is mis-centered as well as too narrow.

theoretical null, $\hat{\pi}_{00}$ is estimated to be only 0.654, while 1548 of the genes now have $\widehat{\text{fdr}}(z_i) \leq 0.20$. Perhaps it is possible that 2464 ($= (1 - \hat{\pi}_{00}) \cdot N$) of the genes display AML/ALL genetic differences, but it seems more likely that there is something inappropriate about the theoretical null. Just what might go wrong is the subject of Section 6.4.

B. Chi-square data

This experiment studied the binding of certain chemical tags at sites within $N = 16\,882$ genes. The number K of sites per gene ranged from three up to several hundred, median $K = 12$. At each site within each gene the number of bound tags was counted. The count was performed under two different experimental conditions, with the goal of the study being to identify genes where the proportion of tags differed between the two conditions. Table 6.1 shows the $K \times 2$ table of counts for the first of the genes, in which $K = 8$.

A z-value z_i was calculated for table$_i$ as follows:

(i) One count was added to each entry of table$_i$.

(ii) S_i, the usual chi-square test statistic for independence, was computed for the augmented table.

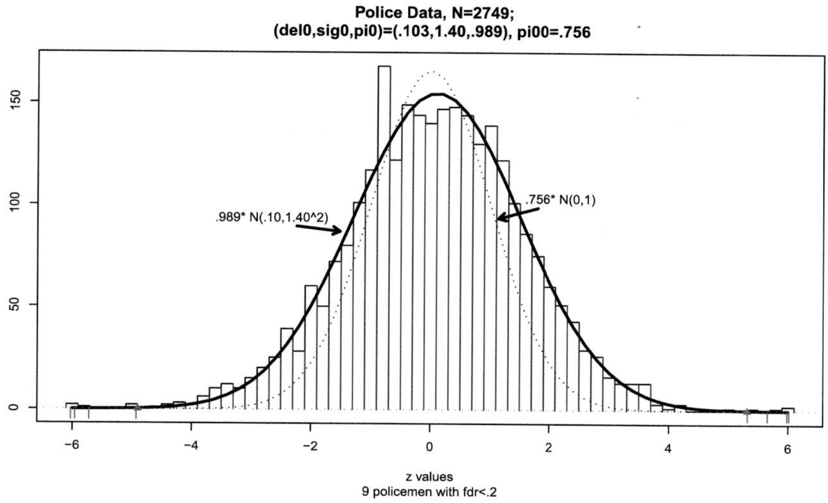

Figure 6.1c z-value histogram for police data as in Figure 6.1a. The theoretical null is about 40% too narrow.

Table 6.1 *First of* $N = 16\,882$ $K \times 2$ *tables for the chi-square data; shown are the number of tags counted at each site under two different experimental conditions.*

Site	1	2	3	4	5	6	7	8
# condition 1	8	8	4	2	1	5	27	9
# condition 2	5	7	1	0	11	4	4	10

(iii) An approximate p-value was calculated,

$$p_i = 1 - F_{K-1}(S_i) \tag{6.2}$$

where F_{K-1} was the cdf of a standard chi-square distribution having $K-1$ degrees of freedom.

(iv) The assigned z-value for table$_i$ was

$$z_i = \Phi^{-1}(1 - p_i) \tag{6.3}$$

with small values of p_i corresponding to large z_i. For table$_1$ in Table 6.1, $p_1 = 0.00132$ and $z_1 = 3.01$.

Exercise 6.1 Verify this calculation.

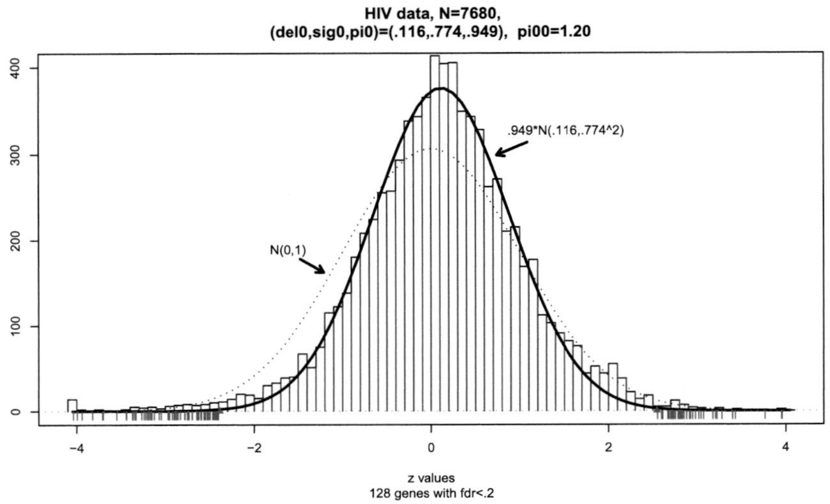

Figure 6.1d z-value histogram for HIV data as in Figure 6.1a. Note reduced scale on x-axis. In this case the theoretical null is too wide.

The ad hoc addition of one count in step (i) greatly stabilized the histogram of z-values.[3] Our methods do not require classical forms for the test statistics such as the standard chi-square definition, but they do depend on being able to approximate the center of the z-value histogram with a normal curve. This leads to questions of comparability and relevance in simultaneous inference, discussed further in Chapter 10.

The empirical null is $\mathcal{N}(0.32, 1.25^2)$ so that, besides being underdispersed, the theoretical null is mis-centered, as apparent in Figure 6.1b. Only ten of the tables had $\widehat{\text{fdr}}(z_i) \leq 0.2$, seven on the right and three on the left. If anything, we would expect the chi-square statistic S_i to be too big instead of too small, but that is not the case here. The three tables on the left with $\widehat{\text{fdr}} \leq 0.2$ all had $K = 3$, the smallest value, raising further doubts about their selection. Chapter 10 returns to this example in its discussion of comparability.

[3] The SAM program of Section 4.5 employs a similar tactic in two-sample comparisons: a small constant is added to the denominator of the usual two-sample t-statistic in order to prevent low-variance cases from dominating the tails of the z-value distribution.

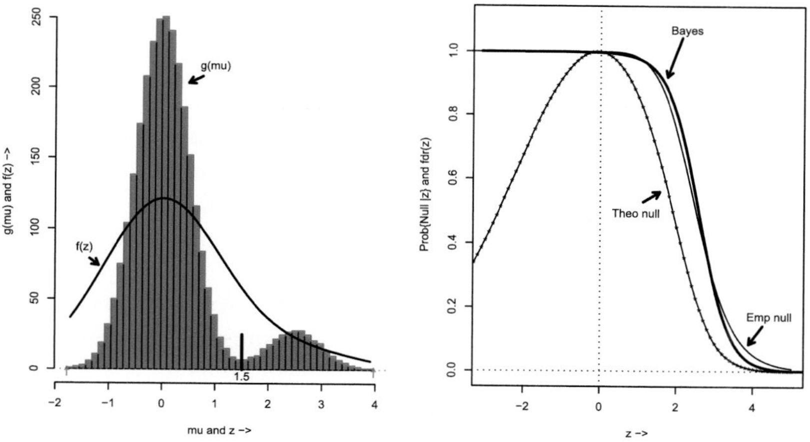

Figure 6.1e *Left panel*: Histogram shows $N = 3000$ draws from prior density $g(\mu)$ (6.6); curve $f(z)$ is corresponding density of observations $z \sim \mathcal{N}(\mu, 1)$. *Right panel*: Heavy curve is Bayes posterior probability $\Pr\{\mu \leq 1.5|z\}$; it is well approximated by the empirical null estimate $\widehat{\text{fdr}}(z)$ (6.9), light curve; $\text{fdr}(z)$ based on the theoretical null $f_0 \sim \mathcal{N}(0, 1)$ is a poor match to the Bayes curve.

C. Police data

A study of possible racial bias in police pedestrian stops was conducted in New York City in 2006. Each of $N = 2749$ officers was assigned a score z_i on the basis of their stop data, with large positive values of z_i being possible evidence of bias. In computing z_i, an ingenious two-stage logistic regression analysis was used to compensate for differences in the time, place, and context of the individual stops.

Let $\boldsymbol{x}_{ij}$ represent the vector of covariates for officer i, stop j. A greatly oversimplified version of the logistic regression model actually used is

$$\text{logit}\left\{\Pr(y_{ij} = 1)\right\} = \beta_i + \boldsymbol{\gamma}'\boldsymbol{x}_{ij} \tag{6.4}$$

where y_{ij} indicates whether or not the stopped person was a member of a defined minority group, β_i is the "officer effect," and $\boldsymbol{\gamma}$ is the vector of logistic regression coefficient for the covariates. The z-score for officer i was

$$z_i = \hat{\beta}_i \Big/ \text{se}\left(\hat{\beta}_i\right) \tag{6.5}$$

where $\hat{\beta}_i$ and $\text{se}(\hat{\beta}_i)$ are the usual estimate and approximate standard error for β_i.

A standard analysis would rely on the theoretical null hypothesis $z_i \dot{\sim} \mathcal{N}(0,1)$. However, the histogram of all $N = 2749$ z_i values is much wider, the empirical null being $\hat{f}_0 \sim \mathcal{N}(0.10, 1.40^2)$. This resulted in nine officers having $\widehat{\text{fdr}}(z_i) \leq 0.20$, only four of whom were on the right (i.e., "racially biased") side. The estimated non-null proportion was $\hat{\pi}_1 = 0.011$, only about 1/5 of which applied to the right side according to the smoothed non-null counts (5.41).

There is a lot at stake here. Relying on the theoretical $\mathcal{N}(0,1)$ null gives $\hat{\pi}_1 = 0.24$, more than 20 times greater and yielding 122 officers having positive z-scores with $\widehat{\text{fdr}}(z_i) \leq 0.20$. The argument for empirical null analysis says we should judge the extreme z-scores by comparison with central variability of the histogram and not according to a theoretical standard. Section 6.4 provides practical arguments for doubting the theoretical standard.

D. HIV data

This was a small study in which $n_2 = 4$ HIV-positive subjects were compared with $n_1 = 4$ healthy controls using cDNA microarrays that measured expression levels for $N = 7680$ genes. Two-sample t-tests (on the logged expressions) yielded z-values z_i as in (2.2)–(2.5) except now with 6 rather than 100 degrees of freedom.

Unlike all of our previous examples (including the prostate and DTI studies), here the central histogram is *less* dispersed than a theoretical $\mathcal{N}(0,1)$ null, with $\hat{f}_0(z) \sim \mathcal{N}(0.12, 0.77^2)$ and $\hat{\pi}_0 = 0.949$. (Underdispersion makes the theoretical null estimate equal the impossible value $\hat{\pi}_{00} = 1.20$.) Using the theoretical null rather than the empirical null reduces the number of genes having $\widehat{\text{fdr}} \leq 0.2$ from 128 to 20.

Figure 6.1e concerns an artificial example involving an overdispersed null distribution, similar to the leukemia, chi-square, and police situations. Pairs (μ_i, z_i) have been independently generated according to the Bayesian hierarchical model (2.47), $\mu \sim g(\cdot)$ and $z|\mu \sim \mathcal{N}(\mu, 1)$; the prior density $g(\mu)$, represented as a histogram in the left panel, is bimodal,

$$g(\mu) = 0.9 \cdot \varphi_{0,0.5}(\mu) + 0.1 \cdot \varphi_{2.5,0.5}(\mu) \tag{6.6}$$

where $\varphi_{a,b}$ represents the density of a $\mathcal{N}(a, b^2)$ distribution. However, the mixture density $f(z) = \int g(\mu)\varphi(z-\mu)d\mu$ is unimodal, reflecting the secondary mode of $g(\mu)$ only by a heavy right tail.

A large majority of the true effects μ generated from (6.6) will have small uninteresting values centered at, but not exactly equaling, zero. The interesting cases, those with large μ, will be centered around 2.5. Having observed z, we would like to predict whether the unobserved μ is interesting or uninteresting. There is no sharp null hypothesis here, but the shape of $g(\mu)$ suggests defining

$$\textit{Uninteresting}: \mu \leq 1.5 \quad \text{and} \quad \textit{Interesting}: \mu > 1.5. \tag{6.7}$$

The heavy curve in the right panel is the Bayes prediction rule for "uninteresting,"

$$\Pr\{\mu \leq 1.5|z\}. \tag{6.8}$$

This assumes that prior density $g(\mu)$ is known to the statistician. But what if it isn't? The curve marked "Emp null" is the estimated local false discovery rate

$$\widehat{\text{fdr}}(z) = \hat{\pi}_0 \hat{f}_0(z)/\hat{f}(z) \tag{6.9}$$

obtained using the central matching method of Section 6.2 which gave empirical null estimates

$$\hat{\pi}_0 = 0.93 \quad \text{and} \quad \hat{f}_0(z) \sim \mathcal{N}(0.02, 1.14^2) \tag{6.10}$$

based on $N = 3000$ observations z_i. We see that it nicely matches the Bayes prediction rule, even though it did not require knowledge of $g(\mu)$ or of the cutoff point 1.5 in (6.7).

The point here is that empirical null false discovery rate methods can deal with "blurry" null hypotheses, in which the uninteresting cases are allowed to deviate somewhat from a sharp theoretical null formulation. (Section 6.4 lists several reasons this can happen.) The empirical null calculation absorbs the blur into the definition of "null". Theoretical or permutation nulls fail in such situations, as shown by the beaded curve in the right panel.

6.2 Empirical Null Estimation

A null distribution is not something one estimates in classic hypothesis testing theory:[4] theoretical calculations provide the null, which the statistician must use for better or worse. Large-scale studies such as the four examples in Section 6.1 can cast severe doubt on the adequacy of the theoretical null. Empirical nulls, illustrated in the figures, use each study's own data to estimate an appropriate null distribution. This sounds circular, but isn't. We

[4] An exception arising in permutation test calculations is discussed in Section 6.5.

will see that a key assumption for empirically estimating the null is that π_0, the proportion of null cases in (2.7), is large, say

$$\pi_0 \geq 0.90 \tag{6.11}$$

as in (2.8), allowing the null distribution opportunity to show itself.

The appropriate choice of null distribution is not a matter of local fdr versus tail area Fdr: both are equally affected by an incorrect choice. Nor is it a matter of parametric versus non-parametric procedures. Replacing t-statistics with Wilcoxon test statistics (each scaled to have mean 0 and variance 1 under the usual null assumptions) gives empirical null $\hat{f}_0(z) \dot{\sim} \mathcal{N}(0.12, 1.72^2)$ for the leukemia data, almost the same as in Figure 6.1a.

The two-groups model (2.7) is unidentifiable: a portion of $f_1(z)$ can be redefined as belonging to $f_0(z)$ with a corresponding increase in π_0. The *zero assumption* (4.44), that the non-null density $f_1(z)$ is zero near $z = 0$, restores identifiability and allows estimation of $f_0(z)$ and π_0 from the central histogram counts.

Exercise 6.2 Suppose $\pi_0 = 0.95$, $f_0 \sim \mathcal{N}(0, 1)$, and f_1 is an equal mixture of $\mathcal{N}(2.5, 1)$ and $\mathcal{N}(-2.5, 1)$. If we redefine the situation to make (4.44) true in $\mathcal{A}_0 = [-1, 1]$, what are the new values of π_0 and $f_0(z)$?

Define

$$f_{\pi 0}(z) = \pi_0 f_0(z) \tag{6.12}$$

so that

$$\text{fdr}(z) = f_{\pi 0}(z)/f(z), \tag{6.13}$$

(5.2). We assume that $f_0(z)$ is normal but not necessarily $\mathcal{N}(0, 1)$, say

$$f_0(z) \sim \mathcal{N}\left(\delta_0, \sigma_0^2\right). \tag{6.14}$$

This yields

$$\log\left(f_{\pi 0}(z)\right) = \left[\log(\pi_0) - \frac{1}{2}\left\{\frac{\delta_0^2}{\sigma_0^2} + \log\left(2\pi\sigma_0^2\right)\right\}\right] + \frac{\delta_0}{\sigma_0^2} z - \frac{1}{2\sigma_0^2} z^2, \tag{6.15}$$

a quadratic function of z.

Central matching estimates $f_0(z)$ and π_0 by assuming that $\log(f(z))$ is quadratic near $z = 0$ (and equal to $f_{\pi 0}(z)$),

$$\log\left(f(z)\right) \doteq \beta_0 + \beta_1 z + \beta_2 z^2, \tag{6.16}$$

estimating $(\beta_0, \beta_1, \beta_2)$ from the histogram counts y_k (5.12) around $z = 0$ and matching coefficients between (6.15) and (6.16): yielding $\sigma_0^2 = -1/(2\beta_2)$ for instance. Note that a different picture of matching appears in Chapter 11.

Exercise 6.3 What are the expressions for δ_0 and π_0?

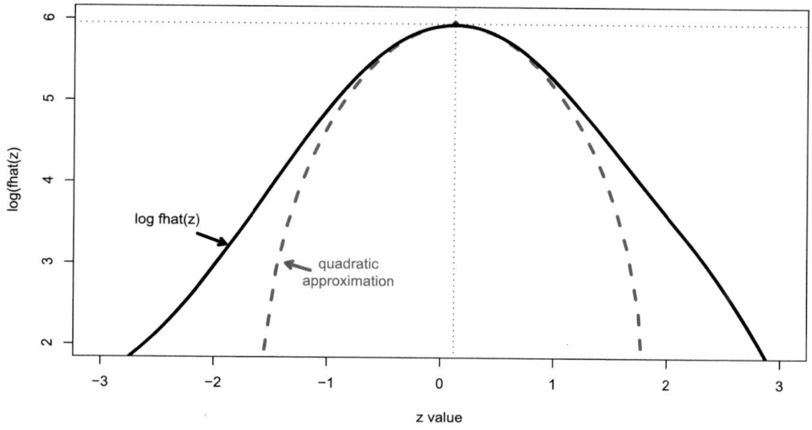

Figure 6.2 Central matching estimation of $f_0(z)$ and π_0 for the HIV data of Section 6.1, (6.15)–(6.16): $(\hat{\delta}_0, \hat{\sigma}_0, \hat{\pi}_0) = (0.12, 0.75, 0.93)$. Chapter 11 gives a different geometrical picture of the matching process.

Figure 6.2 shows central matching applied to the HIV data of Section 6.1. The quadratic approximation $\log \hat{f}_{\pi 0}(z) = \hat{\beta}_0 + \hat{\beta}_1 z + \hat{\beta}_2 z^2$ was computed as the least square fit to $\log \hat{f}(z)$ over the central 50% of the z-values (with the calculation discretized as in Section 5.2), yielding

$$\left(\hat{\delta}_0, \hat{\sigma}_0, \hat{\pi}_0\right) = (0.12, 0.75, 0.93). \tag{6.17}$$

These differ slightly from the values reported in Section 6.1 because those were calculated using the MLE method of Section 6.3.

The zero assumption (4.44) is unlikely to be literally true in actual applications. Nevertheless, central matching tends to produce nearly unbiased estimates of $f_0(z)$, at least under conditions (2.47), (2.49):

$$\begin{gathered} \mu \sim g(\cdot) \quad \text{and} \quad z|\mu \sim \mathcal{N}(\mu, 1), \\ g(\mu) = \pi_0 I_0(\mu) + \pi_1 g_1(\mu). \end{gathered} \tag{6.18}$$

Here $f_0(z) = \varphi(z)$, the $\mathcal{N}(0, 1)$ density, but we are not assuming that the non-null density

$$f_1(z) = \int_{-\infty}^{\infty} g_1(\mu)\varphi(z-\mu)\,d\mu \tag{6.19}$$

satisfies the zero assumption. This introduces some bias into the central matching assessment of δ_0 and σ_0 but, as it turns out, not very much as long as $\pi_1 = 1 - \pi_0$ is small.

With π_0 fixed, let (δ_g, σ_g) indicate the value of (δ_0, σ_0) obtained under model (6.18) by an idealized version of central matching based on $f(z) = \int_{-\infty}^{\infty} g(\mu)\varphi(z-\mu)d\mu$,

$$\delta_g = \arg\max\{f(z)\} \quad \text{and} \quad \sigma_g = \left[-\frac{d^2}{dz^2}\log f(z)\right]_{\delta_g}^{-\frac{1}{2}}. \tag{6.20}$$

Exercise 6.4 What values of $\beta_0, \beta_1, \beta_2$ are implied by (6.20)? In what sense is (6.20) an idealized version of central matching?

We can ask how far (δ_g, σ_g) deviates from the actual parameters $(\delta_0, \sigma_0) = (0, 1)$ for $g(\mu)$ in (6.18). For a given choice of π_0, let

$$\delta_{\max} = \max\{|\delta_g|\} \quad \text{and} \quad \sigma_{\max} = \max\{\sigma_g\}, \tag{6.21}$$

the maxima being over the choice of g_1 in (6.18). Table 6.2 shows the answers. In particular, for $\pi_0 \geq 0.90$ we always have

$$|\delta_g| \leq 0.07 \quad \text{and} \quad \sigma_g \leq 1.04 \tag{6.22}$$

so, no matter how the non-null cases are distributed, the idealized central matching estimates won't be badly biased.

Exercise 6.5 Show that $\sigma_g \geq 1$ under model (6.18).

Table 6.2 *Worst case values* $\delta_{\max}$ *and* $\sigma_{\max}$ (6.21) *as a function of* $\pi_1 = 1 - \pi_0$.

$\pi_1 = 1 - \pi_0$	.05	**.10**	.20	.30
$\sigma_{\max}$	1.02	**1.04**	1.11	1.22
$\delta_{\max}$	.03	**.07**	.15	.27

Figure 6.3 graphs $\delta_{\max}$ and $\sigma_{\max}$ as a function of $\pi_1 = 1 - \pi_0$. In addition to the general model (6.18), which provided the numbers in Table 6.2, $\sigma_{\max}$ is also graphed for the restricted version of (6.18) in which $g_1(\mu)$ is required

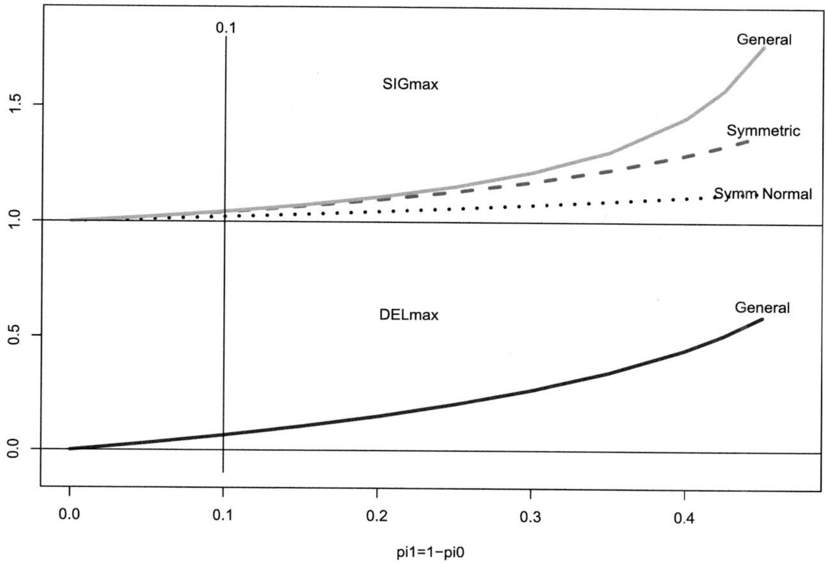

Figure 6.3 $\delta_{\max}$ and $\sigma_{\max}$ (6.21) as a function of $\pi_1 = 1 - \pi_0$; *Symmetric* restricts $g_1(\mu)$ in (6.18) to be symmetric around $\mu = 0$ and similarly *Symm Normal*.

to be symmetric about $\mu = 0$, and also the more restrictive version in which g_1 is both symmetric and normal. The worst case values in Table 6.2 have $g_1(\mu)$ supported on a single point. For example, $\sigma_{\max} = 1.04$ is achieved with $g_1(\mu)$ supported at $\mu_1 = 1.47$.

The default option in `locfdr` is the *MLE method*, discussed in the next section, not central matching. Slight irregularities in the central histogram, as seen in Figure 6.1a, can derail central matching. The MLE method is more stable, but pays the price of possibly increased bias.

Here is a derivation of $\sigma_{\max}$ for the symmetric case in Figure 6.3, the derivations for the other cases being similar. For convenience we consider discrete distributions $g(\mu)$ in (6.18), putting probability π_0 on $\mu = 0$ and π_j on pairs $(-\mu_j, \mu_j)$, $j = 1, 2, \ldots, J$, so the mixture density equals

$$f(z) = \pi_0\varphi(z) + \sum_{j=1}^{J} \pi_j \left[\varphi(z - \mu_j) + \varphi(z + \mu_j)\right] \Big/ 2. \qquad (6.23)$$

Then we can take $\delta_j = 0$ in (6.20) by symmetry (0 being the actual maximum if π_0 exceeds 0.50). We consider π_0 fixed in what follows.

Defining $c_0 = \pi_0/(1-\pi_0)$, $r_j = \pi_j/\pi_0$, and $r_+ = \sum_1^J r_j$, we can express σ_g in (6.20) as

$$\sigma_g = (1-Q)^{-\frac{1}{2}} \qquad \text{where } Q = \frac{\sum_1^J r_j \mu_j^2 e^{-\mu_j^2/2}}{c_0 r_+ + \sum_1^J r_j e^{-\mu_j^2/2}}. \tag{6.24}$$

Actually, $r_+ = 1/c_0$ and $c_0 r_+ = 1$, but the form in (6.24) allows unconstrained maximization of Q (and of σ_g) as a function of $\boldsymbol{r} = (r_1, r_2, \ldots, r_J)$, subject only to $r_j \geq 0$ for $j = 1, 2, \ldots, J$.

Exercise 6.6 Verify (6.24).

Differentiation gives

$$\frac{\partial Q}{\partial r_j} = \frac{1}{\text{den}} \left[\mu_j^2 e^{-\mu_j^2/2} - Q \cdot \left(c_0 + e^{-\mu_j^2/2}\right)\right] \tag{6.25}$$

with den the denominator of Q in (6.24). At a maximizing point $\boldsymbol{r}$ we must have

$$\frac{\partial Q(\boldsymbol{r})}{\partial r_j} \leq 0 \quad \text{with equality if} \quad r_j > 0. \tag{6.26}$$

Defining $R_j = \mu_j^2/(1 + c_0 e^{\mu_j^2/2})$, (6.25)–(6.26) give

$$Q(\boldsymbol{r}) \geq R_j \quad \text{with equality if} \quad r_j > 0. \tag{6.27}$$

At the point where $Q(\boldsymbol{r})$ is maximized, r_j and π_j can only be non-zero if j maximizes R_j.

All of this shows that we need only consider $J = 1$ in (6.23). (In case of ties in (6.27) we can arbitrarily choose one of the maximizing j values.) The maximized value of σ_g is then $\sigma_{\max} = (1 - R_{\max})^{-1/2}$ from (6.24) and (6.27), where

$$R_{\max} = \max_{\mu_1} \left\{\mu_1^2 \Big/ \left(1 + c_0 e^{\mu_1^2/2}\right)\right\}. \tag{6.28}$$

The maximizing argument μ_1 ranges from 1.43 for $\pi_0 = 0.95$ to 1.51 for $\pi_0 = 0.70$. This is considerably less than choices such as $\mu_1 = 2.5$, necessary to give small false discovery rates, at which the values in Figure 6.3 will be conservative upper bounds.

6.3 The MLE Method for Empirical Null Estimation

The *MLE method* takes a more straightforward approach to empirical null estimation: starting with the zero assumption (4.44), we obtain normal theory maximum likelihood estimators $(\hat{\delta}_0, \hat{\sigma}_0, \hat{\pi}_0)$ based on the z_i values in

$\mathcal{A}_0$. These tend to be less variable than central matching estimates, though more prone to bias.

Given the full set of z-values $\boldsymbol{z} = (z_1, z_2, \ldots, z_N)$, let N_0 be the number of z_i in $\mathcal{A}_0$ and $\mathcal{I}_0$ their indices,

$$\mathcal{I}_0 = \{i : z_i \in \mathcal{A}_0\} \quad \text{and} \quad N_0 = \#\mathcal{I}_0 \tag{6.29}$$

and define $\boldsymbol{z}_0$ as the corresponding collection of z-values,

$$\boldsymbol{z}_0 = \{z_i, i \in \mathcal{I}_0\}. \tag{6.30}$$

Also, let $\varphi_{\delta_0,\sigma_0}(z)$ be the $\mathcal{N}(\delta_0, \sigma_0^2)$ density function

$$\varphi_{\delta_0,\sigma_0}(z) = \frac{1}{\sqrt{2\pi\sigma_0^2}} \exp\left\{-\frac{1}{2}\left(\frac{z-\delta_0}{\sigma_0}\right)^2\right\} \tag{6.31}$$

and

$$H_0(\delta_0, \sigma_0) \equiv \int_{\mathcal{A}_0} \varphi_{\delta_0,\sigma_0}(z)\, dz, \tag{6.32}$$

this being the probability that a $\mathcal{N}(\delta_0, \sigma_0^2)$ variate falls in $\mathcal{A}_0$.

We suppose that the N z_i values independently follow the two-groups model (2.7) with $f_0 \sim \mathcal{N}(\delta_0, \sigma_0^2)$ and $f_1(z) = 0$ for $z \in \mathcal{A}_0$. (In terms of Figure 2.3, $\mathcal{A}_0 = \mathcal{Z}$, $N_0 = N_0(\mathcal{A}_0)$, and $N_1(\mathcal{A}_0) = 0$.) Then $\boldsymbol{z}_0$ has density and likelihood function

$$f_{\delta_0,\sigma_0,\pi_0}(\boldsymbol{z}_0) = \left[\binom{N}{N_0}\theta^{N_0}(1-\theta)^{N-N_0}\right]\left[\prod_{\mathcal{I}_0} \frac{\varphi_{\delta_0,\sigma_0}(z_i)}{H_0(\delta_0, \sigma_0)}\right] \tag{6.33}$$

when

$$\theta = \pi_0 H_0(\delta_0, \sigma_0) = \Pr\{z_i \in \mathcal{A}_0\}. \tag{6.34}$$

Notice that $\boldsymbol{z}_0$ provides $N_0 = \#\boldsymbol{z}_0$, distributed as $\text{Bi}(N, \theta)$, while N is a known constant.

Exercise 6.7 Verify (6.33).

Exponential family calculations, described at the end of the section, provide the MLE estimates $(\hat{\delta}_0, \hat{\sigma}_0, \hat{\pi}_0)$. These are usually not overly sensitive to the choice of $\mathcal{A}_0$, which can be made large in order to minimize estimation error. Program `locfdr` centers $\mathcal{A}_0$ at the median of $z_1, z_2, \ldots, z_N$, with half-width about twice the preliminary estimate of σ_0 based on interquartile range. (The multiple is less than 2 if N is very large.)

Table 6.3 reports on the results of a small Monte Carlo experiment:

measurements z_i were obtained from model (2.7) with $\pi_0 = 0.95$, $f_0 \sim \mathcal{N}(0,1)$, $f_1 \sim \mathcal{N}(2.5,1)$, and $N = 3000$. The z_i values were *not* independent, having root mean square correlation $\alpha = 0.10$; see Chapter 8. One hundred simulations gave means and standard deviations for $(\hat{\delta}_0, \hat{\sigma}_0, \hat{\pi}_0)$ obtained by central matching and by the MLE method.

Table 6.3 *MLE and central matching estimates* $(\hat{\delta}_0, \hat{\sigma}_0, \hat{\pi}_0)$ *from 100 simulations with* $\pi_0 = 0.95$, $f_0 \sim \mathcal{N}(0,1)$, $f_1 \sim \mathcal{N}(2.5,1)$, $N = 3000$, *and root mean square correlation* 0.10 *between the z-values. Also shown is correlation between MLE and CM estimates.*

	$\hat{\delta}_0$		$\hat{\sigma}_0$		$\hat{\pi}_0$	
	MLE	CM	MLE	CM	MLE	CM
mean	−.093	−.129	1.004	.984	.975	.963
stdev	.016	.051	.067	.098	.008	.039
corr	.76		.89		.68	

The MLE method yielded smaller standard deviations for all three estimates. It was somewhat more biased, especially for π_0. Note that each 3000-vector $\boldsymbol{z}$ had its mean subtracted before the application of `locfdr` so the true null mean δ_0 was -0.125 not 0.

Besides being computationally more stable than central matching of Figure 6.2, the MLE method benefits from using more of the data for estimation, about 94% of the central z-values in our simulation, compared to 50% for central matching. The latter cannot be much increased without some danger of eccentric results. The upward bias of the MLE $\hat{\pi}_0$ estimate has little effect on $\widehat{\text{fdr}}(z)$ or $\widehat{\text{Fdr}}(z)$, but it can produce overly conservative estimates of power (Section 5.4). Taking a chance on more parametric methods (see Notes) may be necessary here.

Straightforward computations produce maximum likelihood estimates $(\hat{\delta}_0, \hat{\sigma}_0, \hat{\pi}_0)$ in (6.33); $f_{\delta_0,\sigma_0,\pi_0}(\boldsymbol{z}_0)$ is the product of two exponential families[5] which can be solved separately (the two bracketed terms). The binomial term gives

$$\hat{\theta} = N_0/N \tag{6.35}$$

while $\hat{\delta}_0$ and $\hat{\sigma}_0$ are the MLEs from a truncated normal family, obtained by

[5] Appendix A gives a brief review of exponential families.

familiar iterative calculations, finally yielding

$$\hat{\pi}_0 = \hat{\theta}\Big/H_0\left(\hat{\delta}_0, \hat{\sigma}_0\right) \tag{6.36}$$

from (6.34). The log of (6.33) is concave in $(\delta_0, \sigma_0, \pi_0)$, guaranteeing that the MLE solutions are unique.

Exercise 6.8 Show that (6.33) represents a three-parameter exponential family. That is,

$$\log f_{\delta_0,\sigma_0,\pi_0}(\boldsymbol{z}_0) = \eta_1 Y_1 + \eta_2 Y_2 + \eta_3 Y_3 - \psi(\eta_1, \eta_2, \eta_3) + c(\boldsymbol{z}_0) \tag{6.37}$$

where (η_1, η_2, η_3) are functions of $(\delta_0, \sigma_0, \pi_0)$ and (Y_1, Y_2, Y_3) are functions of $\boldsymbol{z}_0$. What are (η_1, η_2, η_3) and (Y_1, Y_2, Y_3)?

6.4 Why the Theoretical Null May Fail

The four examples in Figure 6.1 strongly suggest failure of the theoretical null distribution $f_0(z) \sim \mathcal{N}(0, 1)$. This is somewhat shocking! Theoretical null derivations like that for Student's t-distribution are gems of the statistical literature, as well as pillars of applied practice. Once alerted, however, it isn't difficult to imagine causes of failure for the theoretical null. The difference in large-scale testing is only that we can detect and correct such failures.

Making use of either central matching or the MLE method, false discovery rate estimates (5.5) and (5.6) now become

$$\widehat{\text{fdr}}(z) = \hat{\pi}_0 \hat{f}_0(z)\Big/\hat{f}(z) \quad \text{and} \quad \widehat{\text{Fdr}}(z) = \hat{\pi}_0 \hat{F}_0(z)\Big/\hat{F}(z) \tag{6.38}$$

with $\hat{F}_0(z)$ the left or right cdf of $\hat{f}_0(z)$ as desired. Statement (6.38) is not a universal improvement over (5.5)–(5.6); estimating the null distribution $f_0(z)$ substantially adds to the variability of false discovery rate estimates, as documented in Chapter 7. But using the theoretical null when it is wrong is a recipe for false inference. What follows is a list of practical reasons why the theoretical null might go astray.

(I) *Failed mathematical assumptions* Textbook null hypothesis derivations usually begin from an idealized mathematical framework, e.g., independent and identically distributed (*i.i.d.*) normal components for a two-sample t-statistic. Deviations from the ideal can be easy to spot in large-scale data sets. For example, the logged expression levels x_{ij} for the HIV data have longer-than-normal tails. (Although, by itself, that is not enough to induce the underdispersion seen in the z-value histogram of Figure 6.1d:

repeating the HIV computation for a 7680×8 matrix whose components were randomly selected from the actual matrix gave an almost perfectly $\mathcal{N}(0, 1)$ z-value histogram.)

(II) *Correlation across sampling units* Student's theoretical null distribution for a two-sample t-statistic (2.2) assumes independence across the n sampling units: for instance, across the 72 patient scores for any one gene in the leukemia study. Chapter 8 shows how such independence assumptions can fail in practice. In large-scale studies, even minor experimental defects can manifest themselves as correlation across sampling units. The expression levels x_{ij} for the prostate study of Section 2.1 will be seen to have drifted slightly as the experiment went on (i.e., as j increased), some genes drifting up and others down, inducing minor correlations across microarrays. Other data sets will show more significant correlations, big enough to seriously distort the z-value histogram.

(III) *Correlation across cases* It was argued in Section 4.4 that independence among the z-values is *not* required for valid false discovery rate inference. The hitch is that this is only true if we are using the correct null distribution $f_0(z)$. Section 8.3 discusses the following disconcerting fact: even if the theoretical null distribution $z_i \sim \mathcal{N}(0, 1)$ is valid for all null cases, correlation among the z_i can make $\mathcal{N}(0, 1)$ a misleading choice in likely realizations of $\boldsymbol{z} = (z_1, z_2, \ldots, z_N)$.

Figure 6.4 provides an example. A simulation study was run with $N = 6000$ z-values; 5700 were null cases having $z_i \sim \mathcal{N}(0, 1)$, with root mean square correlation 0.1 among the $5700 \cdot 5699/2$ pairs. The 300 non-null cases followed an exact $\mathcal{N}(2.5, 1)$ distribution, that is,

$$z_i = 2.5 + \Phi^{-1}\left(\frac{i - 0.5}{300}\right) \qquad \text{for } i = 1, 2, \ldots, 300. \tag{6.39}$$

Three quantities were calculated for each of 100 simulations,

$$\text{Fdp}(2.5) = \frac{\#\{\text{null } z_i \geq 2.5\}}{\#\{z_i \geq 2.5\}}, \tag{6.40}$$

the actual false discovery proportion (2.28) for $\mathcal{Z} = [2.5, \infty)$, and both $\widehat{\text{Fdr}}_{\text{theo}}(2.5)$ and $\widehat{\text{Fdr}}_{\text{emp}}(2.5)$, the theoretical and empirical null Fdr estimates (5.6) and (6.38).

The two Fdr estimates are plotted versus Fdp in Figure 6.4. Correlation between the null cases makes Fdp(2.5) highly variable. $\widehat{\text{Fdr}}_{\text{emp}}(2.5)$ follows Fdp(2.5), somewhat noisily, but $\widehat{\text{Fdr}}_{\text{theo}}(2.5)$ moves in the wrong direction, slowly decreasing as Fdp increases. $\widehat{\text{Fdr}}_{\text{emp}}$ is much more vari-

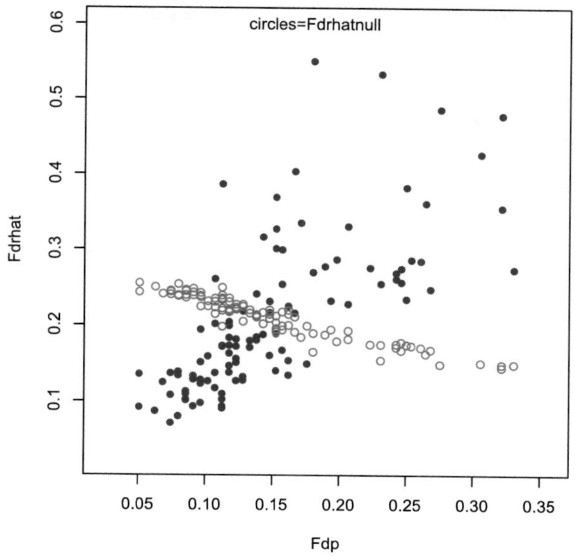

Figure 6.4 Simulation experiment comparing empirical null and theoretical null estimates of Fdr(2.5), plotted against actual false discovery proportion Fdp(2.5), as described in the text. The theoretical null estimates (open circles) decrease as Fdp increases.

able than $\widehat{\text{Fdr}}_{\text{theo}}$ but nevertheless more accurate as an estimate of Fdp, with mean $|\widehat{\text{Fdr}}_{\text{emp}}(2.5) - \text{Fdp}(2.5)| = 0.065$ compared to mean $|\widehat{\text{Fdr}}_{\text{theo}}(2.5) - \text{Fdp}(2.5)| = 0.098$. Even though the theoretical $\mathcal{N}(0, 1)$ null is unconditionally correct here, it is unsatisfactory conditionally.

Exercise 6.9 Explain why $\widehat{\text{Fdr}}_{\text{theo}}(2.5)$ is a decreasing function of Fdp(2.5).

(IV) *Unobserved covariates* Except for the chi-square data, all of our examples are observational studies. The 72 leukemia patients were observed, not assigned, to be in the AML or ALL class. Unobserved covariates such as age, gender, concomitant health conditions, processing variables,[6] etc., may affect the AML/ALL comparison. If the covariates were available we could use linear model techniques to account for them, but if not they tend to broaden the effective null distribution f_0, as discussed next.

Suppose we have a microarray experiment comparing $n/2$ Class A sub-

[6] For example, different areas on a microarray chip are read separately, by devices that may be calibrated differently.

jects with $n/2$ Class B subjects, for N genes. The null genes are assumed to have expression levels

$$x_{ij} = u_{ij} + \frac{I_j}{2}\beta_i \qquad \begin{cases} u_{ij} \sim \mathcal{N}(0,1) \\ \beta_i \sim \mathcal{N}(0,\sigma_\beta^2) \end{cases} \tag{6.41}$$

with $u_{i1}, u_{i2}, \dots, u_{in}, \beta_i$ mutually independent, and

$$I_j = \begin{cases} -1 & j = 1, 2, \dots, n/2 \\ 1 & j = n/2 + 1, \dots, n. \end{cases} \tag{6.42}$$

Here the β_i values are small disturbances caused by unequal effects of unobserved covariates on the two classes. (The β_i may be correlated across genes.)

It is easy to show that the two-sample t-statistic t_i (2.2) comparing Class A with Class B follows a dilated t-distribution with $n-2$ degrees of freedom,

$$t_i \sim \left(1 + \frac{n}{4}\sigma_\beta^2\right)^{\frac{1}{2}} \cdot t_{n-2} \tag{6.43}$$

for the null cases. In other words, the null density is $(1 + n\sigma_\beta^2/4)^{1/2}$ times more dispersed than the usual theoretical null.

Exercise 6.10 (a) Verify (6.43). (b) Suppose that all the genes are null and that $\sigma_\beta = 2/\sqrt{n}$. Show that, for large N, the local fdr, with $\pi_0 = 1$, will be about

$$\text{fdr}(t_i) \doteq \sqrt{2} f_{n-2}(t_i) \Big/ f_{n-2}\left(t_i \big/ \sqrt{2}\right) \tag{6.44}$$

when f_0 is taken to be the usual theoretical null $f_{n-2}(t)$, a Student t-density with $n-2$ degrees of freedom. (c) For $n = 20$, what is the probability that $\text{fdr}(t_i) \leq 0.20$ under (6.44)?

The empirical null fdr estimate $\hat{\pi}_0 \hat{f}_0(z)/\hat{f}(z)$ scales "correctly" using either central matching or the MLE method: if each z_i is multiplied by the same positive constant c, the value of $\widehat{\text{fdr}}$ stays the same for all cases. Another way to say this is that our interest in an outlying z_i is judged relative to the width of the histogram's center, rather than relative to the theoretical $\mathcal{N}(0,1)$ width.

Unobserved covariates are ubiquitous in large-scale studies and are perhaps the most common source of trouble for the theoretical null, a likely culprit for the kind of gross overdispersion seen in Figure 6.1a. Reason (III) above, correlation across cases, produces smaller effects, but is capable of causing the underdispersion seen in Figure 6.1d as well as overdispersion (Chapter 7). Microarray studies are particularly prone to correlation effects, as the examples of Chapter 8 will show. Reason (II), correlation across the supposedly independent sampling units, is surprisingly prevalent as a possible source of overdispersion (Chapter 8). Failed mathematical assumptions, Reason (I), is the only one of the four failure causes that is easily cured by permutation calculations, as discussed in the next section.

Our list of causes is by no means complete. *Filtration*, the data-based preselection of a subset of promising-looking cases for final analysis, can distort both the null and non-null distributions. In a microarray context the investigator might first select only those genes having above-average standard deviations, on the grounds that this hints at differential response across the various experimental conditions. Filtration is a dangerous tactic. Among other dangers, it reduces, in a biased manner, information available for evaluating the appropriate null distribution.

6.5 Permutation Null Distributions

Permutation techniques for assessing a null distribution lie somewhere between the theoretical and empirical methods, but closer to the former than the latter. They are easiest to describe in two-sample situations like the prostate, leukemia, and HIV studies. We have an $N \times n$ data matrix $\boldsymbol{X}$ with the first n_1 columns representing Treatment 1 and the last n_2 columns Treatment 2. These have produced an N-vector $\boldsymbol{z}$ of z-values, perhaps as in (2.2)–(2.5).

In order to compute the permutation null distribution, we randomly permute the columns of $\boldsymbol{X}$ as in (3.39) and recalculate the vector of z-values. Repeating the process B times gives vectors $\boldsymbol{z}^{*1}, \boldsymbol{z}^{*2}, \dots, \boldsymbol{z}^{*B}$, with

$$\boldsymbol{Z}^* = \left(\boldsymbol{z}^{*1}, \boldsymbol{z}^{*2}, \dots, \boldsymbol{z}^{*B}\right) \tag{6.45}$$

representing the full $N \times B$ matrix of permuted values. The usual permutation null is then

$$\hat{f}_0^{\text{perm}} = \text{the empirical distribution of all } N \cdot B \text{ values } z_i^{*b}. \tag{6.46}$$

(Less commonly, we might calculate a separate permutation null for each case i from the B values in the ith row of $\boldsymbol{Z}^*$. This requires B to be very large in order to assess the extreme tail probabilities necessary in large-scale testing, whereas $B = 10$ or 20 is often sufficient for (6.46).)

Figure 6.5 shows QQ plots[7] of $\hat{f}_0^{\text{perm}}$ for the leukemia and HIV studies

[7] A QQ plot of observations $x_1, x_2, \dots, x_m$ plots the ith ordered value $x_{(i)}$ versus the

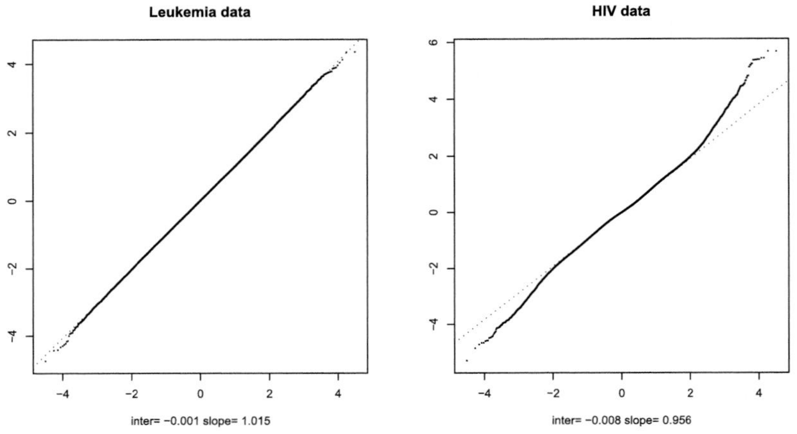

Figure 6.5 *QQ* plots of permutation null distributions for the leukemia and HIV studies, $B = 20$ permutations each. Dotted lines are least squares fits to central 80% of the plot, with intercepts and slopes as indicated; $\hat{f}_0^{\text{perm}} \mathrel{\dot\sim} \mathcal{N}(0, 1)$ for leukemia study but has heavier tails for HIV study.

(z-values as in (2.2)–(2.5)), in both cases based on $B = 20$ permutations. The leukemia plot shows $\hat{f}_0^{\text{perm}}$ nearly $\mathcal{N}(0, 1)$ while $\hat{f}_0^{\text{perm}}(z)$ for the HIV data is roughly $\mathcal{N}(0, 0.96^2)$ near $z = 0$, but with heavier-than-normal tails.

Permutation methods offer a simple and attractive way to deal with mathematically complicated hypothesis testing situations. They are not however a remedy for the theoretical null failures seen in Figure 6.1. There are several points to consider:

- Of the four potential causes of failure raised in Section 6.4, permutation methods deal most effectively with Reason (I), failed mathematical assumptions. The permutation process enforces an i.i.d. structure, so that $\hat{f}_0^{\text{perm}}$ is relevant only in that context. (The second *i*, "identical," may not be realistic, but this usually doesn't seem to cause major problems.) Non-standard statistical definitions — for example, adding a constant to the usual t-statistic denominator, as in SAM — are automatically incorporated into the permutation calculations.
- Permutation methods are no help with Reason (II), correlation across

corresponding normal quantile $\Phi^{-1}[(i - 0.5)/m]$. If the x values come from a $\mathcal{N}(a, b^2)$ distribution, the plot will approximate a straight line with intercept a and slope b.

sampling units, since the sampling process effectively enforces independence.[8]

- They are also of no help with Reason (IV), unobserved covariates. The data vector $(x_{i1}, x_{i2}, \ldots, x_{in})$ in (6.41) has $\beta_i/2$ subtracted from the first $n/2$ observations and added to the last $n/2$. But as far as permutation samples are concerned, the $\pm\beta_i/2$ disturbances are randomly distributed across the n observations. The permutation distribution of the two-sample t-statistic is nearly t_{n-2} under model (6.41), no matter how large β_i may be, rather than equaling the dilated distribution (6.43).
- A virtue of permutation methods is that they preserve correlation between cases, e.g., between genes (since the columns of $\boldsymbol{X}$ are permuted intact). We can estimate the correlation between any pair of z-values from the correlation between the rows of $\boldsymbol{Z}^*$ (6.45), as was done in Section 3.4 and will be seen again in Chapter 8.

  Nevertheless, permutation methods are of no direct assistance with Reason (III), the effects of between-case correlations on the proper choice of f_0. For any given row i of $\boldsymbol{Z}^*$, the values z_i^{*b} do not depend on between-case correlations, and neither does $\hat{f}_0^{\text{perm}}$. Both the leukemia and HIV examples display considerable case-wise correlation (see Chapter 8) but still have $\hat{f}_0^{\text{perm}}$ close to the theoretical null and far from the empirical null.
- In fact, $\hat{f}_0^{\text{perm}}$ will *usually* approximate a $\mathcal{N}(0,1)$ distribution for (2.5), the z-score version of the t-statistic. Fisher's introduction of permutation arguments was intended to justify Student's distribution in just this way. A considerable body of theory in the 1950s showed $\hat{f}_0^{\text{perm}}$ converging quickly to $\mathcal{N}(0,1)$ as n grew large. The results in Figure 6.5 are typical, showing almost perfect convergence in the leukemia example, $n = 72$.
- Permutation and empirical methods can be combined by letting the permutation null replace $\mathcal{N}(0,1)$ in the empirical algorithm. That is, we perform the empirical null calculations on $\Phi^{-1}(F_{\text{perm}}(z_i))$ rather than on the original z-values, with F_{perm} the permutation cdf. Doing so made no difference to the leukemia data but considerably increased $\widehat{\text{fdr}}$ for the HIV data, reducing the number of cases with $\widehat{\text{fdr}} \leq 0.20$ from 128 to 42, now all on the left side.
- Permutation methods are not restricted to two-sample problems. They depend on symmetries of the null hypothesis situation, which may or

[8] It actually enforces a small negative correlation: if x_1^* and x_2^* are any two draws in a permutation sample, then their correlation is $-1/(n-1)$ with respect to the permutation process, regardless of the correlations in the original sample $x_1, x_2, \ldots, x_n$.

may not exist. They would be difficult to find, perhaps impossible, for the chi-square and police examples of Figure 6.1. Bootstrap methods are more flexible but have their own problems in large-scale situations; see Section 7.5.

Notes

The leukemia data was the motivating example in Golub et al. (1999), an early exploration of advanced statistical approaches to microarray experiments. Ridgeway and MacDonald (2009) discuss the police data, providing a much more extensive analysis than the one here. Van't Wout et al. (2003) originated the HIV data, which was used as a main example in Gottardo et al. (2006) as well as Efron (2007b). The chi-square data is part of an ongoing unpublished experiment.

Empirical null methods were introduced in Efron (2004), along with the central matching algorithm. Efron (2007b) suggested the MLE method. Why the theoretical null might fail is discussed in Efron (2008a). Jin and Cai (2007, 2010) present a characteristic function approach to estimating the empirical null, while a more parametric normal mixture method is analyzed in Muralidharan (2010).

The 1935 edition of Fisher's *The Design of Experiments* introduced permutation methods as a justification of Student's t-test. A great deal of subsequent analysis by many authors, including Pitman (1937) and Hoeffding (1952), supported Fisher's contention that permutation methods usually agree closely with Student's distribution.

7

Estimation Accuracy

Empirical Bayes methods blur the line between estimation and hypothesis testing: $\widehat{\text{fdr}}(z_i)$ (5.5) estimates the probability that a null hypothesis is correct, and likewise $\widehat{\text{Fdr}}(z_i)$ (5.6). How accurate are such estimates, like the ones shown in Figure 5.2? In general we would like to assess the accuracy of summary statistics $s(\boldsymbol{z})$ that depend on a vector of z-values $\boldsymbol{z} = (z_1, z_2, \ldots, z_N)$. This would be straightforward if the z_i were independent, using classical methods or perhaps the bootstrap.

Independence is only a dream, though, for most large-scale data sets. Microarray studies are particularly prone to correlation across cases, as will be seen in Chapter 8. Now it seems we are in real trouble, where we must estimate an $N \times N$ matrix of correlations in order to assess a standard deviation for $s(\boldsymbol{z})$. Fortunately, the main result of this chapter shows that this isn't necessary: under reasonable assumptions, good accuracy approximations are available that depend only upon the root mean square of the $N \cdot (N-1)/2$ correlations, a quantity that often can be easily estimated.

As an example of these results, suppose $s(\boldsymbol{z})$ is the right-sided cdf for the leukemia data of Figure 6.1a,

$$\bar{F}(x) = \#\{z_i \geq x\}/N \qquad [N = 7128]. \tag{7.1}$$

Section 7.1 and Section 7.2 show that a good approximation for the variance[1] of $\bar{F}(x)$ is

$$\text{var}\left\{\bar{F}(x)\right\} \doteq \left\{\frac{\bar{F}(x)\cdot\left(1-\bar{F}(x)\right)}{N}\right\} + \left\{\frac{\hat{\alpha}\hat{\sigma}_0^2\hat{f}^{(1)}(x)}{\sqrt{2}}\right\}^2 \tag{7.2}$$

where

- $\hat{\alpha}$ is the estimated root mean square correlation (0.11 for the leukemia data, Chapter 8);

[1] Equation (7.50) gives the full covariance matrix for $\bar{F}(x_k)$, $k = 1, 2, \ldots, K$.

- $\hat{\sigma}_0$ is the estimated central histogram spread (1.68 for the leukemia data, Figure 6.1a);
- $\hat{f}^{(1)}(x)$ is the first derivative of the Poisson regression estimate of the mixture density $f(z)$, as derived in Section 5.2.

The first term in (7.2) is the usual binomial variance, while the second term is a *correlation penalty* accounting for dependence between the z_i values.

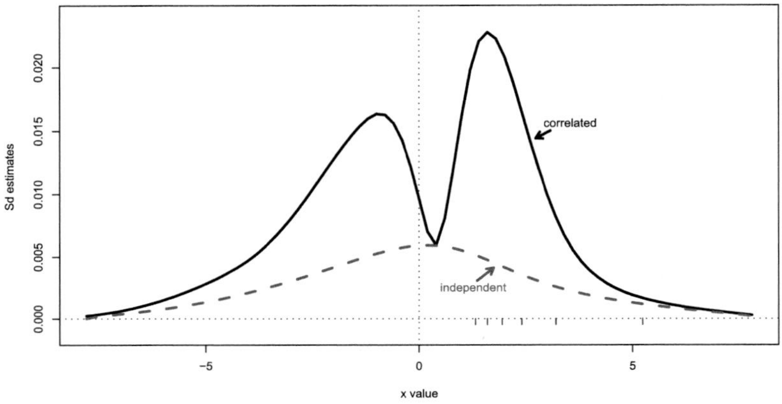

Figure 7.1 Solid curve is estimated standard deviation of right-sided cdf (7.1) for leukemia z-values, from formula (7.2). Dashed curve shows binomial standard deviation from first term of (7.2), ignoring the correlation penalty. Correlation greatly increases the variability of $\bar{F}(x)$ in this case. Hash marks indicate right percentiles used in Table 7.1.

The heavy curve in Figure 7.1 graphs estimated standard deviations $\text{sd}\{\bar{F}(x)\}$ for the leukemia data, the square root of formula (7.2). For comparison, the dashed curve shows the usual binomial standard deviation estimates that would apply if the z_i were independent, the square root of the first term in (7.2). The correlation penalty can be enormous here, multiplying standard deviations by large factors, as shown numerically in Table 7.1.

This chapter derives accuracy formulas like (7.2) for a class of summary statistics $s(\boldsymbol{z})$ pertaining to empirical Bayes calculations, including those seen in Figure 5.3. Our main assumption is that the z_i are individually normal with possibly different means and variances,

$$z_i \sim \mathcal{N}\left(\mu_i, \sigma_i^2\right) \qquad \text{for } i = 1, 2, \ldots, N. \tag{7.3}$$

Table 7.1 *Bold figures are estimated standard deviations for right-sided cdf (7.1), leukemia z-values, formula (7.2);* $\mathrm{sd_{ind}}$ *are binomial standard deviations, ignoring z-value correlations. Standard deviations evaluated at indicated right percentiles of the* $N = 7128$ z_i*'s.*

Percentile	.25	.20	.15	.10	.05	.01
sd	**.021**	**.023**	**.021**	**.016**	**.0082**	**.0017**
$\mathrm{sd_{ind}}$	.005	.005	.004	.004	.0026	.0012
ratio	4.1	4.8	5.0	4.6	3.2	1.4

There is no requirement that they be "z-values" in the hypothesis-testing sense (2.6), though that is our main class of applications.

It will take us some time to justify formula (7.2) and its extensions. The argument proceeds as follows:

- Exact cdf covariance formulas are derived in Section 7.1.
- Simplified approximations are developed in Section 7.2, along with practical data-based estimates.
- Delta-method arguments are used to extend the cdf results to more general summary statistics in Section 7.3.
- Section 7.4 concerns the non-null distribution of z-values, showing model (7.3) to be a good approximation and justifying application of the theory.

The calculations and derivations, some of which are quite involved, emphasize the frequentist side of empirical Bayes theory. Technical details, particularly in Section 7.1 and Section 7.4, can be bypassed for quick reading once the definitions and results are understood, the figures being helpful in this regard.

A natural question is why not use the bootstrap to get accuracy estimates directly? Section 7.5 provides a brief answer: with N much greater than n, the non-parametric bootstrap can behave eccentrically. Figure 7.8 shows this happening in the context of Figure 7.1.

7.1 Exact Covariance Formulas

Given correlated normal observations $z_1, z_2, \dots, z_N$ (7.3), this section derives exact formulas for the mean and covariance of the empirical process $\{\bar{F}(x), -\infty < x < \infty\}$, that is, for the right-sided cdf (7.1) evaluated for all choices of x. Rather than work directly with the cdf, it will be easier to derive first our results for a discretized version of the empirical *density* of the

z_i values; more precisely, we will compute the mean vector and covariance matrix for the histogram counts y_k (5.12). Most statistics $s(\boldsymbol{z})$ of interest to us depend on $\boldsymbol{z}$ only through the vector of counts.

The notation here will be that of (5.11)–(5.14): y_k is the number of z_i values in bin k, $k = 1, 2, \ldots, K$, giving count vector

$$\boldsymbol{y} = (y_1, y_2, \ldots, y_k)'; \tag{7.4}$$

x_k is the centerpoint of bin k, $\boldsymbol{x} = (x_1, x_2, \ldots, x_K)'$; and all bins are of width d.

Exercise 7.1 Let $\boldsymbol{z}_{\text{ord}}$ be the order statistic of $\boldsymbol{z}$, that is, the ordered values $z_{(1)} < z_{(2)} < \cdots < z_{(N)}$ (assuming no ties). Show that, for sufficiently small choice of the bin width d, we can recover $\boldsymbol{z}_{\text{ord}}$ from $\boldsymbol{y}$.

Note All of our previous estimators, such as $\widehat{\text{fdr}}(z)$ and $\widehat{\text{Fdr}}(z)$, depend on $\boldsymbol{z}$ only through $\boldsymbol{z}_{\text{ord}}$ and therefore, effectively, on $\boldsymbol{y}$.

Instead of (7.3), we first make the more restrictive assumption that the z_i are divided into a finite number of classes, with members of the cth class C_c having mean μ_c and standard deviation σ_c,

$$z_i \sim \mathcal{N}\left(\mu_c, \sigma_c^2\right) \qquad \text{for } z_i \in C_c. \tag{7.5}$$

Let N_c be the number of members of C_c, with π_c the proportion,

$$N_c = \#C_c \quad \text{and} \quad \pi_c = N_c/N \tag{7.6}$$

so $\sum_c N_c = N$ and $\sum_c \pi_c = 1$.

It will be convenient to define *adjusted bin centers* $x_{kc} = (x_k - \mu_c)/\sigma_c$ for $k = 1, 2, \ldots, K$, denoting the whole K-vector by

$$\boldsymbol{x}_c = (x_{1c}, x_{2c}, \ldots, x_{Kc})'. \tag{7.7}$$

Similarly, if $h(x)$ is any function of x, we will write

$$\boldsymbol{h}_c = h(\boldsymbol{x}_c) = (h(x_{1c}), h(x_{2c}), \ldots, h(x_{Kc}))' \tag{7.8}$$

for h evaluated at the adjusted bin centers; so, for example, $\boldsymbol{\varphi}_c$ is the standard normal density (2.9) evaluated over $\boldsymbol{x}_c$.

It is easy to calculate the expectation of the count vector $\boldsymbol{y}$ under model (7.5)–(7.6). Let P_{kc} be the probability that z_i from class C_c falls into the kth bin $\mathcal{Z}_k$,

$$P_{kc} = \Pr_c \{z_i \in \mathcal{Z}_k\} \doteq d \cdot \varphi(x_{kc})/\sigma_c. \tag{7.9}$$

This last approximation becomes arbitrarily accurate for d sufficiently small, and we will take it to be exact in what follows. Then

$$E\{\boldsymbol{y}\} = N \sum_c \pi_c \boldsymbol{P}_c = Nd \sum_c \pi_c \boldsymbol{\varphi}_c, \tag{7.10}$$

$\boldsymbol{P}_c = (P_{1c}, P_{2c}, \ldots, P_{Kc})'$ as in (7.8).

Exercise 7.2 Verify (7.10).

The $K \times K$ covariance matrix of the count vector $\boldsymbol{y}$ depends on the $N \times N$ correlation matrix of $\boldsymbol{z}$, but in a reasonably simple way discussed next. Two important definitions are needed to state the first result: there are $M = N(N-1)/2$ correlations $\rho_{ii'}$ between pairs $(z_i, z_{i'})$ of members of $\boldsymbol{z}$, and we denote by $g(\rho)$ the distribution[2] putting weight $1/M$ on each $\rho_{ii'}$. Also, for $\varphi_\rho(u, v)$, the bivariate normal density having zero means, unit standard deviations, and correlation ρ, we define

$$\begin{aligned} \lambda_\rho(u, v) &= \frac{\varphi_\rho(u, v)}{\varphi(u)\varphi(v)} - 1 \\ &= (1-\rho^2)^{-\frac{1}{2}} \exp\left\{\frac{2\rho uv - \rho^2(u^2+v^2)}{2(1-\rho^2)}\right\} - 1 \end{aligned} \tag{7.11}$$

and

$$\lambda(u, v) = \int_{-1}^{1} \lambda_\rho(u, v) g(\rho)\, d\rho. \tag{7.12}$$

Lemma 7.1 *Under the multi-class model* (7.5)–(7.6), *the covariance of the count vector* $\boldsymbol{y}$ (7.4) *has two components,*

$$\mathbf{cov}(\boldsymbol{y}) = \mathbf{cov}_0 + \mathbf{cov}_1 \tag{7.13}$$

where

$$\mathbf{cov}_0 = N \sum_c \pi_c \left\{\mathrm{diag}(\boldsymbol{P}_c) - \boldsymbol{P}_c \boldsymbol{P}_c'\right\} \tag{7.14}$$

and

$$\begin{aligned} \mathbf{cov}_1 = {} & N^2 \sum_c \sum_d \pi_c \pi_d \,\mathrm{diag}(\boldsymbol{P}_c) \boldsymbol{\lambda}_{cd}\, \mathrm{diag}(\boldsymbol{P}_d) \\ & - N \sum_c \mathrm{diag}(\boldsymbol{P}_c) \boldsymbol{\lambda}_{cc}\, \mathrm{diag}(\boldsymbol{P}_c). \end{aligned} \tag{7.15}$$

[2] $g(\rho)$ is a discrete distribution but we will treat it notationally as a density function. Formula (7.17), for example, indicates the average of ρ^j over the M values $\rho_{ii'}$.

Here $\text{diag}(\boldsymbol{P}_c)$ is the $K \times K$ diagonal matrix having diagonal elements P_{kc}, similarly $\text{diag}(\boldsymbol{P}_d)$, while $\boldsymbol{\lambda}_{cd}$ is the $K \times K$ matrix with klth element $\lambda(x_{kc}, x_{ld})$ (7.12). Summations are over all classes.[3] The proof of Lemma 7.1 appears at the end of this section.

Note Equation (7.15) assumes that the correlation distribution $g(\rho)$ is the same across all classes C_c.

The $\mathbf{cov}_0$ term in (7.13)–(7.14) is the sum of the multinomial covariance matrices that would apply if the z_i were mutually independent with fixed numbers drawn from each class; $\mathbf{cov}_1$ is the correlation penalty, almost always increasing $\mathbf{cov}(\boldsymbol{y})$. The N^2 factor in (7.15) makes the correlation penalty more severe as N increases, assuming $g(\rho)$ stays the same.

Expression (7.15) for the correlation penalty can be considerably simplified. *Mehler's identity*[4] for $\lambda_\rho(u, v)$ (7.11) is

$$\lambda_\rho(u, v) = \sum_{j \geq 1} \frac{\rho^j}{j!} h_j(u) h_j(v) \tag{7.16}$$

where h_j is the jth Hermite polynomial. Denoting the jth moment of the correlation distribution $g(\rho)$ by α_j,

$$\alpha_j = \int_{-1}^{1} \rho^j g(\rho)\, d\rho, \tag{7.17}$$

(7.12) becomes

$$\lambda(u, v) = \sum_{j \geq 1} \frac{\alpha_j}{j!} h_j(u) h_j(v) \tag{7.18}$$

so $\boldsymbol{\lambda}_{cd}$ in (7.15) can be expressed in outer product notation as

$$\boldsymbol{\lambda}_{cd} = \sum_{j \geq 1} \frac{\alpha_j}{j!} h_j(\boldsymbol{x}_c) h_j(\boldsymbol{x}_d)'. \tag{7.19}$$

(The outer product of vectors $\boldsymbol{u}$ and $\boldsymbol{v}$ is the matrix having ijth element $u_i v_j$.) Making use of (7.9), taken as exact,

$$\begin{aligned} \text{diag}(\boldsymbol{P}_c) h_j(\boldsymbol{x}_c) &= Nd\, \text{diag}\left(\varphi(\boldsymbol{x}_c)\right) h_j(\boldsymbol{x}_c) / \sigma_c \\ &= (-1)^j Nd\, \varphi_c^{(j)} \big/ \sigma_c \end{aligned} \tag{7.20}$$

where $\varphi_c^{(j)}$ indicates the jth derivative of $\varphi(u)$ evaluated at each component

[3] Note that subscript d is distinct from bin width d, which is a constant.

[4] Equation (7.16) is also known as the *tetrachoric series* and has interesting connections to canonical correlation, the singular value decomposition, Pearson's coefficient of contingency, and correspondence analysis.

of $\boldsymbol{x}_c$. The last line of (7.20) uses the Hermite polynomial relationship $\varphi^{(j)}(u) = (-1)^j \varphi(u) h_j(u)$.

Rearranging (7.15) then gives a simplified formula for $\mathbf{cov}_1$.

Lemma 7.2 *Defining*

$$\bar{\phi}^{(j)} = \sum_c \pi_c \varphi_c^{(j)} \big/ \sigma_c, \tag{7.21}$$

formula (7.15) *for the correlation penalty becomes*

$$\mathbf{cov}_1 = N^2 d^2 \left\{ \sum_{j \geq 1} \frac{\alpha_j}{j!} \bar{\phi}^{(j)} \bar{\phi}^{(j)\prime} - \frac{1}{N} \sum_{j \geq 1} \frac{\alpha_j}{j!} \left(\sum_c \pi_c \varphi_c^{(j)} \varphi_c^{(j)\prime} \big/ \sigma_c^2 \right) \right\}. \tag{7.22}$$

Returning to right-sided cdfs (7.1), let $\boldsymbol{B}$ be the $K \times K$ matrix

$$\boldsymbol{B}_{kk'} = \begin{cases} 1 & \text{if } k \leq k' \\ 0 & \text{if } k > k' \end{cases} \tag{7.23}$$

so

$$\bar{\boldsymbol{F}} = \frac{1}{N} \boldsymbol{B} \boldsymbol{y} \tag{7.24}$$

is a K-vector with kth component the proportion of z_i values in bins indexed $\geq k$,

$$\bar{F}_k = \#\{z_i \geq x_k - d/2\}/N \qquad (k = 1, 2, \ldots, K). \tag{7.25}$$

($\boldsymbol{B}$ would be transposed if we were dealing with left-sided cdfs.)

The expectation of $\bar{\boldsymbol{F}}$ is easy to obtain from (7.10),

$$\begin{aligned} E\left\{\bar{F}_k\right\} &= \sum_c \pi_c \left[\sum_{k' \geq k} d\varphi \left(\frac{x_{k'} - \mu_c}{\sigma_c} \right) \bigg/ \sigma_c \right] \doteq \sum_c \pi_c \int_{x_{kc}}^{\infty} \varphi(u)\, du \\ &= \sum_c \pi_c \Phi^+(x_{kc}) \end{aligned} \tag{7.26}$$

where $\Phi^+(u) \equiv 1 - \Phi(u)$. Now that we are working with tail areas rather than densities, we can let the bin width $d \to 0$, making (7.26) exact.

$\bar{\boldsymbol{F}}$ has covariance matrix $\boldsymbol{B}\mathbf{cov}(\boldsymbol{y})\boldsymbol{B}'/N^2$. Applying integration calculations like that in (7.26) to Lemma 7.1 gives a covariance decomposition for $\bar{\boldsymbol{F}}$.

Theorem 7.3 *Under the multi-class model* (7.5)–(7.6),

$$\mathbf{Cov}\left(\bar{\boldsymbol{F}}\right) = \mathbf{Cov}_0 + \mathbf{Cov}_1 \tag{7.27}$$

where $\mathbf{Cov}_0$ *has klth entry*

$$\frac{1}{N}\sum_{c}\pi_c\left\{\Phi^+\left(\max(x_{kc},x_{lc})\right)-\Phi^+(x_{kc})\Phi^+(x_{lc})\right\} \tag{7.28}$$

and

$$\mathbf{Cov}_1=\sum_j\frac{\alpha_j}{j!}\bar{\varphi}^{(j-1)}\bar{\varphi}^{(j-1)'}-\frac{1}{N}\sum_j\frac{\alpha_j}{j!}\left\{\sum_c\pi_c\bar{\varphi}_c^{(j-1)}\bar{\varphi}_c^{(j-1)'}\right\}. \tag{7.29}$$

Here π_c is from (7.6), x_{kc} and x_{lc} are from (7.7), α_j is as in (7.17), and

$$\bar{\varphi}^{(j-1)}=\sum_c\pi_c\varphi_c^{(j-1)}. \tag{7.30}$$

(Note the distinction between $\bar{\varphi}$ and $\bar{\phi}$ (7.21).)

Proof of Lemma 7.1 Let $I_k(i)$ denote the indicator function of the event $z_i\in\mathcal{Z}_k$ (5.11) so that the number of z_i values from class C_c in $\mathcal{Z}_k$ is

$$y_{kc}=\sum_{\mathbf{c}}I_k(i), \tag{7.31}$$

the boldface subscript indicating summation over the members of C_c. We first compute $E\{y_{kc}y_{ld}\}$ for bins k and l, $k\neq l$, and classes c and d,

$$\begin{aligned}E\{y_{kc}y_{ld}\}&=E\left\{\sum_{\mathbf{c}}\sum_{\mathbf{d}}I_k(i)I_l(j)\right\}\\&=d^2\sum_{\mathbf{c}}\sum_{\mathbf{d}}\frac{\varphi_{\rho_{ij}}(x_{kc},x_{ld})(1-\chi_{ij})}{\sigma_c\sigma_d}\end{aligned} \tag{7.32}$$

following notation (7.6)–(7.10), with χ_{ij} the indicator function of event $i=j$ (which can only occur if $c=d$). This reduces to

$$E\{y_{kc}y_{ld}\}=N^2d^2\pi_c(\pi_d-\chi_{cd}/N)\frac{\int_{-1}^{1}\varphi_\rho(x_{kc},x_{ld})g(\rho)\,d\rho}{\sigma_c\sigma_d} \tag{7.33}$$

under the assumption that the same correlation distribution $g(\rho)$ applies across all class combinations. Since $y_k=\sum_c y_{kc}$, we obtain

$$E\{y_ky_l\}=N^2d^2\sum_c\sum_d\pi_c(\pi_d-\chi_{cd}/N)\frac{\int_{-1}^{1}\varphi_\rho(x_{kc},x_{ld})g(\rho)\,d\rho}{\sigma_c\sigma_d}, \tag{7.34}$$

the non-bold subscripts indicating summation over classes. Subtracting

$$E\{y_k\}E\{y_l\}=N^2d^2\sum_c\sum_d\frac{\varphi(x_{kc})\varphi(x_{ld})}{\sigma_c\sigma_d} \tag{7.35}$$

from (7.34) results, after some rearrangement, in

$$\begin{aligned}\operatorname{cov}(y_k, y_l) = N^2 d^2 \sum_c \sum_d \frac{\varphi(x_{kc})\varphi(x_{ld})}{\sigma_c \sigma_d} \\ \times \left\{ \pi_c \left(\pi_d - \frac{\chi_{cd}}{N} \right) \int_{-1}^{1} \left(\frac{\varphi_\rho(x_{kc}, x_{ld})}{\varphi(x_{kc})\varphi(x_{ld})} - 1 \right) g(\rho)\, d\rho \right\} \\ - N d^2 \sum_c \pi_c \frac{\varphi(x_{kc})\varphi(x_{ld})}{\sigma_c \sigma_d}. \end{aligned} \tag{7.36}$$

Using $P_{kc} = d \cdot \varphi(x_{kc})/\sigma_c$ as in (7.9), expression (7.36) is seen to equal the klth element of $\mathbf{cov}(\boldsymbol{y})$ in Lemma 7.1, when $k \neq l$. The case $k = l$ proceeds in the same way, the only difference being that $Nd\pi_c \chi_{cd} \varphi(x_{kc})/\sigma_c$ must be added to formula (7.32), again in agreement with $\mathbf{cov}(\boldsymbol{y})$ in Lemma 7.1. □

7.2 Rms Approximations

The exact covariance expressions of Section 7.1 are useful for theoretical and simulation calculations, but less so for actual applications. This section develops simplified versions called *rms approximations* (for "root mean square") that facilitate the kind of data-based estimates seen in (7.2).

Convenient approximations to $\mathbf{Cov}_1$ and $\mathbf{cov}_1$ are based on three simplifications:

1 The second terms in (7.28) and in (7.22) are of order $1/N$ compared to the first terms, and can be ignored for N large.

2 Common standardization methods for data matrix $\boldsymbol{X}$, such as subtracting off the mean of each column as we did in our microarray examples, make α_1, the first coefficient in Mehler's identity (7.18), exactly or nearly zero; this leaves α_2 as the lead.

3 With ρ confined to $[-1, 1]$, the higher-order moments α_j of $g(\rho)$, $j \geq 3$, decrease quickly to zero if α_2 is not too large.

Making all three simplifications reduces (7.28) and (7.22) to the rms approximations

$$\mathbf{Cov}_1 \doteq \alpha^2 \bar{\varphi}^{(1)} \bar{\varphi}^{(1)\prime} \big/ 2 \tag{7.37}$$

and

$$\mathbf{cov}_1 \doteq (Nd\alpha)^2 \bar{\phi}^{(2)} \bar{\phi}^{(2)\prime} \big/ 2 \tag{7.38}$$

where α is the *root mean square correlation*

$$\alpha = \alpha_2^{1/2} = \left[\int_{-1}^{1} \rho^2 g(\rho)\, d\rho\right]^{\frac{1}{2}}. \tag{7.39}$$

Here $\bar{\varphi}^{(1)}$ is the average first derivative in (7.30),

$$\bar{\varphi}^{(1)} = \sum_c \pi_c \varphi_c^{(1)} \tag{7.40}$$

and likewise

$$\bar{\phi}^{(2)} = \sum_c \pi_c \bar{\varphi}^{(2)} \big/ \sigma_c \tag{7.41}$$

from (7.21).

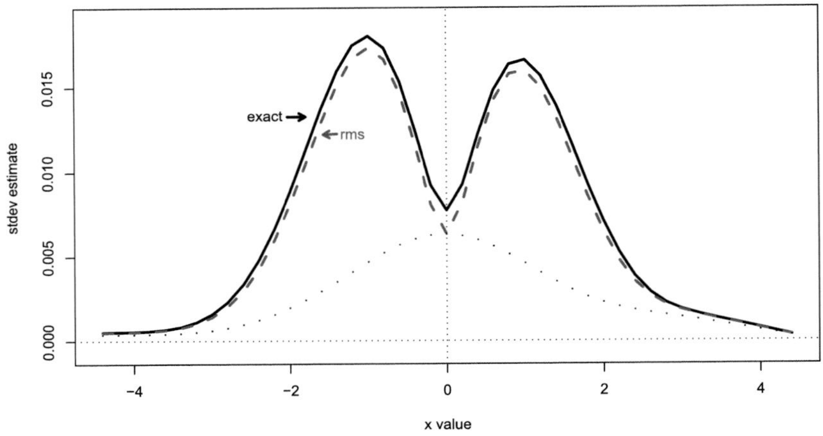

Figure 7.2 Comparison of exact formula for sd$\{\bar{F}(x)\}$ from Theorem 7.3 (solid curve) with the rms approximation (dashed) for $N = 6000$, $\alpha = 0.10$, in the two-class situation (7.42). Dots indicate binomial standard deviation, ignoring correlation penalty.

Figure 7.2 compares exact and approximate standard deviations of $\bar{F}(x)$ (7.1), for the situation where $N = 6000$, $\alpha = 0.10$, and there are two classes (7.5), (7.6) having

$$\begin{aligned} &(\mu_0, \sigma_0) = (-0.125, 1) \qquad \pi_0 = 0.95 \\ \text{and} \quad &(\mu_1, \sigma_1) = (2.38, 1) \qquad \pi_1 = 0.05. \end{aligned} \tag{7.42}$$

(If we begin with $\mu_0 = 0$ and $\mu_1 = 2.5$ and recenter z by subtracting $\bar{z}$, we get μ_0 and μ_1 as in (7.42).) The solid curve traces the exact standard deviations, i.e., square roots of the diagonal elements of $\mathbf{Cov}(\bar{F})$ (7.27); the dashed curve shows standard deviations when substituting the rms approximation (7.37) for $\mathbf{Cov}_1$ in (7.27). We see that the rms approximation performs quite satisfactorily.

Practical application of the rms approximations requires us to estimate the rms correlation α (7.39) and $\bar{\varphi}^{(1)}$ (7.40) or $\bar{\phi}^{(2)}$ (7.41). Chapter 8 shows that an estimate $\hat{\alpha}$ is easy to obtain from data matrices $\boldsymbol{X}$ such as that for the leukemia study. Here we will describe the estimation of $\bar{\varphi}^{(1)}$ or $\bar{\phi}^{(2)}$. These look difficult since they depend on the class parameters (μ_c, σ_c, π_c) in (7.5)–(7.6) but, at least under some assumptions, a simple estimate is available.

The marginal density $f(z)$ under model (7.5)–(7.6) is

$$f(z) = \sum_c \pi_c \varphi\left(\frac{z-\mu_c}{\sigma_c}\right)\bigg/\sigma_c \tag{7.43}$$

so the first and second derivatives of $f(z)$ are

$$\begin{aligned} f^{(1)}(z) &= \sum_c \pi_c \varphi^{(1)}\left(\frac{z-\mu_c}{\sigma_c}\right)\bigg/\sigma_c^2 \\ \text{and} \quad f^{(2)}(z) &= \sum_c \pi_c \varphi^{(2)}\left(\frac{z-\mu_c}{\sigma_c}\right)\bigg/\sigma_c^3. \end{aligned} \tag{7.44}$$

Suppose we make the *homogeneity assumption* that all σ_c values are the same, say $\sigma_c = \sigma_0$. Comparison with definitions (7.30) and (7.21) then gives

$$\bar{\varphi}^{(1)} = \sigma_0^2 \boldsymbol{f}^{(1)} \quad \text{and} \quad \bar{\phi}^{(2)} = \sigma_0^2 \boldsymbol{f}^{(2)} \tag{7.45}$$

where $\boldsymbol{f}^{(j)} = f^{(j)}(\boldsymbol{x})$, the jth derivative evaluated at the bin centerpoints. It is worth noting that the class structure (7.5)–(7.6) has disappeared entirely in (7.45), leaving $z_i \sim \mathcal{N}(\mu_i, \sigma_0^2)$ for $i = 1, 2, \ldots, N$ as the basic assumption.

Exercise 7.3 Verify (7.45).

Substituting (7.45) into the rms approximations (7.37)–(7.38) yields convenient expressions for the correlation penalties,

$$\mathbf{Cov}_1 \doteq \frac{(\sigma_0^2\alpha)^2}{2}\boldsymbol{f}^{(1)}\boldsymbol{f}^{(1)\prime} \quad \text{and} \quad \mathbf{cov}_1 \doteq \frac{(Nd\sigma_0^2\alpha)^2}{2}\boldsymbol{f}^{(2)}\boldsymbol{f}^{(2)\prime}. \tag{7.46}$$

This finally returns us to the correlation penalty used in equation (7.2), $(\hat{\alpha}\hat{\sigma}_0^2 \hat{f}^{(1)}(x)/\sqrt{2})^2$, where $f^{(1)}(z)$ has been estimated by differentiation of the Poisson regression estimate $\hat{f}(z)$ of Section 5.2.

Table 7.2 *Estimated 3-class model for the leukemia data.*

	Left	Center	Right
μ_c	−4.2	.09	5.4
σ_c	1.16	1.68	1.05
π_c	.054	.930	.016

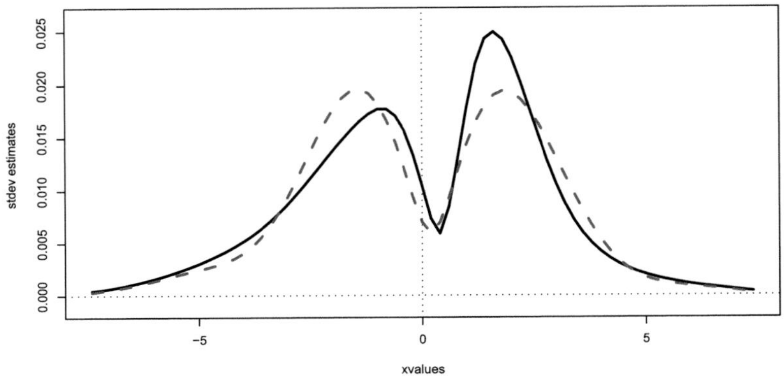

Figure 7.3 Leukemia data; comparison of standard deviation estimates for empirical cdf (7.1), using formula (7.2) (solid curve) or 3-class model based on Table 7.2 (dashed curve).

Suppose we distrust the homogeneity assumption. The smoothed non-null counts $\hat{y}_{1k}$ (5.42) allow direct, though rather crude, estimates of the class parameters (μ_c, σ_c, π_c). The equivalent of Figure 5.5 for the leukemia data, along with the MLE estimates $(\hat{\mu}_0, \hat{\sigma}_0, \hat{\pi}_0)$, suggests a 3-class model with the parameter values given in Table 7.2. The dashed curve in Figure 7.3 shows standard deviations for $\bar{F}(x)$ if $\bar{\varphi}^{(1)}$ in (7.29) is estimated from (7.40) using the parameter values in Table 7.2. It differs only modestly from the results based on the homogeneity assumption (7.45), as used in formula (7.2).

Exercise 7.4 How were the left and right class parameters in Table 7.2 calculated from the smooth counts $\hat{y}_{1k}$?

The independence terms as stated in Theorem 7.3 and Lemma 7.1, $\mathbf{Cov}_0$ and $\mathbf{cov}_0$, depend on the class structure (7.5)–(7.6). They assume that the class sample sizes N_c are fixed constants so that $\mathbf{cov}_0$ is a mixture of multinomial covariance matrices. A more realistic assumption might be that $N_1, N_2, \ldots, N_C$ are a multinomial sample of size N, sampled with probabilities $\pi_1, \pi_2, \ldots, \pi_C$, in which case $\mathbf{cov}_0$ becomes the single multinomial covariance

$$\mathbf{cov}_0 = N\left\{\text{diag}(d \cdot \boldsymbol{f}) - d^2 \boldsymbol{f}\boldsymbol{f}'\right\}, \tag{7.47}$$

$\boldsymbol{f} = f(\boldsymbol{x})$ (7.43). Here $d \cdot \boldsymbol{f}$ is the vector of bin probabilities

$$d \cdot \boldsymbol{f} = \sum_c \pi_c \boldsymbol{P}_c \tag{7.48}$$

as in (7.9)–(7.10).

Exercise 7.5 (a) Verify (7.48). (b) Show that $\mathbf{cov}_0$ (7.47) is greater than or equal to $\mathbf{cov}_0$ (7.14) in the sense that the difference is a positive semi-definite matrix.

Formula (7.47) has the advantage of not depending explicitly on the class structure. In the same way, we can replace $\mathbf{Cov}_0$ (7.28) with a single multinomial expression.

To summarize, the rms approximations and the homogeneity assumption together yield these convenient estimates for the covariance matrices of the counts and their cdf values:

$$\widehat{\mathbf{cov}}\{\boldsymbol{y}\} = N\left\{\text{diag}\left(d \cdot \hat{\boldsymbol{f}}\right) - d^2 \hat{\boldsymbol{f}}\hat{\boldsymbol{f}}'\right\} + \frac{\left(\hat{\alpha} N d \hat{\sigma}_0^2\right)^2}{2} \hat{\boldsymbol{f}}^{(2)} \hat{\boldsymbol{f}}^{(2)\prime} \tag{7.49}$$

and

$$\widehat{\mathbf{Cov}}\left\{\bar{\boldsymbol{F}}\right\} = \frac{1}{N}\left\{\left(\bar{F}_{\max(k,l)} - \bar{F}_k \bar{F}_l\right)\right\} + \frac{\left(\hat{\alpha}\hat{\sigma}_0^2\right)^2}{2} \hat{\boldsymbol{f}}^{(1)} \hat{\boldsymbol{f}}^{(1)\prime}. \tag{7.50}$$

Here $\bar{\boldsymbol{F}} = (\bar{F}(x_1), \bar{F}(x_2), \ldots, \bar{F}(x_K))'$, the leading matrix having elements as indicated in (7.50). Expression (7.2) is obtained from the diagonal elements of (7.50).

Exercise 7.6 Show that the "independence" matrices in (7.49)–(7.50) are related by $\widehat{\mathbf{Cov}}_0 = \boldsymbol{B}\widehat{\mathbf{cov}}_0\boldsymbol{B}'$, with $\boldsymbol{B}$ as in (7.23).

Formula (7.2) for the standard deviation of $\bar{F}(x)$ was tested in a simulation experiment: for each of 100 replications, a 6000×80 matrix $\boldsymbol{X}$ was constructed such that two-sample t-statistics comparing the first and last 40 columns gave z-values satisfying model (7.42), with $N = 6000$ and $\alpha = 0.10$ (see Section 8.2). The solid curve in Figure 7.4 graphs the mean

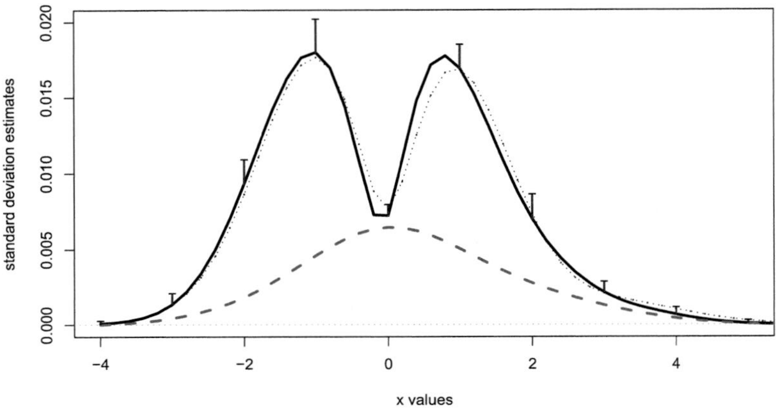

Figure 7.4 Simulation experiment testing formula (7.2), 100 replications of model (7.42), $N = 6000$, $\alpha = 0.10$, generated as described in Chapter 8. *Solid curve* is mean of 100 values $\widehat{\text{sd}}_k$, square root of (7.2) at $x = x_k$; error bars indicate empirical standard deviations of $\widehat{\text{sd}}_k$ at $x = -4, -3, \ldots, 4$. *Dotted curve*: true sd_k from Theorem 7.3. *Dashed curve* is mean of $\widehat{\text{sd}}_0$, square root of first term in (7.2).

of the 100 values of $\widehat{\text{sd}}_k$ (square roots of (7.2) for $x = x_k$) with error bars indicating the empirical standard deviations. The dotted curve shows true sd_k values from Theorem 7.3. We see that formula (7.2) gave nearly unbiased results except at the extreme right, with reasonably low variability.

7.3 Accuracy Calculations for General Statistics

We would like to assess the accuracy of estimates such as $\widehat{\text{fdr}}(z)$ and $\widehat{\text{Fdr}}(z)$. To this end, standard *delta-method* (i.e., Taylor series) techniques allow us to extend the covariance estimates of Section 7.2 to statistics that can be expressed as smoothly differentiable functions of the count vector $\boldsymbol{y}$ (7.4).

Suppose that $Q(\boldsymbol{y})$ is a q-dimensional function of the K-vector $\boldsymbol{y}$ (considered as varying continuously) such that a small change $d\boldsymbol{y}$ produces change dQ in Q according to the local linear approximation

$$dQ = \hat{\boldsymbol{D}} d\boldsymbol{y} \tag{7.51}$$

where $\hat{\boldsymbol{D}}$ is the $q \times K$ matrix $dQ/d\boldsymbol{y}$, with jkth element

$$\hat{\boldsymbol{D}}_{jk} = \left.\frac{\partial Q_j}{\partial y_k}\right|_{\boldsymbol{y}}. \tag{7.52}$$

If $\widehat{\mathbf{cov}}(\boldsymbol{y})$ is a covariance estimate for $\boldsymbol{y}$, perhaps (7.49), then the usual delta-method estimate of $\text{cov}(Q)$ is the $q \times q$ matrix

$$\widehat{\text{cov}}(Q) = \hat{\boldsymbol{D}}\,\widehat{\mathbf{cov}}(\boldsymbol{y})\hat{\boldsymbol{D}}'. \tag{7.53}$$

In a theoretical context where $\mathbf{cov}(\boldsymbol{y})$ is known we could use instead

$$\mathbf{cov}(Q) \doteq \boldsymbol{D}\,\mathbf{cov}(\boldsymbol{y})\boldsymbol{D}', \tag{7.54}$$

now with the derivative matrix $\boldsymbol{D}$ evaluated at the expectation of $\boldsymbol{y}$. Sometimes $\boldsymbol{D}$ or $\hat{\boldsymbol{D}}$ is called the *influence function* of $Q(\boldsymbol{y})$.

As a first example, consider assessing the accuracy of $\widehat{\text{fdr}}(z) = \pi_0 f_0(z)/\hat{f}(z)$ (5.5), except with π_0 considered known. Notation (7.8) will be useful here, where, if $h(z)$ is any function of z, then $\boldsymbol{h} = h(\boldsymbol{x}) = (h(x_1), h(x_2), \ldots, h(x_K))'$ is $h(z)$ evaluated at the bin centers $\boldsymbol{x}$. Thus, for example, $\widehat{\mathbf{fdr}} = \pi_0 \boldsymbol{f}_0/\hat{\boldsymbol{f}}$ is the K-vector[5] of values $\widehat{\text{fdr}}(x_k)$.

It is convenient to work first with the logarithm of $\widehat{\mathbf{fdr}}$, say $\widehat{\mathbf{lfdr}} = \log(\widehat{\mathbf{fdr}})$,

$$\widehat{\mathbf{lfdr}} = \log(\pi_0) + \boldsymbol{l}_0 - \hat{\boldsymbol{l}}. \tag{7.55}$$

Since $\log(\pi_0)$ and $\boldsymbol{l}_0 = \log(\boldsymbol{f}_0)$ are constants, the influence function $d\widehat{\mathbf{lfdr}}/d\boldsymbol{y}$ equals $-d\boldsymbol{l}/d\boldsymbol{y}$.

The influence function $d\hat{\boldsymbol{l}}/d\boldsymbol{y}$ is well known when $\hat{\boldsymbol{f}}$ is the Poisson regression estimate of Section 5.2. Let M be the $K \times m$ structure matrix for the regression. (In the implementation of (5.10), M has kth row $(1, x_k, x_k^2, \ldots, x_k^J)$, $m = J + 1$.) Also let $\boldsymbol{\nu} = Nd\boldsymbol{f}$ be the expectation of $\boldsymbol{y}$ (5.14) with[6] $\hat{\boldsymbol{\nu}} = Nd\hat{\boldsymbol{f}}$ and

$$\hat{G}_{m\times m} = M' \operatorname{diag}(\hat{\boldsymbol{\nu}})\, M, \tag{7.56}$$

$\operatorname{diag}(\hat{\boldsymbol{\nu}})$ indicating a $K \times K$ diagonal matrix with diagonal elements $\hat{\nu}_k$. A standard generalized linear model (GLM) argument discussed at the end of this section gives

$$d\hat{\boldsymbol{l}}\big/d\boldsymbol{y} = M\hat{G}M' = -d\widehat{\mathbf{lfdr}}\big/d\boldsymbol{y}. \tag{7.57}$$

[5] Following conventions of computer language R, the ratio of vectors is the vector of ratios, etc., and $\log(\pi_0)$ is added to each component at (7.55).

[6] The estimates $\hat{\boldsymbol{f}}$ and $\hat{\boldsymbol{f}}_0$ returned by `locfdr` sum to N rather than to $1/d$ for easy graphical comparison with $\boldsymbol{y}$.

Applying (7.53) yields an estimated covariance matrix for $\widehat{\mathbf{lfdr}}$,

$$\widehat{\text{cov}}\left(\widehat{\mathbf{lfdr}}\right) = \left(M\hat{G}M'\right) \widehat{\mathbf{cov}}(\boldsymbol{y}) \left(M\hat{G}M'\right) \tag{7.58}$$

where $\widehat{\mathbf{cov}}(\boldsymbol{y})$ can be taken from (7.49). If we are in the independence situation $\hat{\alpha} = 0$, then we can write (7.49) as

$$\widehat{\mathbf{cov}}(\boldsymbol{y}) = \text{diag}(\hat{\boldsymbol{\nu}}) - \hat{\boldsymbol{\nu}}\hat{\boldsymbol{\nu}}/N. \tag{7.59}$$

Ignoring the last term, (7.58) gives a nice result,

$$\widehat{\mathbf{cov}}\left(\widehat{\mathbf{lfdr}}\right) \doteq M\hat{G}M' \,\text{diag}(\hat{\nu}) M\hat{G}M' = M\hat{G}M'. \tag{7.60}$$

In our examples, however, the correlation penalty was usually too big to ignore.

Corresponding to $\widehat{\mathbf{lfdr}}$ (7.55) is the right-sided log tail area false discovery rate

$$\widehat{\mathbf{lFdr}} = \log(\pi_0) + \boldsymbol{L}_0 - \hat{\boldsymbol{L}} \tag{7.61}$$

where $\hat{\boldsymbol{L}} = \log(\hat{F}(\boldsymbol{x}))$ and $\boldsymbol{L}_0 = \log(F_0(\boldsymbol{x}))$, as in (5.6), except with π_0 known. The $K \times K$ derivative matrix $d\hat{\mathbf{L}}/d\hat{\mathbf{l}}$ has elements

$$\frac{\partial \hat{L}_j}{\partial \hat{l}_k} = \begin{cases} \hat{f}_k/\hat{F}_j & \text{if } k \geq j \\ 0 & \text{if } k < j. \end{cases} \tag{7.62}$$

Calling this matrix $\hat{B}$, the chain rule of differentiation gives

$$d\hat{\mathbf{L}}\big/d\boldsymbol{y} = \hat{B}M\hat{G}M' \tag{7.63}$$

and

$$\widehat{\mathbf{Cov}}\left(\widehat{\mathbf{lFdr}}\right) = \left(\hat{B}M\hat{G}M'\right) \widehat{\mathbf{Cov}}(\boldsymbol{y}) \left(M\hat{G}M'\hat{B}'\right). \tag{7.64}$$

Exercise 7.7 Verify (7.62) and (7.64).

Figure 7.5 compares the accuracy of $\widehat{\text{fdr}}(z)$ with $\widehat{\text{Fdr}}(z)$ for the two-class model (5.18) with $N = 6000$ and $\alpha = 0, 0.1$, or 0.2. The solid curves graph the standard deviation of $\log \widehat{\text{fdr}}(z)$ obtained from (7.58), (7.49), plotted versus the right percentiles of the mixture distribution $f(z)$. The corresponding standard deviations for $\log \widehat{\text{Fdr}}(z)$, obtained from (7.64), (7.50), appear as the dashed curves. Remembering that standard deviation on the log scale nearly equals the coefficient of variation, Exercise 5.3, we can see the CVs increasing from around 5% to over 20% as the correlation parameter α goes from 0 to 0.2.

The dotted curves in Figure 7.2 depict the accuracy of the non-parametric

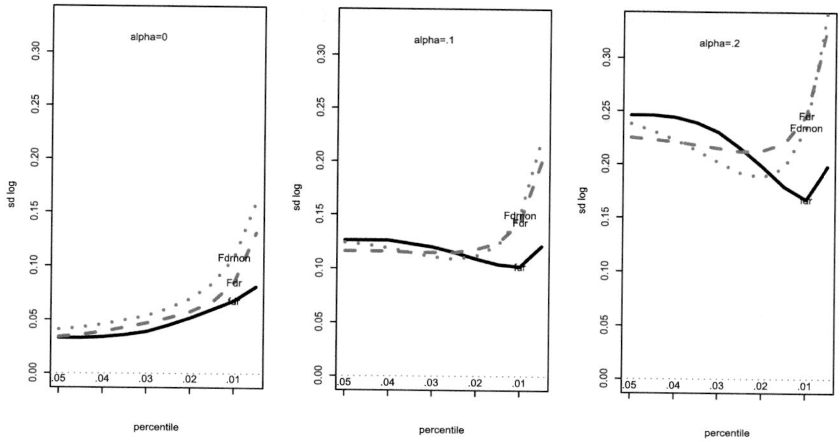

Figure 7.5 *Solid curves* show standard deviation of $\log \widehat{\text{fdr}}(z)$ as a function of z at the upper percentiles of $f(z)$ for two-class model (5.18), $N = 6000$, $\alpha = 0$, 0.1, and 0.2; *dashed curves* show same for $\log \widehat{\text{Fdr}}(z)$; *dotted curves* denote non-parametric version $\log \overline{\text{Fdr}}(z)$. Accuracy decreases sharply with increasing correlation α. The local fdr estimate is more accurate at the extreme percentiles.

tail-area Fdr estimate $\overline{\text{Fdr}}(z) = \pi_0 F_0(z)/\bar{F}(z)$ (7.1), obtained using (7.50) and

$$\frac{d \log \bar{F}_j}{dy_k} = \begin{cases} 1/(N\bar{F}_j) & \text{if } k \geq j \\ 0 & \text{if } k < j. \end{cases} \tag{7.65}$$

All three methods perform about equally well except at the extreme right, where $\widehat{\text{fdr}}(z)$ is noticeably more accurate.

Exercise 7.8 Verify (7.65).

All of our calculations have so far assumed that π_0 and $f_0(z)$ are known. Having π_0 unknown barely degrades accuracy, but f_0 is a different matter. Next we will calculate the influence function of $\widehat{\text{fdr}}(z) = \hat{\pi}_0 \hat{f}_0(z)/\hat{f}(z)$ when $\hat{\pi}_0$ and $\hat{f}_0(z)$ are estimated by central matching, as in Figure 6.2.

Suppose that central matching is accomplished by a least squares quadratic fit of the counts y_k to the bin centers x_k, over a central subset of K_0 bins having index set $\boldsymbol{k}_0$. Define M_0 as the $K \times 3$ matrix with kth row $(1, x_k, x_k^2)$

and let $\tilde{M}$ and $\tilde{M}_0$ be the submatrices whose columns are defined by $\boldsymbol{k}_0$,

$$\tilde{M} = M[\boldsymbol{k}_0] \quad \text{and} \quad \tilde{M}_0 = M_0[\boldsymbol{k}_0] \tag{7.66}$$

of dimensions $K_0 \times m$ and $K_0 \times 3$.

Theorem 7.4 *The influence function of* $\log \widehat{\mathbf{fdr}}$ *with respect to* $\boldsymbol{y}$ *when using central matching is*

$$\frac{d\widehat{\mathbf{lfdr}}}{d\boldsymbol{y}} = A\hat{G}M' \tag{7.67}$$

where

$$A = M_0\tilde{G}_0^{-1}\tilde{M}_0'\tilde{M} - M \qquad \left[\tilde{G}_0 = \tilde{M}_0'\tilde{M}_0\right]. \tag{7.68}$$

Proof of Theorem 7.4 appears at the end of this section.

Note The theorem applies as stated to other models for the null: if $f_0(z)$ is assumed known, with only π_0 unknown, then M_0 becomes the $K \times 1$ matrix of 1's; for a cubic null, the kth row of M_0 is $(1, x_k, x_k^2, x_k^3)$.

Theorem 7.4 yields the delta-method estimate

$$\widehat{\mathbf{cov}}\left(\widehat{\mathbf{lfdr}}\right) = \left(A\hat{G}M\right)\widehat{\mathbf{cov}}(\boldsymbol{y})\left(M'\hat{G}A'\right). \tag{7.69}$$

As in (7.60), the independence estimate $\widehat{\mathbf{cov}}(\boldsymbol{y}) \doteq \text{diag}(\hat{\boldsymbol{\nu}})$ gives the compact expression

$$\widehat{\mathbf{cov}}\left(\widehat{\mathbf{lfdr}}\right) \doteq A\hat{G}A'. \tag{7.70}$$

Comparing (7.67) with (7.57), the latter amounts to taking the first term in A (7.68) to be zero, which would be appropriate if we had an infinite amount of prior information on π_0 and $f_0(z)$.

Table 7.3 compares sd$\{\log \widehat{\text{fdr}}(z)\}$ using the theoretical null (7.58) versus the empirical null (7.70). The standard errors are *much* larger for the empirical null. Partly, however, that reflects a limitation of our analysis, which focuses on unconditional standard deviations. Some of the variability in the empirical null case is "good variability," in that $\widehat{\text{fdr}}(z)$ is correctly adjusting for the true false discovery proportion, as in Figure 6.4. The central matching estimates $(\hat{\delta}_0, \hat{\sigma}_0)$ are acting as approximate ancillary statistics, to use classical terminology, which adapt the estimate $\widehat{\text{fdr}}(z)$ to differing situations. At the present level of development, it is difficult to separate out the good and bad variability components of the empirical null estimate.

Table 7.3 *Comparison of* sd$\{\log \widehat{\mathrm{fdr}}(z)\}$ *for* $N = 6000$, $\alpha = 0.1$ (5.18), *as in middle panel of Figure 7.5. Top row uses central matching* (7.70)*; bottom row assumes* π_0, $f_0(z)$ *known* (7.58)*. Some of the extra variability in the empirical case comes from correctly adjusting* $\widehat{\mathrm{fdr}}(z)$ *to differing situations, as in Figure 6.4.*

Percentile	.05	.04	.03	.02	.01
sd empirical null	**.18**	**.26**	**.36**	**.54**	**.83**
sd theoretical null	.13	.13	.12	.11	.10

Proof of Theorem 7.4 In a Poisson GLM, the MLE of the expectation vector $\boldsymbol{\nu} = \exp(M\beta)$ satisfies

$$M'\left[\boldsymbol{y} - e^{M\hat{\beta}}\right] = 0. \tag{7.71}$$

A small change $d\boldsymbol{y}$ in $\boldsymbol{y}$ therefore produces change $d\hat{\beta}$ satisfying

$$M'\,d\boldsymbol{y} = M' \operatorname{diag}(\hat{\boldsymbol{\nu}}) M\, d\hat{\beta}. \tag{7.72}$$

Exercise 7.9 Verify (7.72) and use it to prove (7.57).

Let $\hat{\boldsymbol{l}}_{00} = \log(\hat{\pi}_0 \hat{\boldsymbol{f}}_0)$, the log numerator of $\widehat{\mathbf{fdr}}$, which we are modeling in central matching by

$$\hat{\boldsymbol{l}}_{00} = M_0 \hat{\gamma} \tag{7.73}$$

where $\hat{\gamma}$ is an m_0-dimensional vector and $m_0 = 3$ as before (7.66). We are fitting $\hat{\boldsymbol{l}}_{00}$ by least squares to the central portion $\tilde{\boldsymbol{l}} = \hat{\boldsymbol{l}}[\boldsymbol{k}_0]$ of $\hat{\boldsymbol{l}} = \log \hat{\boldsymbol{f}}$, so ordinary least squares theory gives

$$d\hat{\gamma} = \tilde{G}_0^{-1} M_0'\, d\tilde{\boldsymbol{l}} \quad \text{and} \quad d\hat{\boldsymbol{l}}_{00} = M_0 \tilde{G}_0^{-1} M_0'\, d\tilde{\boldsymbol{l}}. \tag{7.74}$$

Then (7.57) yields

$$d\tilde{\boldsymbol{l}} = d\hat{\boldsymbol{l}}[\boldsymbol{k}_0] = \tilde{M}\hat{G}M'\,d\boldsymbol{y} \tag{7.75}$$

and

$$d\hat{\boldsymbol{l}}_{00} = M_0 \tilde{G}_0^{-1} \tilde{M}_0' \tilde{M} \hat{G} M'\, d\boldsymbol{y} \tag{7.76}$$

from (7.74). Finally,

$$d\widehat{\mathbf{lfdr}} = d\left(\hat{\boldsymbol{l}}_{00} - \hat{\boldsymbol{l}}\right) = \left(M_0 \hat{G}_0 \tilde{M}_0' \tilde{M} - M\right) \hat{G} M'\, d\boldsymbol{y}, \tag{7.77}$$

verifying (7.67). □

7.4 The Non-Null Distribution of z-Values

The main results of this chapter, summarized in (7.49)–(7.50), depend on normality, $z_i \sim \mathcal{N}(\mu_i, \sigma_i^2)$ (7.3). In all of our examples, the z_i were actually z-values, i.e., standardized test statistics, obtained for instance as in (2.2)–(2.6). By definition, a z-value is a statistic having a $\mathcal{N}(0,1)$ distribution under a null hypothesis H_0 of interest. But will it still be normal under non-null conditions? In other words, is it legitimate to apply results like (7.2) to the leukemia data, as we did in Figure 7.1?

The results of this section are reassuring. They show that, under repeated sampling, the non-null distribution of a typical z-value will have standard deviation of order $O(n^{-1/2})$ and non-normality $O_p(n^{-1})$ as we move away from the null hypothesis: so $z_i \sim \mathcal{N}(\mu_i, \sigma_i^2)$ holds to a good approximation, though we can expect σ_i^2 to differ noticeably from 1 when μ_i is far from 0.

Figure 7.6 illustrates the phenomenon for the Student-t situation (2.2)–(2.6). Here we have

$$z = \Phi^{-1}\left(F_\nu(t)\right) \qquad \text{with } t \sim t_\nu(\delta), \tag{7.78}$$

the notation $t_\nu(\delta)$ indicating a non-central t variable with ν degrees of freedom and non-centrality parameter δ (not δ^2),

$$t = \frac{Z+\delta}{S} \qquad \left[Z \sim \mathcal{N}(0,1) \text{ independent of } S^2 \sim \chi_\nu^2/\nu\right]. \tag{7.79}$$

F_ν is the cdf of a *central* t_ν distribution as in (2.5), where $\nu = 100$.

Table 7.4 *Non-central t example $t \sim t_\nu(\delta)$ for $\nu = 20$, $\delta = 0,1,2,3,4,5$; moment parameters of $z = \Phi^{-1}(F_\nu(t))$ (7.78) indicate near-normality even for δ far from 0.*

δ	0	1	2	3	4	5
mean	0	.98	1.89	2.71	3.41	4.01
sd	1	.98	.92	.85	.77	.71
skew	0	−.07	−.11	−.11	−.10	−.07
kurt	0	.02	.06	.08	.09	.07

Figure 7.6 shows the density of z from (7.78) for $\nu = 20$ and non-centrality $\delta = 0,1,2,3,4,5$. By definition, the density is $\mathcal{N}(0,1)$ for $\delta = 0$. At $\delta = 5$, the density almost exactly matches that of a $\mathcal{N}(4.01, 0.71^2)$ distribution. Mean, standard deviation, skewness, and kurtosis for each choice of δ are given in Table 7.4, bolstering faith in the approximation $z \sim \mathcal{N}(\mu, \sigma^2)$.

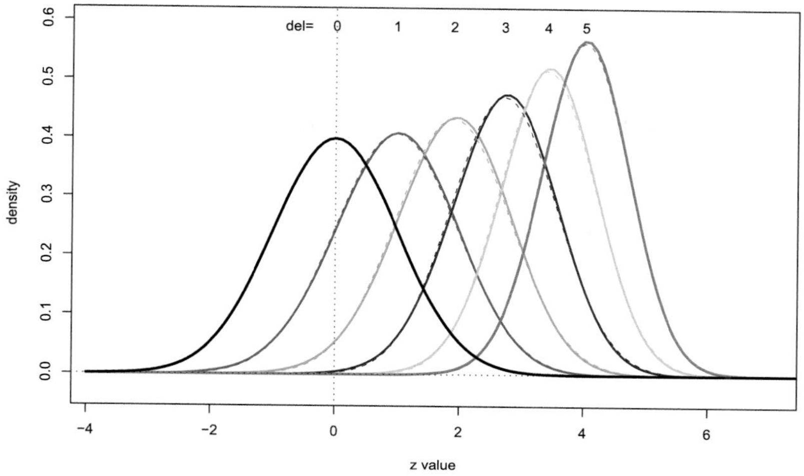

Figure 7.6 z-value density for non-central t distribution (7.78)–(7.79), for degrees of freedom $\nu = 20$ and non-centrality $\delta = 0, 1, 2, 3, 4, 5$. Dashed curves are normal densities matched in mean and standard deviation; see Table 7.4. Negative δ gives mirror image results.

Exercise 7.10 A test statistic t has possible densities $\{f_\theta(t), \theta \in \Theta\}$ with corresponding cdfs $F_\theta(t)$ and we wish to test $H_0 : \theta = \theta_0$. Show that the z-value statistic $z = \Phi^{-1}\{F_{\theta_0}(t)\}$ has densities

$$g_\theta(z) = \varphi(z) f_\theta(t) / f_{\theta_0}(t) \tag{7.80}$$

with $\varphi(z)$ the standard normal density (2.9).

T-statistics are ubiquitous in statistical applications. Suppose that in addition to the $N \times n$ expression matrix $\boldsymbol{X}$ in a microarray experiment, we observe a primary response variable u_j (u_j the treatment/control indicator in a two-sample study) and covariates $w_{j1}, w_{j2}, \dots, w_{jp}$ for the jth of n subjects. Given the observed expression levels $x_{i1}, x_{i2}, \dots, x_{in}$ for gene i, we could calculate t_i, the usual t-value for u_j as a function of x_{ij} after adjusting for the covariates in a linear model. Then, as in (7.78),

$$z_i = \Phi^{-1}\left(F_{n-p-1}(t_i)\right) \tag{7.81}$$

would be a z-value under the usual Gaussian assumption, with behavior like that in Table 7.4.

As a second example where exact calculations are possible, suppose we observe an i.i.d. sample $x_1, x_2, \ldots, x_n$ from an exponential distribution of unknown scale θ,

$$f_\theta(x) = \frac{1}{\theta} e^{-x/\theta} \qquad [x \geq 0]. \tag{7.82}$$

An exact z-value Z_0 for testing $H_0 : \theta = \theta_0$ is based on the mean $\bar{x}$,

$$Z_0 = \Phi^{-1}\left(G_n\left(n\bar{x}/\theta_0\right)\right) \tag{7.83}$$

where G_n is the cdf of a gamma distribution with n degrees of freedom.

Exercise 7.11 Using the fact that $\bar{x}$ is distributed as θ/n times a gamma variable with n degrees of freedom, show that $Z_0 \sim \mathcal{N}(0, 1)$ if $\theta = \theta_0$.

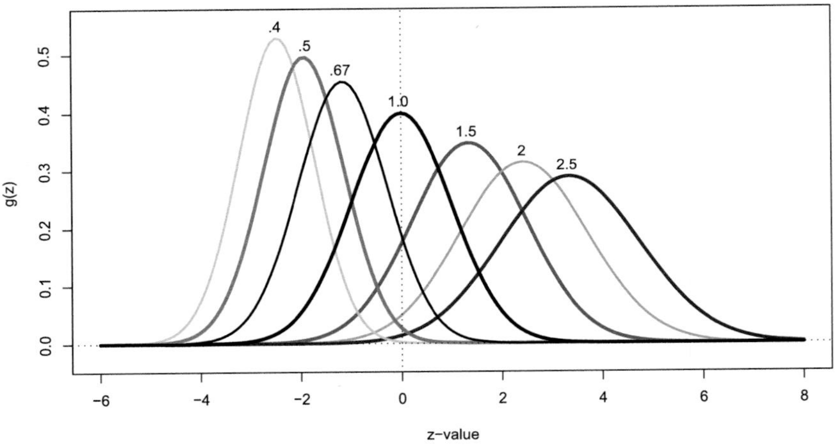

Figure 7.7 z-value density curves for gamma example, Z_0 as in (7.83) with $n = 10$, $\theta_0 = 1$. True value θ indicated above curves. Normality for non-null cases is nearly perfect here; see Table 7.5.

Figure 7.7 displays the density of Z_0 for different choices of θ when $n = 10$ and $\theta_0 = 1$. Normality is excellent: there is no visible discrepancy between the density curves and their matching normal equivalents (having means and standard deviations as in Table 7.5).

What follows is a heuristic argument showing why we can expect non-null normality for z-values obtained from an independent and identically distributed (i.i.d.) sample.[7] Suppose that $x_1, x_2, \ldots, x_n$ is such a sample,

[7] The argument is more technical than the previous material, but yields a simple and useful result at (7.102).

Table 7.5 *Mean, standard deviation, skewness, and kurtosis for* Z_0 (7.83); *the gamma example with* $n = 10$ *and* $\theta_0 = 1$. *For true* θ *in* (7.82) *as indicated.*

θ	.4	.5	.67	1	1.5	2.0	2.5
mean	−2.49	−1.94	−1.19	0	1.36	2.45	3.38
stdev	.76	.81	.88	1	1.15	1.27	1.38
skew	−.05	−.04	−.02	0	.02	.04	.04
kurt	.01	.01	.00	0	.00	−.01	−.04

obtained from cdf F_θ, a member of a one-parameter family

$$\mathcal{F} = \{F_\theta, \theta \in \Theta\} \tag{7.84}$$

having its moment parameters {mean, standard deviation, skewness, kurtosis}, denoted by

$$\{\mu_\theta, \sigma_\theta, \gamma_\theta, \delta_\theta\}, \tag{7.85}$$

defined differentiably in θ.

Under the null hypothesis $H_0 : \theta = 0$, which we can write as

$$H_0 : x \sim \{\mu_0, \sigma_0, \gamma_0, \delta_0\} \tag{7.86}$$

the standardized variate

$$Y_0 = \sqrt{n}\left(\frac{\bar{x} - \mu_0}{\sigma_0}\right) \qquad \left[\bar{x} = \sum_{i=1}^{n} y_i/n\right] \tag{7.87}$$

satisfies

$$H_0 : Y_0 \sim \left\{0, 1, \frac{\gamma_0}{\sqrt{n}}, \frac{\delta_0}{n}\right\}. \tag{7.88}$$

Normality can be improved to second order by means of a Cornish–Fisher transformation,

$$Z_0 = Y_0 - \frac{\gamma_0}{6\sqrt{n}}\left(Y_0^2 - 1\right) \tag{7.89}$$

which reduces the skewness in (7.88) from $O(n^{-1/2})$ to $O(n^{-1})$,

$$H_0 : Z_0 \sim \{0, 1, 0, 0\} + O\left(n^{-1}\right). \tag{7.90}$$

We can interpret (7.90) as saying that Z_0 is a *second-order z-value*,

$$H_0 : Z_0 \sim \mathcal{N}(0, 1) + O_p\left(n^{-1}\right) \tag{7.91}$$

e.g., *a test statistic giving standard normal p-values accurate to* $O(n^{-1})$.

Suppose now that H_0 is false and instead H_1 is true, with $x_1, x_2, \ldots, x_n$ i.i.d. according to

$$H_1 : x \sim \{\mu_1, \sigma_1, \gamma_1, \delta_1\} \tag{7.92}$$

rather than (7.86). Setting

$$Y_1 = \sqrt{n}\left(\frac{\bar{x} - \mu_1}{\sigma_1}\right) \quad \text{and} \quad Z_1 = Y_1 - \frac{\gamma_1}{6\sqrt{n}}\left(Y_1^2 - 1\right) \tag{7.93}$$

makes Z_1 second-order normal under H_1,

$$H_1 : Z_1 \sim \mathcal{N}(0, 1) + O_p\left(n^{-1}\right). \tag{7.94}$$

We wish to calculate the distribution of Z_0 (7.89) under H_1. Define

$$c = \sigma_1/\sigma_0, \quad d = \sqrt{n}(\mu_1 - \mu_0)/\sigma_0, \quad \text{and} \quad g_0 = \gamma_0 \big/ \left(6\sqrt{n}\right). \tag{7.95}$$

Some simple algebra yields the following relationship between Z_0 and Z_1.

Lemma 7.5 *Under definitions* (7.89), (7.93), *and* (7.95),

$$Z_0 = M + SZ_1 + g_0\left\{\left(\frac{\gamma_1}{\gamma_0}S - c^2\right)\left(Y_1^2 - 1\right) + \left(1 - c^2\right)\right\} \tag{7.96}$$

where

$$M = d \cdot (1 - dg_0) \quad \textit{and} \quad S = c \cdot (1 - 2dg_0). \tag{7.97}$$

The asymptotic relationships claimed at the start of this section are easily derived from Lemma 7.5. We consider a sequence of alternatives θ_n approaching the null hypothesis value θ_0 at rate $n^{-1/2}$,

$$\theta_n - \theta_0 = O\left(n^{-1/2}\right). \tag{7.98}$$

The parameter $d = \sqrt{n}(\mu_{\theta_n} - \mu_0)/\sigma_0$ defined in (7.95) is then of order $O(1)$, as is

$$M = d(1 - dg_0) = d\left(1 - d\gamma_0 \big/ \left(6\sqrt{n}\right)\right) \tag{7.99}$$

while standard Taylor series calculations give

$$c = 1 + \frac{\dot{\sigma}_0}{\dot{\mu}_0}\frac{d}{\sqrt{n}} + O\left(n^{-1}\right) \quad \text{and} \quad S = 1 + \left(\frac{\dot{\sigma}_0}{\dot{\mu}_0} - \frac{\gamma_0}{3}\right)\frac{d}{\sqrt{n}} + O\left(n^{-1}\right), \tag{7.100}$$

the dot indicating differentiation with respect to θ.

Theorem 7.6 *Under model* (7.84), (7.98), *and the assumptions of Lemma 7.5,*

$$Z_0 \sim \mathcal{N}\left(M, S^2\right) + O_p\left(n^{-1}\right) \tag{7.101}$$

with M and S as given in (7.99)–(7.100). *Moreover,*

$$\left.\frac{dS}{dM}\right|_{\theta_0} = \frac{1}{\sqrt{n}}\left(\left.\frac{d\sigma}{d\mu}\right|_{\theta_0} - \frac{\gamma_0}{3}\right) + O\left(n^{-1}\right). \tag{7.102}$$

Proof The proof of Theorem 7.6 uses Lemma 7.5, with θ_n determining H_1 in (7.96). Both $1 - c^2$ and $(\gamma_1/\gamma_0)S - c^2$ are of order $O(n^{-1/2})$; the former from (7.100) and the latter using $\gamma_1/\gamma_0 = 1 + (\dot{\gamma}_0/\gamma_0)(\theta_n - \theta_0) + O(n^{-1})$. Since $Y_1^2 - 1$ is $O_p(1)$, this makes the bracketed term in (7.96) $O_p(n^{-1/2})$; multiplying by $g_0 = \gamma_0/(6\sqrt{n})$ reduces it to $O_p(n^{-1})$, and (7.101) follows from (7.94). Differentiating M and S in (7.99)–(7.100) with respect to d verifies (7.102). □

Theorem 7.6 supports our claim that, under non-null alternatives, the null hypothesis normality of Z_0 degrades more slowly than its unit standard deviation, the comparison being $O_p(n^{-1})$ versus $O(n^{-1/2})$.

Result (7.102) simplifies further if $\mathcal{F}$ is a one-parameter exponential family, i.e., a family with densities proportional to $\exp\{\theta x\} \cdot f_0(x)$. Then $d\sigma/d\mu = \gamma/2$, so that (M, S) in (7.101) satisfy

$$\left.\frac{dS}{dM}\right|_{\theta_0} = \frac{\gamma_0}{6\sqrt{n}} + O\left(n^{-1}\right). \tag{7.103}$$

For the gamma family, $\gamma_0 = 2$, giving $\gamma_0/(6\sqrt{n}) = 0.1054$ for $n = 10$. This matches to better than three decimal places the observed change in standard deviation with respect to mean, as illustrated in Figure 7.7.

Moving beyond one-parameter families, suppose $\mathcal{F}$ is a p-parameter exponential family, having densities proportional to $\exp\{\eta_1 x_1 + \eta_2' x_2\} f_0(x_1, x_2)$, where η_1 and x_1 are real-valued while η_2 and x_2 are $(p-1)$-dimensional vectors, but where we are only interested in η_1 and not the nuisance vector η_2. The *conditional* distribution of x_1 given x_2 is then a one-parameter exponential family with natural parameter η_1, which puts us back in the context of Theorem 7.6.

The non-central t family does not meet the conditions of Lemma 7.5 or Theorem 7.6: (7.78) is symmetric in δ around zero, causing γ_0 in (7.98) to equal 0 and likewise the derivative in (7.102). Nevertheless, as Figure 7.5 shows, it does exhibit impressive non-null normality. In fact, my experience has been that z-values in general follow near-normal non-null distributions, making the crucial assumption (7.3) a reasonable starting point for their analysis.

7.5 Bootstrap Methods

It would be very nice to avoid all the normal theory parametric intricacies of the past sections, and obtain our accuracy estimates directly from a non-parametric bootstrap. Figure 7.8 tries this for the situation illustrated in Figure 7.1. Here the data matrix X has $N = 7128$ rows (genes) and $n = 72$ columns (patients), with $n_1 = 45$ ALL and $n_2 = 25$ AML patients as in Figure 6.1a.

Bootstrap data sets X^* were obtained by randomly selecting 45 of the first 45 X columns *with replacement*, and likewise 27 of the last 27 columns (under the assumption that the patients, though not the genes, were sampled independently). X^* gave N-vector $\boldsymbol{z}^*$ as for the actual data, and then $\bar{F}(x)^*$ (7.1). $B = 50$ replications of this whole process yielded values $\bar{F}(x)^{*1}, \bar{F}(x)^{*2}, \ldots, \bar{F}(x)^{*50}$, whose empirical standard deviation $\widehat{\text{sd}}(x)$ was the non-parametric bootstrap standard error estimate for $\bar{F}(x)$.

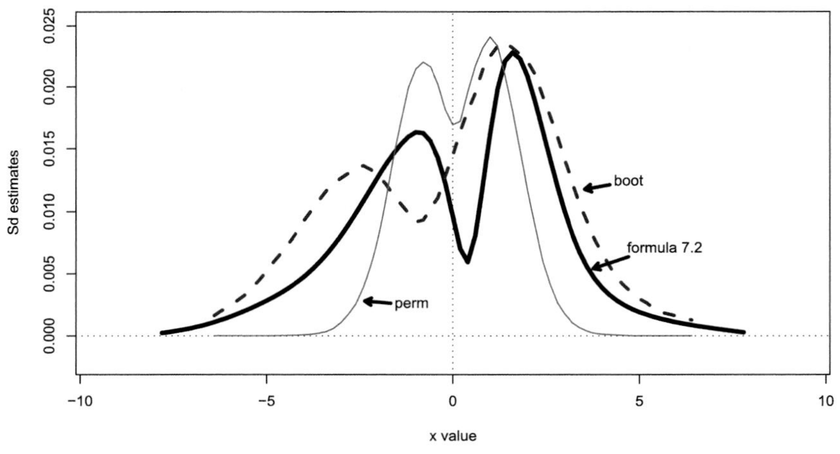

Figure 7.8 Standard deviation estimates for leukemia data cdf as in Figure 7.1. Non-parametric bootstrap (dashed line) gives a dilated version of sd(x) from parametric formula (7.2). Permutation $\widehat{\text{sd}}(x)$ curve (thin line) is too narrow.

The dashed curve in Figure 7.8 traces $\widehat{\text{sd}}(x)$ for $-8 < x < 8$. It looks like a dilated version of the curve from (7.2), with both peaks noticeably widened. This turns out to be an unfortunately dependable feature of non-parametric bootstrapping in an $N \gg n$ situation: for each i, the bootstrap replications $z_i^{*1}, z_i^{*2}, \ldots, z_i^{*50}$ roughly follow a normal distribution centered at z_i, say

$$z_i^* \dot{\sim} \mathcal{N}(z_i, \hat{v}_i) \qquad i = 1, 2, \ldots, N. \tag{7.104}$$

This widens the original z-value histogram of Figure 6.1a by adding the $\hat{v}_i$ component of variance, and subsequently widening the $\widehat{\mathrm{sd}}(x)$ curve.

Figure 7.8 also shows a permutation $\widehat{\mathrm{sd}}(x)$ curve, where now the $\boldsymbol{z}^*$ vectors are obtained by permuting all 72 columns of X. There are two reasons that this curve is much narrower than that from formula (7.2). The less important one here is that permutations nullify any true differences between ALL and AML patients (in contrast to the two-sample bootstrapping above). More crucially, permutations tend to enforce a $\mathcal{N}(0, 1)$ null distribution, Figure 6.5, much narrower than the $\mathcal{N}(0.09, 1.68^2)$ null seen in Figure 6.1a.

Column-wise bootstrapping and permuting both preserve within-column correlations. (Their correctly bimodal $\widehat{\mathrm{sd}}(x)$ curves in Figure 7.8 reflect this.) Trouble for the non-parametric bootstrap stems from a simpler cause: the empirical distribution of n points in N dimensions is a poor estimator of the true distribution when N is much larger than n. Parametric bootstrapping is less prone to failure, but of course that assumes knowledge of a reasonably accurate parametric model for X.

Notes

Most of this chapter's material is taken from Efron (2007a) and Efron (2010). Other good references for large-scale correlation effects include Dudoit et al. (2004); Owen (2005); Qiu et al. (2005a,b); Clarke and Hall (2009); and Desai et al. (2010). Schwartzman and Lin (2009) use Mehler's expansion to pursue higher-order accuracy formulas. The influence function calculations of Section 7.3 appear in Efron (2007b, Sect. 5) as part of a more extensive array of accuracy formulas.

Lancaster (1958) provides an insightful discussion of Mehler's formula, published in 1866! There are a variety of sources for Cornish–Fisher expansions, including Chapter 2 of Johnson and Kotz (1970) and, for much greater detail, Section 2.2 of Hall (1992).

The quantity $\gamma_0/6\sqrt{n}$ (7.103) is called the *acceleration* in Efron (1987), interpreted as "the rate of change of standard deviation with respect to expectation on a normalized scale," which agrees nicely with its use in (7.103).

Efron and Tibshirani (1993) provide a full treatment of the bootstrap, with emphasis on the importance of estimating the underlying probability distribution of the sampling units, as in their Figure 8.3. Effective bootstrapping usually requires $n \gg N$, just what we don't have here.

8

Correlation Questions

Correlation played a central role in the accuracy calculations of Chapter 7, with the estimated root mean square correlation $\hat{\alpha}$ being the key quantity in approximations (7.2), (7.49), and (7.50). This chapter takes up several questions concerning correlation in large-scale studies, including the calculation of $\hat{\alpha}$, which often turns out to be quite straightforward. The setting will be like that for our microarray examples: we observe an $N \times n$ matrix X as in (2.1); the columns of X correspond to individual subjects in Section 2.1, hopefully sampled independently of each other, while we expect the gene expressions within a column to be correlated.

Independence of the columns of X is a tacit assumption of all theoretical and permutation null hypothesis testing methods (calculations like (2.2)–(2.6) make no sense otherwise), though not of empirical null procedures. Section 8.3 considers tests for column-wise independence. We begin in Section 8.1 with a general result relating the row-wise and column-wise correlations of X, leading directly to the estimator $\hat{\alpha}$.

8.1 Row and Column Correlations

We will be interested in both the row-wise and column-wise correlations of the the $N \times n$ matrix[1] $X = (x_{ij})$. It simplifies notation to assume that X has been "demeaned" by the subtraction of row and column means, so that

$$\sum_{i=1}^{N} x_{ij} = \sum_{j=1}^{n} x_{ij} = 0 \qquad \text{for } i = 1, 2, \ldots, N \text{ and } j = 1, 2, \ldots, n. \tag{8.1}$$

[1] The number of rows of X is often denoted by p rather than N in the microarray literature, following (in transposed fashion) an analogy with the familiar ANOVA $n \times p$ structure matrix. This can cause confusion with p-values and probabilities. Our use of "big N" suggests the usual microarray situation $N \gg n$.

Sometimes we will go further and assume *double standardization*: that in addition to (8.1) we have

$$\sum_{i=1}^{N} x_{ij}^2 = N \quad \text{and} \quad \sum_{j=1}^{n} x_{ij}^2 = n$$
$$\text{for} \quad i = 1, 2, \ldots, N \quad \text{and} \quad j = 1, 2, \ldots, n. \tag{8.2}$$

Double standardization is convenient for our discussion because it makes sample covariances into sample correlations. How to achieve (8.2) is discussed at the end of this section.

With means removed as in (8.1), the sample covariance between columns j and j' of X is

$$\hat{\Delta}_{jj'} = \sum_i x_{ij} x_{ij'} / N \tag{8.3}$$

(*not* reducing the denominator by 1) and similarly

$$\hat{\sigma}_{ii'} = \sum_j x_{ij} x_{i'j} / n \tag{8.4}$$

for the sample covariance between rows i and i'. The $n \times n$ column sample covariance matrix $\hat{\boldsymbol{\Delta}}$ and the $N \times N$ row sample covariance matrix $\hat{\boldsymbol{\Sigma}}$ are

$$\hat{\boldsymbol{\Delta}} = X'X/N \quad \text{and} \quad \hat{\boldsymbol{\Sigma}} = XX'/n. \tag{8.5}$$

Double standardization makes the diagonal elements $\hat{\Delta}_{jj}$ and $\hat{\sigma}_{ii}$ equal 1. Then the sample correlation $\hat{\rho}_{ii'}$ between rows i and i' of X is

$$\hat{\rho}_{ii'} = \hat{\sigma}_{ii'} / (\hat{\sigma}_{ii} \hat{\sigma}_{i'i'})^{1/2} = \hat{\sigma}_{ii'}. \tag{8.6}$$

Our main result is stated and proved in terms of the *singular value decomposition* (svd) of X,

$$\underset{N \times n}{X} = \underset{N \times K}{U} \; \underset{K \times K}{\boldsymbol{d}} \; \underset{K \times n}{V'} \tag{8.7}$$

where K is the rank of X, $\boldsymbol{d}$ is the diagonal matrix of ordered *singular values* $d_1 \geq d_2 \geq \cdots \geq d_K > 0$, and U and V are orthonormal matrices of sizes $N \times K$ and $n \times K$,

$$U'U = V'V = I_K, \tag{8.8}$$

with I_K the $K \times K$ identity. The squares of the singular values,

$$e_1 \geq e_2 \geq \cdots \geq e_K > 0 \qquad [e_k = d_k^2] \tag{8.9}$$

are the eigenvalues of $N\hat{\boldsymbol{\Delta}} = V\boldsymbol{d}^2V'$ and also of $n\hat{\boldsymbol{\Sigma}} = U\boldsymbol{d}^2U'$.

The singular value decomposition reveals an interesting correspondence between the row and column correlations of X.

Theorem 8.1 *If X has row and column means equal to zero* (8.1), *then the n^2 elements of $\hat{\mathbf{\Delta}}$ and the N^2 elements of $\hat{\mathbf{\Sigma}}$ both have mean zero and empirical variance*

$$c_2 = \sum_{k=1}^{K} e_k^2 \Big/ (Nn)^2. \tag{8.10}$$

Proof Letting 1_n denote the vector of n 1's, the sum of the entries of $\hat{\mathbf{\Delta}}$ is

$$1_n' X' X 1_n / N = 0 \tag{8.11}$$

according to (8.1), while the mean of the squared entries is

$$\frac{\sum_{j=1}^{n} \sum_{j'=1}^{n} \hat{\Delta}_{jj'}^2}{n^2} = \frac{\operatorname{tr}\left((X'X)^2\right)}{(Nn)^2} = \frac{\operatorname{tr}\left(V\boldsymbol{d}^4 V'\right)}{(Nn)^2} = c_2. \tag{8.12}$$

Replacing $X'X$ with XX' yields the same result for $\hat{\mathbf{\Sigma}}$. □

Exercise 8.1 Here tr indicates the trace of a square matrix, the sum of its diagonal elements. Verify (8.12). *Hint*: Use the trace commutative property $\operatorname{tr}(AB) = \operatorname{tr}(BA)$.

The covariances $\hat{\Delta}_{jj'}$ and $\hat{\sigma}_{ii'}$ become correlations under double standardization, as in (8.6). Then Theorem 8.1 says that the row and column sample correlations have the same means and empirical standard deviations. This sounds definitely counter-intuitive in a typical microarray setting, where the genes (rows) are assumed correlated and arrays (columns) independent.

Figure 8.1 shows an example. Here the data is from the *Cardio Study*, a microarray experiment comparing $n = 63$ subjects, $n_1 = 44$ healthy controls and $n_2 = 19$ cardiovascular patients, each measured on $N = 20\,426$ genes. The matrix X used in Figure 8.1 is the doubly standardized expression data for the healthy controls, so X is $20\,426 \times 44$ and satisfies (8.1)–(8.2).

The solid histogram shows all 44^2 column sample correlations, while the line histogram is a random sample of 10 000 row correlations (out of the more than 400 million values $\hat{\rho}_{ii'}$!). In this case c_2 (8.8) equals 0.283^2, so both histograms have mean zero and standard deviation 0.283.

Exercise 8.2 What is a computationally efficient way to calculate c_2 (without using the SVD)? *Hint*: The trace of a square symmetric matrix is the sum of its eigenvalues.

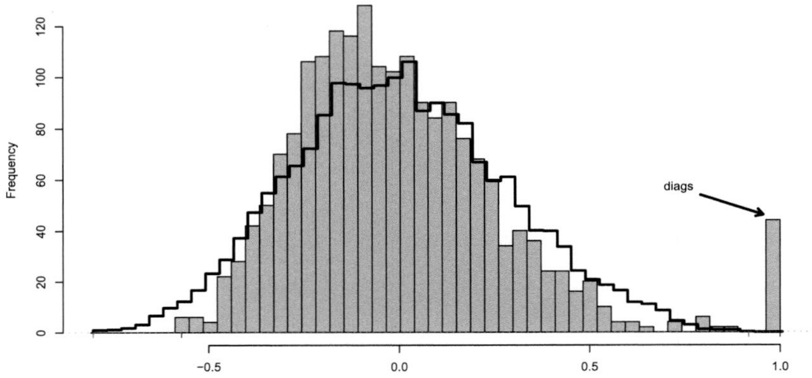

Figure 8.1 *Solid histogram* shows the 44^2 column sample correlations of the doubly standardized expression matrix X for the healthy cardio controls. *Line histogram* depicts the sample of 10 000 of the $20\,426^2$ row correlations. Both histograms have mean 0 and standard deviation 0.283.

The 44 diagonal elements of $\hat{\mathbf{\Delta}}$, the column correlation matrix, protrudes as a prominent spike at 1. (We cannot see the spike of 20 426 elements of $\hat{\mathbf{\Sigma}}$, the row correlation matrix, because they form such a small fraction of all $20\,426^2$.) It is easy to remove the diagonal 1's from consideration.

Corollary 8.2 *In the doubly standardized situation, the off-diagonal elements of the column correlation matrix* $\hat{\mathbf{\Delta}}$ *have empirical mean and variance*

$$\hat{\mu} = -\frac{1}{n-1} \quad \textit{and} \quad \hat{\alpha}^2 = \frac{n}{n-1}\left(c_2 - \frac{1}{n-1}\right) \tag{8.13}$$

with c_2 *as in* (8.10).

Exercise 8.3 (a) Verify (8.13). (b) Show that $c_2 \geq 1/K$. *Hint*: Under double standardization, $\bar{e}$, the average of the eigenvalues (8.9), equals Nn/K.

Standardization

A matrix X is *column standardized* by individually subtracting the mean and dividing by the standard deviation of each column, and similarly for row standardization. Table 8.1 shows the effect of successive row and column standardizations on the original $20\,426 \times 44$ matrix of healthy cardio subjects. Here "Col" is the empirical standard deviation of the 946 column-wise correlations $\hat{\Delta}_{jj'}$ $j < j'$; "Eig" is $\hat{\alpha}$ in (8.13); and "Row" is the empirical standard deviation of a 1% sample of the row correlations $\hat{\rho}_{ii'}$, adjusted for overdispersion as in (8.18) of the next section. The sampling error of the Row entries is about ±0.0034.

Table 8.1 *Estimates of the rms correlation $\hat{\alpha}$ after successive column and row standardization of the* 20 426 × 44 *matrix of healthy cardio subjects, as explained in the text. Sampling error of Row entries* ±0.0034.

	Col	Row	Eig
demeaned	.252	.286	.000
col	.252	.249	.251
row	.242	.255	.246
col	.242	.241	.242
row	.241	.246	.235
col	.241	.244	.241
row	.241	.245	.234
col	.241	.238	.241

	Col	Row	Eig
demeaned	.252	.286	.000
row	.241	.283	.279
col	.241	.251	.240
row	.240	.247	.241
col	.240	.247	.240
row	.241	.240	.235
col	.241	.237	.240
row	.241	.233	.233

The doubly standardized matrix X used in Figure 8.1 was obtained after five successive column/row standardizations of the original cardio matrix. This was excessive: the figure looked almost the same after two iterations. Other microarray examples converged equally rapidly, and it can be shown that convergence is guaranteed except in some special small-size counterexamples.

Microarray analyses usually begin with some form of column-wise standardization, designed to negate "brightness" differences among the n arrays. In the same spirit, row standardization helps prevent incidental gene differences (such as very large or very small expression level variabilities) from obscuring the effects of interest. This isn't necessary for z-values based on t statistics, which are "self-standardizing," but can be important in other contexts, as demonstrated in Chapter 10.

8.2 Estimating the Root Mean Square Correlation

For $n = 44$ and $c_2 = 0.283^2$, formula (8.13) yields

$$\left(\hat{\mu}, \hat{\alpha}^2\right) = \left(-0.023, 0.241^2\right), \qquad (8.14)$$

reducing our estimated standard deviation for the column correlations from 0.283 to 0.241. The corresponding diagonal-removing correction for the row correlations (replacing n by N in (8.13)) is negligible for $N = 20\,426$. However, c_2 overestimates the variance of the row correlations for another reason: with only $n = 44$ bivariate points available to estimate each correlation, random error adds a considerable component of variance to the $\hat{\rho}_{ii'}$ histogram in Figure 8.1. Once this overdispersion is corrected for, $\hat{\alpha}$ in (8.13) turns out to be a reasonable estimator for α, the root mean square correlation (7.39).

We assume that the columns $\boldsymbol{x}$ of $X = (\boldsymbol{x}_1, \boldsymbol{x}_2, \ldots, \boldsymbol{x}_n)$ are independent and identically distributed N-vectors, with correlation $\rho_{ii'}$ between entries x_i and $x_{i'}$. The rms correlation is

$$\alpha = \left[\sum_{i<i'} \rho_{ii'}^2 \Big/ \binom{N}{2} \right]^{\frac{1}{2}} \tag{8.15}$$

which is another notation for (7.39).

Define $\bar{\alpha}$ to be the observed row-wise root mean square correlation,

$$\bar{\alpha} = \left[\sum_{i<i'} \hat{\rho}_{ii'}^2 \Big/ \binom{N}{2} \right]^{\frac{1}{2}}. \tag{8.16}$$

In the doubly standardized case,

$$\bar{\alpha}^2 = \frac{Nc_2 - 1}{N - 1} \doteq c_2 \tag{8.17}$$

but definition (8.16) can be evaluated directly, with or without standardization. (Program `alpha`, Appendix B, approximates α by randomly selecting a subset I of the rows of X, evaluating $\hat{\rho}_{ii'}$ for i and i' in I, and calculating the corresponding root mean square $\hat{\rho}_{ii'}$ value.) Our preferred estimator for the rms correlation, the one used in Figure 7.1, is

$$\hat{\alpha} = \left[\frac{n}{n-1} \left(\bar{\alpha}^2 - \frac{1}{n-1} \right) \right]^{\frac{1}{2}}. \tag{8.18}$$

Under double standardization this is almost the same as $\hat{\alpha}$ in (8.13). Section 8.4 justifies (8.18) squared as a nearly unbiased estimator of the true α^2 in multivariate normal models, a result supported more generally by numerical simulations. It gave $\hat{\alpha} = 0.245$ for the cardio example.

Table 8.2 shows $\hat{\alpha}$ estimates for the leukemia data of Figure 6.1a, applied separately to the two patient classes and also to the entire 7128×72 expression matrix. The original matrices were first standardized by having

Table 8.2 $\hat{\alpha}$ (8.18) *for the leukemia data of Figure 6.1a.*

	AML	ALL	Together
standardized	**.120**	**.104**	**.117**
doubly standardized	.110	.088	.103

each column linearly transformed to mean 0 and variance 1; $\hat{\alpha}$ values for the doubly standardized versions, described below, are seen to be somewhat smaller. The "together" calculation is biased upward since genes that actually express differently in the two patient classes will appear to be more highly correlated, though that effect doesn't seem crucial here. In general, we should remove any suspected "signal" from the rows of X (such as a regression on a known covariate) before estimating the row-wise correlations.

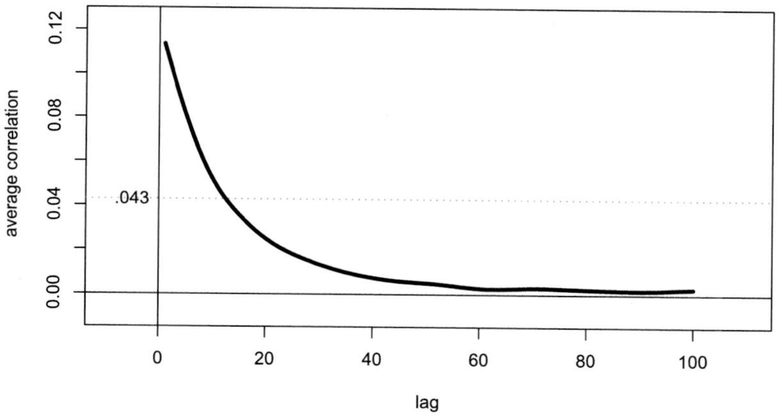

Figure 8.2 Average serial correlations for the snp counts along $N = 26\,874$ sites on chromosome 9 for 313 healthy Caucasian men; rms correlation $\hat{\alpha} = 0.043$.

Figure 8.2 concerns a single nucleotide polymorphism (snp) study that we shall see more of in Chapter 10: $n = 313$ healthy Caucasian men had their genetic bases — A, G, T, or C — read at $N = 26\,874$ sites on both copies of chromosome number 9. The $N \times n$ matrix X had entries x_{ij} equaling the number of bases disagreeing with the modal choice at that site, $x_{ij} = 0$, 1, or 2, i.e., the number of "snps." Program `alpha` gave $\hat{\alpha} = 0.043$, much smaller than in the cardio or leukemia studies. Figure 8.2 shows the average lagged correlation $\hat{\rho}_{i,i+j}$ for lags $j = 1, 2, \ldots, 100$, as we move along chromosome 9. Nearby locations are seen to be more positively correlated. However, most of the rms value $\hat{\alpha} = 0.043$ is not accounted for by serial correlation.

Exercise 8.4 Why do we know this last statement is true?

Table 8.3 *Rms correlation estimates $\hat{\alpha}$ for prostate data, Section 2.1, and HIV data, Section 6.1.*

	Controls	Patients	Together
prostate data	.014	.016	.033
HIV data	.191	.209	.290

Program `alpha` was used to estimate α for the prostate and HIV microarray studies of Section 2.1 and Section 6.1. Table 8.3 shows $\hat{\alpha}$ to be near zero for the prostate study, and quite large for the HIV data (about as large as $\hat{\alpha} = 0.245$ for the cardio study).

Chapter 7 showed how correlation can drastically degrade estimation accuracy, a point reiterated in Section 8.4. Microarrays are still a relatively new technology, and improved biomedical methods in exposure, registration, and background control may reduce correlation problems. Purely statistical improvements can also reduce correlations, perhaps by more extensive standardization techniques. However, none of this will help if microarray correlations are inherent in the way genes interact at the DNA level, rather than a limitation of current methodology. My own experience has shown microarray studies in general to be the worst correlation offenders; imaging experiments often appear to be highly correlated, as in Figure 2.4, but the correlation is local rather than long-range, yielding small values of the rms correlation.

Simulating correlated z-values

A great advantage of the theory in Chapter 7 is that the rms accuracy formulas depend only on the rms correlation α and not on the entire $N \times N$ covariance matrix Σ for the N-vector of z-values $\boldsymbol{z}$. However, simulations like that for Figure 7.4 require us to generate vectors $\boldsymbol{z}$ whose Σ matrix has the desired α. The program `simz` uses a simple construction to do so; the N

cases are divided into J blocks of length $H = N/J$ (with default $J = 5$); independent $\mathcal{N}(0, 1)$ variates U_j and V_{hj} are generated for $j = 1, 2, \ldots, J$ and $h = 1, 2, \ldots, H_j$, and the N z-values are set equal to $(\gamma U_j + V_{hj})/(1 + \gamma^2)^{1/2}$. The constant γ is chosen to make the rms correlation of the demeaned variates $z_i - \bar{z}$ equal the desired value of α.

For the simulation of Figure 7.4, with $N = 6000$ and $\alpha = 0.10$, the correlation matrix of each column of X had five blocks of 1200 "genes" each, correlation 0.20 within blocks and -0.05 across blocks. The 80 columns were generated independently in this way. A constant δ was added to the first 300 genes of the last 40 columns in order to simulate the two-class model (7.42).

Exercise 8.5 X was demeaned for the simulation in Figure 7.4. What was the value of the constant δ?

8.3 Are a Set of Microarrays Independent of Each Other?

Most of the statistical microarray literature assumes that the columns of X, the individual arrays, are sampled independently of each other, even though the gene expressions within each array may be highly correlated. This was the assumption in Section 2.1, for example, justifying the t-test and the theoretical null hypothesis (2.6). Simple permutation methods (not related to permutation null hypotheses) offer a quick check on the assumption of column-wise independence, sometimes with surprising results.

The scatterplot in Figure 8.3 shows the $N = 20\,426$ points $(x_{i,31}, x_{i,32})$ for the cardio data matrix. It seems to indicate a very strong positive correlation between arrays 31 and 32, with $\hat{\Delta}_{31,32} = 0.805$ (8.3). (Since X is doubly standardized, each row has mean 0, ruling out so-called "ecological correlation," due only to differing gene-wise response levels.) This seems like a smoking gun argument against column-wise independence, but the gunsmoke isn't as convincing as it looks, for reasons discussed in Section 8.4.

A simple test for column-wise independence is based on the first eigenvector v_1 of the $n \times n$ column sample covariance matrix $\hat{\boldsymbol{\Delta}}$ (8.5), v_1 also being the first column of V in (8.7). The left panel of Figure 8.4 shows the components of v_1 for the cardio matrix, plotted versus the column number $1, 2, \ldots, 44$. Suppose that the columns of the original expression matrix before standardization were independent and identically distributed vectors. This implies that the columns of X are exchangeable, and that all orderings of the components of v_1 are equally likely.

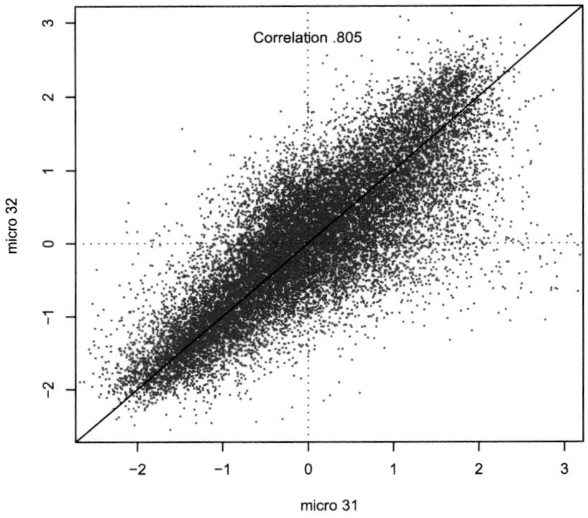

Figure 8.3 Scatterplot of microarrays 31 and 32 of the cardio data, $(x_{i,31}, x_{i,32})$ for $i = 1, 2, \ldots, N = 20\,426$. It seems to indicate strong correlation: $\hat{\Delta}_{31,32} = 0.805$.

This is not what the figure shows: the components seem to increase from left to right, with a noticeable block of large values at arrays 27–32. Time trends and block effects are signs of experimental instabilities that cause correlation across arrays. Uncontrolled changes in the development process, for example, can raise or lower expression measurements for whole sets of genes on different areas of the microarray chip, correlating the subsequent block of arrays and raising the corresponding entries of v_1.

Let $S(v_1)$ be a real-valued statistic that measures structure, for instance the linear regression coefficient of the components of v_1 versus the array index. Comparing $S(v_1)$ with a set of permuted values

$$\left\{S^{*b} = S\left(v^{*b}\right),\ b = 1, 2, \ldots, B\right\} \tag{8.19}$$

where v^{*b} is a random permutation of the components of v_1, provides a quick test of the i.i.d. null hypothesis. Taking $S(v_1)$ to be the absolute value of the above-mentioned linear regression statistic, $B = 2000$ permutations yielded only eight values S^{*b} exceeding $S(v_1)$. This gave p-value $8/2000 = 0.004$, strongly significant evidence against the null hypothesis.

Permutation testing was also applied using the *block statistic*

$$S(v_1) = v_1' \boldsymbol{B} v_1 \tag{8.20}$$

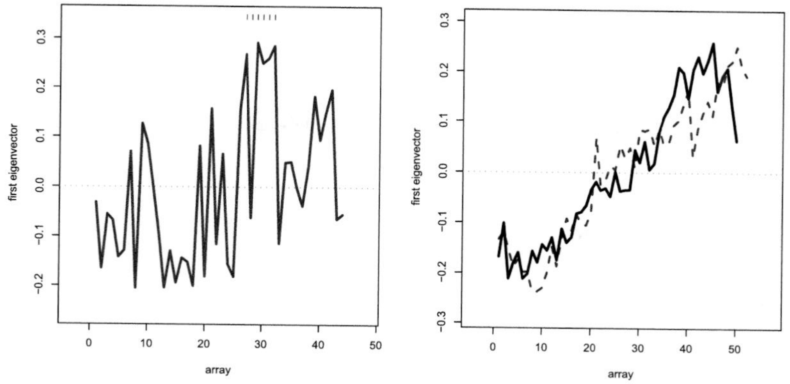

Figure 8.4 *Left panel*: Components of the first eigenvector plotted versus array index $1, 2, \ldots, 44$, for the cardio data matrix; dashes emphasize the block of large components for arrays 27–32. *Right panel*: First eigenvector for control subject (solid line) and cancer patient (dashed line) matrices, prostate cancer study of Section 2.1; there was a drift in expression levels as the study progressed.

where $\boldsymbol{B}$ is the $n \times n$ matrix

$$\boldsymbol{B} = \sum_h \beta_h \beta_h'. \tag{8.21}$$

The sum in (8.21) is over all vectors β_h of the form

$$\beta_h = (0, 0, \ldots, 0, 1, 1, \ldots, 1, 0, 0, \ldots, 0)' \tag{8.22}$$

with the 1's forming blocks of length between 2 and 10 inclusive. $B = 5000$ permutations of the cardio vector v_1 gave only three v^{*b} values exceeding $S(v_1)$, p-value 0.0006, still stronger evidence against the i.i.d. null hypothesis.

The right panel of Figure 8.4 concerns the prostate data of Section 2.1. The first eigenvectors v_1 of $\hat{\boldsymbol{\Delta}}$, computed separately for the 50 control subjects and the 52 cancer patients (with the two matrices individually doubly standardized), are plotted versus array number. Both vectors increase nearly linearly from left to right, indicating a systematic drift in expression levels as the study progressed. Some genes drifted up, others down, the average drift equaling zero because of standardization. The resulting correlation effect is small (or else $\hat{\alpha}$ would be larger than Table 8.3 indicates) but genuine: both the slope and block tests decisively rejected independence across the columns.

Exercise 8.6 Show that if X is doubly standardized, the average linear regression coefficient $\hat{\beta}_i$ of x_{ij} on $j = 1, 2, \ldots, n$ equals 0.

Table 8.4 *Correlation matrix and first eigenvector v_1 for the eight BRCA2 cases, $N = 3226$ genes. The last four arrays are strongly positively correlated.*

	1	2	3	4	5	6	7	8
1	1.00	.07	.09	.26	−.14	−.16	−.19	−.12
2	.07	1.00	.17	−.02	−.08	−.10	−.04	−.10
3	.09	.17	1.00	−.08	.01	−.06	−.10	−.03
4	.26	−.02	−.08	1.00	−.09	−.06	−.08	−.10
5	−.14	−.08	.01	−.09	1.00	.24	.34	.43
6	−.16	−.10	−.06	−.06	.24	1.00	.35	.35
7	−.19	−.04	−.10	−.08	.34	.35	1.00	.31
8	−.12	−.10	−.03	−.10	.43	.35	.31	1.00
v_1	**.30**	**.15**	**.10**	**.19**	**−.45**	**−.44**	**−.46**	**−.48**

Sometimes a pattern of column-wise correlation is obvious to the eye. Table 8.4 concerns the *BRCA data*, a microarray experiment concerning BRCA mutations, two known genetic factors that increase the risk of breast cancer. In this study, tumors from seven breast cancer patients having the BRCA1 mutation were compared against eight with BRCA2, on microarrays measuring $N = 3226$ genes. The table shows the 8×8 correlation matrix for the BRCA2 subjects, column standardized. We see strong positive correlations in the second group of four arrays, perhaps indicating an uncontrolled change in the development process. The first-four/last-four effect is particularly striking in the first eigenvector v_1. A block test (now using all blocks of length 2 through 4) rejected independence at p-value 0.025, which is close to the most extreme result possible with only eight columns.

Two objections can be raised to our permutation tests: (1) they are really testing i.i.d., not independence; (2) non-independence might not manifest itself in the order of v_1 (particularly if the order of the microarrays has been shuffled in some unknown way).

Column-wise standardization makes the column distributions more similar, mitigating objection (1). Going further, "quantile standardization" — say replacing each column's entries by normal scores — makes the marginals exactly the same. The cardio data was reanalyzed using normal scores, with almost identical results.

Objection (2) is more worrisome from the point of view of statistical power. The order in which the arrays were obtained *should* be available to the statistician, and should be analyzed to expose possible trends like those in Figure 8.4. It would be desirable, nevertheless, to have independence tests that do not depend on order, that is, test statistics invariant under column-wise permutations.

This turns out to be problematic, for reasons having to do with Theorem 8.1. An obvious test statistic for column-wise independence, and one that doesn't depend on the order of the columns, is

$$S = \sum_{j<j'} \hat{\Delta}_{jj'}^2 \Big/ \binom{n}{2} \tag{8.23}$$

the average squared off-diagonal column correlation. But if X is doubly standardized then S is a monotone increasing function of $\hat{\alpha}$ (8.18), the estimated rms row correlation: so large values of S can always be blamed on large row-wise correlations.

Exercise 8.7 Express S as a function of $\hat{\alpha}$.

In other words, the spread of the solid histogram in Figure 8.1 can always be attributed to the spread of the line histogram. This doesn't rule out all histogram tests: the histogram of the 946 off-diagonal values $\hat{\Delta}_{jj'}$ appears to be long-tailed to the right. Applying `locfdr` yielded five values $\hat{\Delta}_{jj'}$ having $\widehat{\text{fdr}}$ (6.38) less than 0.20, those exceeding 0.743. (Null density estimates $(\hat{\delta}_0, \hat{\sigma}_0, \hat{\pi}_0) = (-0.04, 0.236, 998)$ from the MLE method.) All five were from the block 27–32 emphasized in Figure 8.4.

The threshold 0.743 for significance seems remarkably high given that each $\hat{\Delta}_{jj'}$ is based on $N = 20\,426$ pairs $(x_{ij}, x_{ij'})$, as seen in Figure 8.3. However, the pairs are not independent of each other because of gene-wise correlations. We will see in Section 8.4 that the "effective sample size" for estimating $\Delta_{jj'}$ is *much* smaller than N.

8.4 Multivariate Normal Calculations

Classic statistical theory depended heavily on normal distributions. Normality is a risky assumption in large-scale studies but, risky or not, it is a useful tool for analyzing our methodology. This section gives a brief review of the ideas involved.

A random $N \times n$ matrix X is said to have *matrix normal distribution*

$$\underset{N\times n}{X} \sim \mathcal{N}_{N\times n}\Big(\underset{N\times n}{M}, \underset{N\times N}{\Sigma} \otimes \underset{n\times n}{\Delta}\Big) \tag{8.24}$$

if the Nn entries x_{ij} are jointly normal, with means and covariances

$$E\{x_{ij}\} = m_{ij} \quad \text{and} \quad \operatorname{cov}(x_{ij}, x_{i'j'}) = \sigma_{ii'}\Delta_{jj'}. \tag{8.25}$$

The notation $\otimes$ indicates a *Kronecker product*[2] as defined by (8.25). If $\Delta = I_n$, the $n \times n$ identity matrix, then

$$\operatorname{cov}(x_{ij}, x_{i'j}) = \begin{cases} \sigma_{ii'} & \text{if } j = j' \\ 0 & \text{if } j \neq j'; \end{cases} \tag{8.26}$$

that is, the columns of X are independent of each other, each with covariance matrix Σ, which was the assumption leading to our estimator $\hat{\alpha}$ in Section 8.2. If Δ is not diagonal, X has both its rows and columns correlated, a possibility we want to explore further.

Linear transformations are particularly convenient for matrix normal distributions. If

$$\underset{a\times b}{Y} = \underset{a\times N}{A}\ \underset{N\times n}{X}\ \underset{n\times b}{B'} \tag{8.27}$$

then

$$Y \sim \mathcal{N}_{a\times b}(AMB', A\Sigma A' \otimes B\Delta B'). \tag{8.28}$$

(The mean and covariance formulas in (8.28) depend only on (8.25) and do not require normality.)

Demeaning is an example of (8.27). Let $B = I_n - 1_n 1_n'/n$ and $A = I_N - 1_N 1_N'/N$, where 1_n is a column vector of n 1's. Then Y is the demeaned version of X, i.e., X with its row and column means subtracted off.

Exercise 8.8 (a) Verify this statement. (b) Show that Y has mean $M^Y = 0$ and covariance $\Sigma^Y \otimes \Delta^Y$, where

$$\sigma^Y_{ii'} = \sigma_{ii'} - \sigma_{i\cdot} - \sigma_{\cdot i'} + \sigma_{\cdot\cdot} \quad \text{and} \quad \Delta^Y_{jj'} = \Delta_{jj'} - \Delta_{j\cdot} - \Delta_{\cdot j'} + \Delta_{\cdot\cdot} \tag{8.29}$$

where the dots indicate averaging over the missing indices. (c) Show that demeaning reduces the rms correlations of both rows and columns.

[2] Many references reverse the notation to $\Delta \otimes \Sigma$, this being the convention for the R function `kronecker`, but both notations agree on (8.25).

In what follows we will assume that X has been both demeaned and doubly standardized, in the population sense that $M = 0$ and Σ and Δ have all diagonal elements 1,

$$m_{ij} = 0, \quad \sigma_{ii} = 1, \quad \text{and} \quad \Delta_{jj} = 1. \tag{8.30}$$

An unbiased estimator of α^2

The matrix normal model leads directly to an unbiased estimator of the squared rms correlation α^2 (8.15), *assuming that the columns of X are independent*, which, under (8.30), is equivalent to the column covariance matrix equaling the identity

$$\boldsymbol{\Delta} = I_n. \tag{8.31}$$

As in (8.16), let $\bar{\alpha}$ be the observed rms row-wise correlation.

Theorem 8.3 *Under the matrix normal model* (8.24), (8.30), (8.31),

$$\tilde{\alpha}_n^2 = \frac{n}{n+1}\left(\bar{\alpha}^2 - \frac{1}{n}\right) \tag{8.32}$$

is an unbiased estimator of α^2.

The proof appears at the end of this section. It depends on taking $\hat{\boldsymbol{\Sigma}}$ (8.5) to have a Wishart distribution with n degrees of freedom. Under the usual normal assumptions, demeaning reduces the degrees of freedom to $n - 1$, suggesting the estimator

$$\tilde{\alpha}_{n-1}^2 = \frac{n-1}{n}\left(\bar{\alpha}^2 - \frac{1}{n-1}\right). \tag{8.33}$$

A comparison with our preferred estimator $\hat{\alpha}$ in (8.18) shows that they are proportional, with

$$\hat{\alpha}/\tilde{\alpha}_{n-1} = n/(n-1) \doteq 1 + 1/n. \tag{8.34}$$

Usually this small factor is neglible. Jensen's inequality says that $\tilde{\alpha}_{n-1}$ will be biased downward as an estimator of α, and simulations indicate that the slight increase in $\hat{\alpha}$ helps mitigate this bias.

Effective sample size

The $N = 20\,426$ points in Figure 8.3 appear strongly correlated, giving $\hat{\Delta}_{31,32} = 0.805$. There would be no questioning correlation if the points (x_{i32}, x_{i33}) were independent across cases $i = 1, 2, \ldots, N$. But they are not independent, and their dependence dramatically lowers the effective sample size.

Assume again the matrix normal model (8.24), (8.30): $X \sim \mathcal{N}_{N\times n}(0, \Sigma \otimes \Delta)$ with the diagonal elements of both Σ and Δ equaling 1. Traditional multivariate analysis[3] deals with the situation $\Sigma = I_N$, i.e., with the rows of X independent, say

$$\boldsymbol{x}_i \overset{\text{iid}}{\sim} \mathcal{N}_n(0, \boldsymbol{\Delta}), \qquad i = 1, 2, \ldots, N. \tag{8.35}$$

This gives $\hat{\boldsymbol{\Delta}}$ a scaled Wishart distribution, unbiased for $\boldsymbol{\Delta}$, whose n^2 means and $n^2 \times n^2$ covariances we can denote as

$$\underset{n\times n}{\hat{\boldsymbol{\Delta}}} \sim \left(\underset{n\times n}{\boldsymbol{\Delta}}, \underset{n^2\times n^2}{\boldsymbol{\Delta}^{(2)}} \Big/ N\right) \qquad \left[\Delta^{(2)}_{jk,lh} = \Delta_{jl}\Delta_{kh} + \Delta_{jh}\Delta_{kl}\right]. \tag{8.36}$$

So, for example, $\text{cov}(\hat{\Delta}_{11}, \hat{\Delta}_{23}) = 2\Delta_{12}\Delta_{13}/N$.

Correlation between the rows of X doesn't change the fact that $\hat{\boldsymbol{\Delta}}$ is unbiased for $\boldsymbol{\Delta}$, but it increases the variability of $\hat{\boldsymbol{\Delta}}$.

Theorem 8.4 *Under the matrix normal model* (8.24), (8.30)*, the column covariance estimator* $\hat{\boldsymbol{\Delta}} = X'X/N$ *has mean and covariance*

$$\hat{\boldsymbol{\Delta}} \sim \left(\boldsymbol{\Delta}, \boldsymbol{\Delta}^{(2)} \Big/ N_{\text{eff}}\right) \tag{8.37}$$

where N_{eff} *is the "effective sample size"*

$$N_{\text{eff}} = N \Big/ \left[1 + (N-1)\alpha^2\right] \tag{8.38}$$

where α *is the rms correlation* (8.15) *and* $\boldsymbol{\Delta}^{(2)}$ *is as in* (8.36).

In other words, $\hat{\boldsymbol{\Delta}}$ has the accuracy obtained from N_{eff} independent draws (8.35). For $N = 20\,426$ and $\alpha = \hat{\alpha} = 0.245$, the cardio estimate, we get $N_{\text{eff}} = 16.7$ (!). One has to imagine Figure 8.3 with only 17 points instead of 20 426 for a realistic impression of its accuracy.

Exercise 8.9 Show that the estimated standard error of $\hat{\Delta}_{31,32} = 0.805$ is at least 0.24, assuming $\alpha = 0.245$.

There is a tricky point to consider here: $\hat{\alpha}$ (8.18) is *not* a dependable estimate of α (8.15) if the columns of X are highly correlated. Theorem 8.1 works both ways: substantial column-wise correlations can induce an unwarranted impression of row-wise correlation. For the cardio study, in which column-wise correlation is strongly suspected, α may be considerably smaller than $\hat{\alpha}$.

It should be noted that formula (8.38) applies specifically to the estimation of $\boldsymbol{\Delta}$ and that other estimation problems would have different effective

[3] Most texts transpose the set-up so that the data matrix is $n \times N$, usually denoted $p \times n$, with the *columns* i.i.d.

sample sizes, perhaps not so conveniently depending on only the rms correlation α. The proof of Theorem 8.4 appears at the end of this section.

Correlation of t-values

Theorem 8.3 showed that if the columns of X were i.i.d. normal vectors, say

$$\boldsymbol{x}_j \overset{\text{iid}}{\sim} \mathcal{N}_N(\boldsymbol{\mu}, \Sigma) \qquad j = 1, 2, \dots, n, \tag{8.39}$$

we could construct an unbiased estimator for Σ's mean squared correlation α^2. Our real interest, however, lies not in the individual columns but with summary statistics across the columns, like t_i and z_i in Section 2.1. Fortunately, the covariance matrix of summary statistics is often itself nearly Σ, so that $\tilde{\alpha}$ or $\hat{\alpha}$ remain relevant to them. Here we will give a specific result for t_i, the one-sample t-statistic.

For row i of X in (8.39) define

$$D_i = \sqrt{n}\bar{x}_i, \quad \delta_i = \sqrt{n}\mu_i, \quad \text{and} \quad S_i = \sum_{j=1}^{n} \left(x_{ij} - \bar{x}_i\right)^2 \Big/ (n-1), \tag{8.40}$$

$\bar{x}_i = \sum_j x_{ij}/n$, so that the t-statistic is

$$t_i = D_i / S_i^{1/2}. \tag{8.41}$$

Theorem 8.5 *The covariance between t_i and $t_{i'}$ under model* (8.39) *is*

$$\operatorname{cov}(t_i, t_{i'}) = \frac{n+1}{n-1}\begin{pmatrix} 1 & c_0\rho_{ii'} \\ c_0\rho_{ii'} & 1 \end{pmatrix} + \frac{1}{2(n-1)}\begin{pmatrix} \delta_i^2 & \delta_i\delta_{i'}\rho_{ii'}^2 \\ \delta_i\delta_{i'}\rho_{ii'}^2 & \delta_{i'}^2 \end{pmatrix} + O\left(\frac{1}{n^2}\right) \tag{8.42}$$

where

$$c_0 = 1 - \left(1 - \rho_{ii'}^2\right) \Big/ (2(n+1)). \tag{8.43}$$

In the null case, $\delta_i = \delta_{i'} = 0$,

$$\operatorname{cor}(t_i, t_{i'}) = c_0\rho_{ii'} + O\left(1/n^2\right). \tag{8.44}$$

(The proof is sketched at the end of this section.)

The effect of Theorem 8.4 is to make the t-statistic correlations just slightly smaller than those for the column entries. With $n = 20$ and $\rho_{ii'} = 0.4$, for example, we get $\operatorname{cor}(t_i, t_{i'}) \doteq 0.392$. Numerical experimentation indicates the same kind of effect even in non-null cases, not surprisingly since the second term in (8.42) is of order only $O(1/n)$.

Proof of Theorem 8.3 With $\sigma_{ii'} \equiv 1$, we have $\rho_{ii'} = \sigma_{ii'}$ and $\bar{\alpha}^2$ (8.16) equaling

$$\bar{\alpha}^2 = \frac{\sum_i \sum_{i'} \hat{\sigma}^2_{ii'} - N}{N(N-1)} \tag{8.45}$$

compared to

$$\alpha^2 = \frac{\sum_i \sum_{i'} \sigma^2_{ii'} - N}{N(N-1)}. \tag{8.46}$$

A standard Wishart result, as in (8.36), says that $\hat{\sigma}^2_{ii'}$ has mean and variance

$$\hat{\sigma}^2_{ii'} \sim \left(\sigma^2_{ii'}, \frac{1+\sigma^2_{ii'}}{n}\right) \tag{8.47}$$

so that

$$E\left\{\sum_i \sum_{i'} \hat{\sigma}^2_i\right\} = \frac{n+1}{n} \sum_i \sum_{i'} \sigma^2_{ii'} + \frac{N^2}{n}. \tag{8.48}$$

□

Exercise 8.10 Complete the proof of Theorem 8.3.

Proof of Theorem 8.4 The covariance calculation for $\hat{\mathbf{\Delta}} = X'X/N$ involves the expansion

$$\begin{aligned} \hat{\Delta}_{jk}\hat{\Delta}_{lh} &= \left(\sum_i X_{ij}X_{ik}/N\right)\left(\sum_{i'} x_{i'l}X_{i'h}/N\right) \\ &= \frac{1}{N^2}\left(\sum_i X_{ij}X_{ik}X_{il}X_{ih} + \sum_{i\neq i'} X_{ij}X_{ik}X_{i'l}X_{i'h}\right). \end{aligned} \tag{8.49}$$

Using the formula

$$E\{Z_1Z_2Z_3Z_4\} = \gamma_{12}\gamma_{34} + \gamma_{13}\gamma_{24} + \gamma_{14}\gamma_{23} \tag{8.50}$$

for a normal vector $(Z_1, Z_2, Z_3, Z_4)'$ with means zero and covariances γ_{ij}, (8.25) gives

$$E\left\{\sum_i X_{ij}X_{ik}X_{il}X_{ih}\right\} = N\left[\Delta_{jk}\Delta_{lh} + \Delta_{jl}\Delta_{kh} + \Delta_{jh}\Delta_{kl}\right] \tag{8.51}$$

(using $\sigma_{ii'} = 1$) and

$$\begin{aligned} E\left\{\sum_{i\neq i'} X_{ij}X_{ik}X_{i'l}X_{i'h}\right\} &= N(N-1)\Delta_{jk}\Delta_{kh} \\ &\quad + (\Delta_{jl}\Delta_{kh} + \Delta_{jh}\Delta_{kl})\sum_{i\neq i'} \sigma^2_{ii'}. \end{aligned} \tag{8.52}$$

Then (8.49) yields

$$E\left\{\hat{\Delta}_{jk}\hat{\Delta}_{lh}\right\} = \Delta_{jk}\Delta_{lh} + (\Delta_{jl}\Delta_{kh} + \Delta_{jh}\Delta_{kl})\left(\frac{1+(N-1)\alpha^2}{N}\right) \tag{8.53}$$

giving

$$\text{cov}(\Delta_{jk}, \Delta_{lh}) = (\Delta_{jh}\Delta_{kl} + \Delta_{jh}\Delta_{kl})/N_{\text{eff}} \tag{8.54}$$

as in (8.37). □

Proof of Theorem 8.5 Letting $D = (D_i, D_{i'})'$, $\delta = (\delta_i, \delta_{i'})$, and $S = (S_i, S_{i'})'$, we have

$$D \sim \mathcal{N}_2\left(\delta, \begin{pmatrix} 1 & \rho_{ii'} \\ \rho_{ii'} & 1 \end{pmatrix}\right) \quad \text{and} \quad S \sim \left(\begin{pmatrix} 1 \\ 1 \end{pmatrix}, \frac{2}{n-1}\begin{pmatrix} 1 & \rho_{ii'}^2 \\ \rho_{ii'}^2 & 1 \end{pmatrix}\right) \tag{8.55}$$

with D and S independent. The notation indicates the mean and covariance of S, obtained from (8.36). Also defining $t = (t_i, t_{i'})'$ and $U = (S_i^{-1/2}, S_{i'}^{-1/2})'$,

$$t = D \cdot U, \tag{8.56}$$

where the dot indicates component-wise multiplication. A standard two-term Taylor expansion gives approximate mean and variance for U,

$$U \mathrel{\dot\sim} \left(\left(1 + \frac{3}{4(n-1)}\right)\begin{pmatrix} 1 \\ 1 \end{pmatrix}, \frac{1}{2(n-1)}\begin{pmatrix} 1 & \rho_{ii'}^2 \\ \rho_{ii'}^2 & 1 \end{pmatrix}\right). \tag{8.57}$$

Because D and U are independent in (8.56), we can use the general identity

$$\text{cov}(t) = \text{cov}(D) \cdot E\{UU'\} + (E\{D\}E\{D\}') \cdot \text{cov}(U) \tag{8.58}$$

to finish the proof. □

Exercise 8.11 (a) Derive (8.57). (b) Verify (8.58). (c) Complete the proof.

8.5 Count Correlations

Correlations among the components of $\boldsymbol{z}$, the vector of z-values, lead to correlations in the count vector $\boldsymbol{y}$ (7.4). This is more than a theoretical curiosity since it directly affects the interpretation of false discovery rates (Reason (III), Section 6.4).

Here we will trace the effects of correlation on counts of the null cases, those z-values having $z_i \sim \mathcal{N}(0, 1)$. Figure 8.5 used program `simz`, Section 8.2, to generate 2000 simulated $\boldsymbol{z}$ vectors, each of length $N = 4000$ and rms correlation $\alpha = 0.20$, all N cases null. The figure plots Y_1 versus Y_0, where

$$Y_0 = \#\{|z_i| \le 1\} \quad \text{and} \quad Y_1 = \#\{z_1 \ge 2\}. \tag{8.59}$$

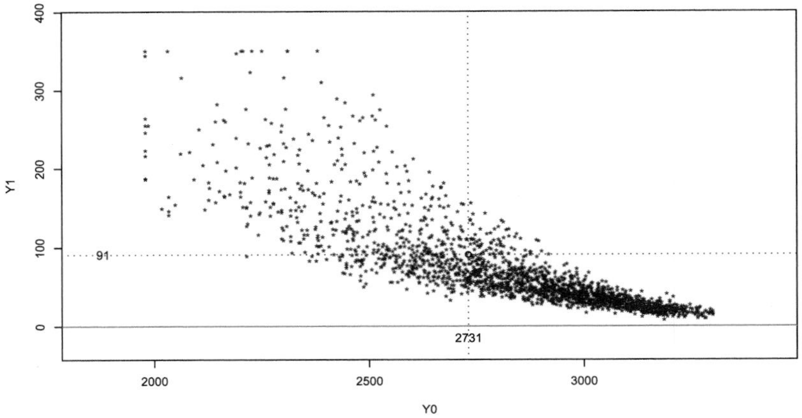

Figure 8.5 Tail counts $Y_1 = \#\{z_i \geq 2\}$ plotted versus central counts $Y_0 = \#\{|z_i| \leq 1\}$ for 2000 simulated $\boldsymbol{z}$ vectors with $N = 4000$, $\alpha = 0.20$. Large values of Y_0 give small values of Y_1, while small Y_0 gives large Y_1.

Table 8.5 *Standard deviations and correlation for Y_0 and Y_1* (8.59) *for the situation in Figure 8.5 ($\alpha = 0.20$) and also for independent z-values ($\alpha = 0$).*

	$\alpha = .20$		$\alpha = 0$	
	Y_0	Y_1	Y_0	Y_1
standard deviation	272.0	57.3	30.1	8.89
correlation	**−.82**		**−.26**	

There is a very strong negative correlation: if the number of central counts Y_0 much exceeds its expectation 2731, then the number of tail counts Y_1 is less than its expectation 91, and conversely. Table 8.5 shows that the negative correlation is far stronger than in the situation where the z-values are independent.

In an actual large-scale analysis, we might be interested in the false discovery rate for region $\mathcal{Z} = [2, \infty)$, for which Y_1 would be the unobserved numerator $N_0(\mathcal{Z})$ in the false discovery proportion (2.28). The fact that central counts help predict tail counts motivates use of the empirical null distributions, Section 6.2. What goes wrong if we ignore this effect is illustrated in Figure 6.4.

The striking correlation seen in Figure 8.5 is predicted by the rms approximation (7.38). Combined with Lemma 7.1, this suggests that the null count vector $\boldsymbol{y}$ is distributed approximately as

$$\boldsymbol{y} = \boldsymbol{y}_0 + Nd\boldsymbol{W}A. \tag{8.60}$$

Here $\boldsymbol{y}_0$ is a multinomial vector with expectation $\boldsymbol{\nu}_0 = Nd(\ldots, \varphi(x_k), \ldots)'$ and covariance diag $(\boldsymbol{\nu}_0) - \boldsymbol{\nu}_0\boldsymbol{\nu}_0'/N$, this being $\mathbf{cov}_0$ in (7.14); $\boldsymbol{W}$ is the "wing-shaped function"

$$W_k = \frac{\varphi''(x_k)}{\sqrt{2}} = \varphi(x_k)\frac{x_k^2 - 1}{\sqrt{2}} \qquad \text{for } k = 1, 2, \ldots, K, \tag{8.61}$$

illustrated in Figure 8.6; and A is a random variable, independent of $\boldsymbol{y}_0$, with mean and variance

$$A \sim \left(0, \alpha^2\right). \tag{8.62}$$

Exercise 8.12 Show that $\boldsymbol{y}$ in (8.60) has the mean and covariance suggested by Lemma 7.1 and (7.38).

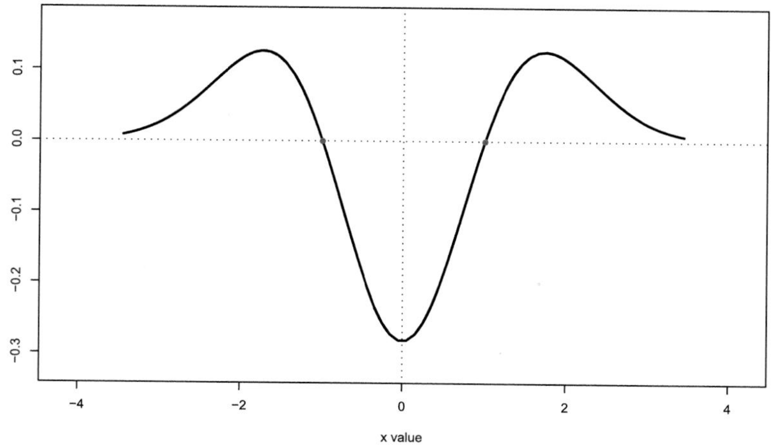

Figure 8.6 Wing-shaped function $\varphi(x)(x^2 - 1)/\sqrt{2}$.

A positive value of A in (8.60) is seen to depress the count vector $\boldsymbol{y}$ for x in the interval $[-1, 1]$ and to increase it outside that interval: in other words, it decreases the central count Y_0 and increases tail counts such as

Y_1. Negative A has the opposite effects. This accounts nicely for Figure 8.5. The magnitude of the negative correlation effects depends on α in (8.62), which controls the size of A's excursions from 0.

Notes

Large-scale correlation problems have attracted considerable current interest, as indicated in the notes for Chapter 7. Most of this chapter's material comes from Efron (2009a), where tests for column-wise independence are investigated more thoroughly, and Efron (2007a). Muralidharan (2010) devised the nice unbiasedness result of Theorem 8.3. Olshen and Rajaratnam (2009) show, by advanced methods, that the double standardization argument converges, except for small, artificially contructed matrices X.

The BRCA data comes from Hedenfalk et al. (2001). Ashley et al. (2006), working out of the laboratory of T. Quertermous, analyze the cardio data (Agilent two-color hybridization arrays), while the snp data (unpublished) is also from the Quertermous team. Program `simz` is referenced in Appendix B.

9

Sets of Cases (Enrichment)

Microarray experiments, through a combination of insufficient data per gene and the difficulties of large-scale simultaneous inference, often yield disappointing results. In search of greater detection power, *enrichment analysis* considers the combined outcomes of biologically determined sets of genes, for example the set of all the genes in a predefined genetic pathway. If all 20 z-values in a hypothetical pathway were positive, we might assign significance to the pathway's effect, whether or not any of the individual z_i were deemed non-null. We will consider enrichment methods in this chapter, and some of the theory, which of course applies just as well to similar situations outside the microarray context.

Our main example concerns the *p53 data*, partially illustrated in Figure 9.1; p53 is a transcription factor, that is, a gene that controls the activity of other genes. Mutations in p53 have been implicated in cancer development. A National Cancer Institute microarray study compared 33 mutated cell lines with 17 in which p53 was unmutated. There were $N = 10\,100$ gene expressions measured for each cell line, yielding a $10\,100 \times 50$ matrix X of expression measurements. Z-values based on two-sample t-tests were computed for each gene, as in (2.1)–(2.5), comparing mutated with unmutated cell lines. Figure 9.1 displays the 10 100 z_i values.

The results are disappointing. The histogram looks like a slightly short-tailed normal density, with MLE empirical null estimate $\mathcal{N}(0.06, 1.11^2)$. One gene, "BAX", stands out at $z = 4.83$, but the other 10 099 genes have uninteresting fdr and Fdr values, even using a $\mathcal{N}(0, 1)$ null.

This is where enrichment analysis comes to at least a partial rescue. A collection of 522 potentially interesting gene sets were identified from pre-existing[1] catalogs relating to p53 function. The sets ranged in size from two to 358 genes, median 17, with considerable overlap among the sets.

[1] Here we are dealing only with pre-selected gene sets. "Unsupervised learning," in which the sets are chosen on the basis of the observed data — say by a clustering algorithm — lies beyond the scope of the theory.

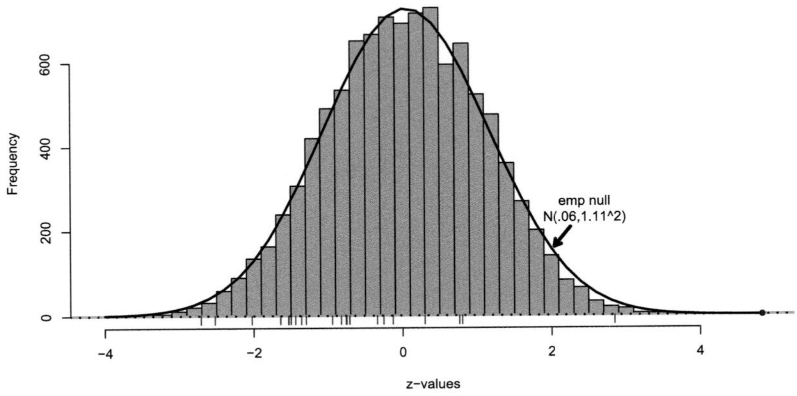

Figure 9.1 Histogram of $N = 10\,100$ z-values from the p53 study. Curve shows the empirical null density $\mathcal{N}(0.06, 1.11^2)$. False discovery rate tests label all genes null, except for BAX gene at $z = 4.83$. Enrichment analysis, however, will yield 8 non-null gene sets. Hash marks indicate the 22 z-values for gene set *ras pathway*.

We will see that eight of the 522 sets indicate non-null enrichment, even taking account of having 522 candidate sets to consider. One of the eight is the *ras pathway*, whose 22 z_i values, indicated by hash marks in Figure 9.1, are mostly < 0.

9.1 Randomization and Permutation

We have an $N \times n$ data matrix X and a corresponding N-vector of scores $\boldsymbol{z}$, with z_i determined from the ith row of X. The p53 study, with $N = 10\,100$, takes z_i to be the two-sample statistic (2.1)–(2.5), but our calculations will not require any specific form for the z-values. To begin with, we consider a single gene set[2] $\mathcal{S}$ comprising m z-values,

$$\boldsymbol{z}_{\mathcal{S}} = \{z_i, i \in \mathcal{S}\} \qquad [\#\mathcal{S} = m]; \tag{9.1}$$

$m = 22$ for the *ras pathway* in Figure 9.1. Our goal is to assign a p-value $p_{\mathcal{S}}$ to the null hypothesis H_0 of "no enrichment" for $\mathcal{S}$. Later we will use false discovery rate methods to take into account simultaneous testing on a catalog of possible gene sets.

[2] Microarray terminology is convenient here, but of course the theory and methods apply more generally.

Let $S(\boldsymbol{z}_S)$ be a score we attach to the z-values in $\mathcal{S}$. We will mainly deal with simple averages,

$$S(\boldsymbol{z}_S) = \sum_{i \in \mathcal{S}} s(z_i)/m \tag{9.2}$$

where $s(z)$ is some function like z or z^2, but more complicated scoring functions are also of interest. A popular scoring choice is the *GSEA statistic* (gene set enrichment analysis), essentially the maximum absolute difference[3] between the cumulative distribution functions of the m z-values in $\mathcal{S}$ and the $N - m$ z-values not in $\mathcal{S}$. A p-value p_S is obtained by calculating resampled versions of S and seeing what proportion of them exceed $S(z_S)$.

The problem is that there are two quite plausible but different resampling schemes:

Column permutations The columns of X are randomly permuted giving recalculated z-value vectors $\boldsymbol{z}^{*1}, \boldsymbol{z}^{*2}, \ldots, \boldsymbol{x}^{*B}$ as in Section 6.5, corresponding scores

$$S^{*b} = S\left(\boldsymbol{z}_S^{*b}\right) \qquad b = 1, 2, \ldots, B, \tag{9.3}$$

and p-value

$$p_S^* = \#\left\{S^{*b} > S\right\}\big/B. \tag{9.4}$$

Row randomization Random subsets of size m are selected from $\{1, 2, \ldots, N\}$, say $\boldsymbol{i}^{\dagger 1}, \boldsymbol{i}^{\dagger 2}, \ldots, \boldsymbol{i}^{\dagger B}$, giving corresponding m-vectors and scores

$$\boldsymbol{z}^{\dagger b} = \left\{z_i, i \in \boldsymbol{i}^{\dagger b}\right\}, \qquad S^{\dagger b} = S\left(\boldsymbol{z}^{\dagger b}\right), \tag{9.5}$$

and p-value

$$p_S^\dagger = \#\left\{S^{\dagger b} > S\right\}\big/B. \tag{9.6}$$

Exercise 9.1 In what way does a single permutation z-vector $\boldsymbol{z}^*$ require more computation than a single randomization vector $\boldsymbol{z}^\dagger$?

Row randomization has the appealing feature of operating *conditionally* on the observed vector $\boldsymbol{z}$: it tests the null hypothesis that S is no bigger than what might be obtained by a random selection of m elements from $\boldsymbol{z}$.

Its defect concerns correlation between the genes. Suppose the N values $s_i = s(z_i)$ in (9.2) have empirical mean and standard deviation

$$s \sim (m_s, \mathrm{sd}_s). \tag{9.7}$$

[3] This difference is a version of the Kolmogorov–Smirnov test statistic.

Then a row-randomized version of (9.2),

$$S^{\dagger} = \sum_{i \in S^{\dagger}} s(z_i)/m \tag{9.8}$$

will have mean and standard deviation

$$S^{\dagger} \sim (\mu^{\dagger}, \sigma^{\dagger}) = \left(m_s, \mathrm{sd}_s \big/ \sqrt{m}\right) \tag{9.9}$$

just as if the genes were independent of each other.

Exercise 9.2 Formula (9.9) correctly gives $(\mu^{\dagger}, \sigma^{\dagger})$ if we allow the subsets $\boldsymbol{i}^{\dagger}$ to be chosen randomly and with replacement from $\{1, 2, \ldots, N\}$. What if we insist that they be drawn *without* replacement?

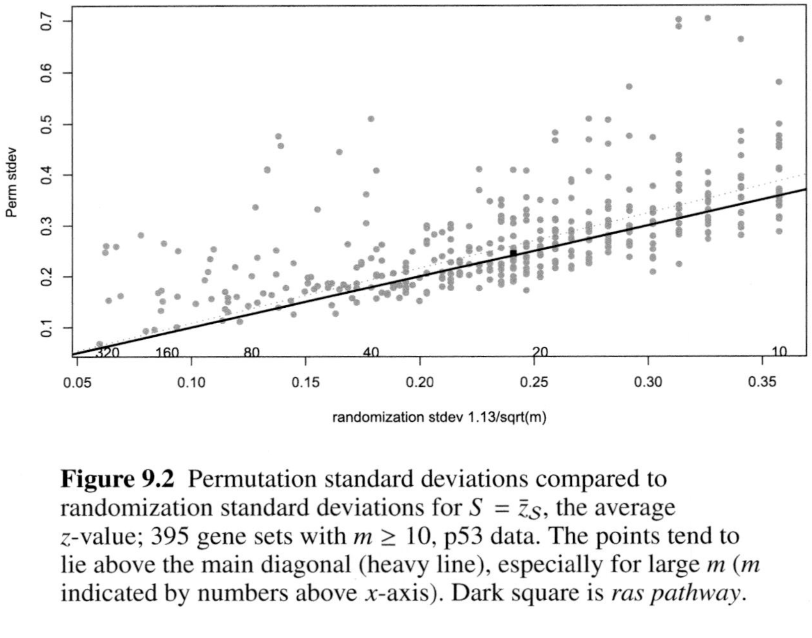

Figure 9.2 Permutation standard deviations compared to randomization standard deviations for $S = \bar{z}_S$, the average z-value; 395 gene sets with $m \geq 10$, p53 data. The points tend to lie above the main diagonal (heavy line), especially for large m (m indicated by numbers above x-axis). Dark square is *ras pathway*.

Taking $s(z_i) = z_i$ for the p53 data gives mean and standard deviation[4]

$$s \sim (0.00, 1.13). \tag{9.10}$$

Figure 9.2 compares $\sigma_S^{\dagger} = 1.13/\sqrt{m}$ with the column permutation standard

[4] The catalog of 522 gene sets includes, with repetitions, 15 059 genes, of which 4486 are distinct. The values in (9.10) are those for the list of 15 059 rather than (0.04, 1.06) for the original 10 100 values; (9.10) seems more appropriate for the comparison made in Figure 9.2.

deviation σ_S^* obtained from $B = 1000$ values S^{*b} (9.3) (for the 395 gene sets $\mathcal{S}$ having $m \geq 10$ members). The ratio $\sigma_S^*/\sigma_S^\dagger$ has median 1.08 (light line) but is considerably greater for large m, with median 1.50 for sets $\mathcal{S}$ with $m \geq 50$.

Positive correlations between the gene expressions in $\mathcal{S}$ make the permutation standard deviation σ_S^* of S greater than the row randomization value $\sigma_s = \text{sd}_s / \sqrt{m}$. The randomization p-value (9.6) will then be misleadingly small, at least compared with p_S^*, the usual permutation value.

Exercise 9.3 Let $\boldsymbol{s}_S^*$ be the m-vector of values $s(z_i^*)$, $i \in \mathcal{S}$, corresponding to a permutation vector $\boldsymbol{z}_S^*$, as in (9.3). If $\Sigma_{\mathcal{S}}$ is the $m \times m$ covariance matrix of $\boldsymbol{s}_S^*$ under permutation, show that $\sigma_S^*/\sigma_S^\dagger = \bar{\Sigma}_S^{1/2} / \text{sd}_s$, with $\bar{\Sigma}_{\mathcal{S}}$ the average of the m^2 elements of $\Sigma_{\mathcal{S}}$. (So $\Sigma_{\mathcal{S}} = I$ gives $\sigma_S^* = \sigma_S^\dagger / \text{sd}_s$, but positive off-diagonal elements make $\sigma_S^* > \sigma_S^\dagger / \text{sd}_s$.)

Permutation tests for enrichment have a serious weakness of their own, as demonstrated by the following example. The BRCA microarray data of Section 8.4 involved $n = 15$ breast cancer patients, seven with BRCA1 mutations, eight with BRCA2, on $N = 3226$ genes; z-values for each gene were calculated as in (2.1)–(2.5), comparing the two BRCA categories. The z-value histogram was highly overdispersed, empirical null estimate $\mathcal{N}(-0.04, 1.52^2)$; Fdr analysis yielded no non-null genes.

An artificial catalog of 129 gene sets $\mathcal{S}$, each of size $m = 25$, was constructed by random selection from all N genes, making it highly unlikely that any set was actually enriched. Enrichment was tested using $S = \sum_{\mathcal{S}} |z_i|/m$, that is, (9.2) with $s(z) = |z|$. Figure 9.3 shows that the permutation values S^* greatly underestimated the actual 129 S values. The permutation p-values (9.4), $B = 400$, were very small for most of the sets $\mathcal{S}$: a standard Fdr analysis (BH(0.1) as in Section 4.2) labeled 113 of the 129 as "enriched," even though we know that all of them were chosen at random.

It is easy to spot the trouble here: the mean and standard deviation of the $N = 3226$ values $|z_i|$ is

$$(m_s, \text{sd}_s) = (1.17, 0.82) \tag{9.11}$$

compared to the mean and standard deviation (m_s^*, sd_s^*) for all $N \cdot B$ permutation values $|z_i^{*b}|$,

$$(m_s^*, \text{sd}_s^*) = (0.81, 0.59). \tag{9.12}$$

This makes a typical permuted score $S^* = \sum_{\mathcal{S}} |z_i^*|/m$ smaller than the actual score $S = \sum_{\mathcal{S}} |z_i|/m$.

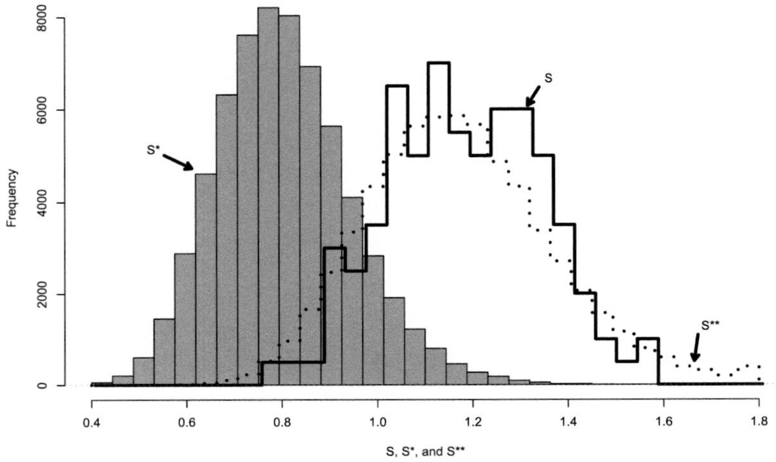

Figure 9.3 *Line histogram* shows scores $S = \sum_{\mathcal{S}} |z_i|/m$ for 129 randomly selected gene sets from BRCA data, all $m = 25$. *Solid histogram* is permutation scores $S^{*b} = \sum_{\mathcal{S}} |z_i^{*b}|/m$, $b = 1, 2, \dots,$ 400. *Dotted histogram* shows restandardized scores S^{**} (9.13).

The values reported in (9.12) are just about what we would get if the z_i^* replicates followed a theoretical $\mathcal{N}(0, 1)$ null distribution. This is a dependable feature of permutation t-tests, Section 6.5, which are immune to the overdispersion seen in the BRCA z-values. The actual scores $S = \sum_{\mathcal{S}} |z_i|/m$ are not immune, and are shifted to the right in Figure 9.3.

Row randomization handles overdispersion correctly, but ignores correlation; column permutation ignores overdispersion but correctly accounts for correlation. *Restandardization* is a somewhat ad hoc attempt to deal correctly with both problems. Given a permutation score S^{*b} (9.3), we compute the *restandardized score*

$$S^{**b} = m_s + \frac{\text{sd}_s}{\text{sd}_s^*} \left(S^{*b} - m_s^* \right) \tag{9.13}$$

where (m_s, sd_s) are from (9.7) while (m_s^*, sd_s^*) are the permutation mean and standard deviation of $s(z_i^{*b})$ over all B permutations and N genes as in (9.13). The restandardized p-value is then

$$p_{\mathcal{S}}^{**} = \# \left\{ S^{**b} > S \right\} \Big/ B. \tag{9.14}$$

The dotted curve in Figure 9.3 is the histogram of all $129 \cdot B$ restandardized scores S^{**}. It reasonably matches the actual S histogram; a BH(0.1) analysis of the 129 p_S^{**} values now correctly declares all the sets to be unenriched.

Another way to write (9.14) is

$$p_S^{**} = \#\left\{\frac{S^{*b} - m_s^*}{\text{sd}_s^*} > \frac{S - m_s}{\text{sd}_s}\right\} \Big/ B. \tag{9.15}$$

If we change $s(z)$ to the *standardized* function

$$t(z) = (s(z) - m_s) \,/\, \text{sd}_s \tag{9.16}$$

with (m_s, sd_s) thought of as fixed constants, e.g., as in (9.11), then the corresponding scoring function is

$$T(\boldsymbol{z}) = \sum_{\mathcal{S}} t(z_i)/m = (S(\boldsymbol{z}) - m_s) \,/\, \text{sd}_s \tag{9.17}$$

and (9.15) becomes

$$p_S^{**} = \#\left\{T^{*b} > T\right\} \Big/ B. \tag{9.18}$$

In this sense, restandardization amounts to calculating the usual permutation p-value for a row-standardized version of S.

Exercise 9.4 Verify (9.18). *Hint*: $(m_t, \text{sd}_t) = (0, 1)$.

The need for restandardization resembles the arguments for the empirical null in Chapter 6. Theoretically (9.11) should match (9.12) in the BRCA example, but factors like those of Section 6.4 have spread out the distribution of $|z_i|$. These factors don't affect the permutation distribution, and some form of standardization is necessary before we can trust permutation p-values. This doesn't have to be of form (3.16). The GSEA algorithm achieves a similar effect by comparing the cdf of the z-values in $\mathcal{S}$ to the cdf of all the others rather than to a theoretical $\mathcal{N}(0, 1)$ cdf.

Restandardization can be shown to yield reasonable inferences in a variety of situations:

- if $\mathcal{S}$ was selected randomly, as in the row randomization model;
- if the theoretical null $z \sim \mathcal{N}(0, 1)$ agrees with the empirical distribution of the N z-values;
- if the z_i are uncorrelated.

The method is not perfect, as examples can be constructed to show. Neither the randomization nor permutation models perfectly describes how gene sets $\mathcal{S}$ come under consideration in practice, making some sort of compromise formula a necessity. Section 9.3 provides some theoretical support for formula (9.13), in an admittedly specialized formulation.

9.2 Efficient Choice of a Scoring Function

The previous section concerned the proper choice of a null hypothesis for enrichment testing, with the control of Type I error (false rejection) in mind. We are also interested in power, of course. This requires us to specify alternatives to null selection, which then leads to recipes for efficient scoring functions $S(\boldsymbol{z})$.

Row randomization (9.6) tests the null hypothesis that set $\mathcal{S}$ was chosen by random selection from the N cases. The *Poisson selection model* allows non-random selection. It starts with independent Poisson variates

$$I_i \stackrel{\text{ind}}{\sim} \text{Poi}(\nu_i) \qquad \nu_i = \alpha e^{\beta' s_i}/\tau_\beta \tag{9.19}$$

for $i = 1, 2, \ldots, N$, where $s_i = s(z_i)$ is some *feature function*, a J-dimensional function that we believe extracts the features of z_i indicative of enrichment, e.g., $s(z) = (z, z^2)'$; β is an unknown J-dimensional parameter vector, and

$$\tau_\beta = \sum_{i=1}^{N} e^{\beta' s_i}; \tag{9.20}$$

α is an unknown scalar parameter satisfying

$$\alpha = \sum_{i=1}^{N} \nu_i \tag{9.21}$$

according to (9.19)–(9.20).

In what follows, ν_i will be small and the I_i almost all zeroes or ones, mostly zeroes. It will be notationally convenient to assume that they are *all* 0 or 1. Then we define the selected gene set $\mathcal{S}$ as composed of those cases having $I_i = 1$,

$$\mathcal{S} = \{i : I_i = 1\} \tag{9.22}$$

so $\mathcal{S}$ has

$$m = \sum_{1}^{N} I_i \tag{9.23}$$

members. The Poisson assumptions (9.19) give $m \sim \text{Poi}(\sum \nu_i)$, or by (9.21),

$$m \sim \text{Poi}(\alpha). \tag{9.24}$$

It is easy to calculate the probability $g_{\alpha,\beta}(\mathcal{S})$ of selecting any particular gene set $\mathcal{S}$ under model (9.19),

$$g_{\alpha,\beta}(\mathcal{S}) = \left(\frac{e^{-\alpha}\alpha^m}{m!}\right)\left(m!e^{m[\beta'\bar{s}_{\mathcal{S}} - \log(\tau_\beta)]}\right) \tag{9.25}$$

where

$$\bar{s}_{\mathcal{S}} = \sum_{\mathcal{S}} s_i/m, \tag{9.26}$$

called $S(\boldsymbol{z}_{\mathcal{S}})$ in Section 9.1. This is a product of two exponential family likelihoods, yielding maximum likelihood estimates $\hat{\alpha} = m$ and $\hat{\beta}$ satisfying

$$\sum_{i=1}^{N} s_i e^{\hat{\beta}' s_i} \bigg/ \sum_{i=1}^{N} e^{\hat{\beta}' s_i} = \bar{s}_{\mathcal{S}}. \tag{9.27}$$

Exercise 9.5 Verify (9.25) and (9.27). Show that the conditional distribution of $\mathcal{S}$ given its size m is

$$\begin{aligned} g_\beta(\mathcal{S}|m) &= m! e^{m[\beta' \bar{s}_{\mathcal{S}} - \log(\tau_\beta)]} \\ &= m! e^{m[\beta'(\bar{s}_{\mathcal{S}} - \bar{s}) - (\log(\tau_\beta) - \beta' \bar{s})]} \end{aligned} \tag{9.28}$$

where $\bar{s} = \sum_1^N s_i/N$.

Parameter vector $\beta = \mathbf{0}$ corresponds to the row randomization null hypothesis that $\mathcal{S}$ was selected at random. Non-zero β "tilts" the selection toward gene sets having larger values of $\beta'(\bar{s}_{\mathcal{S}} - \bar{s})$, as seen in (9.28). An efficient test rejects $H_0 : \beta = 0$, i.e., that $\mathcal{S}$ was randomly selected, if $\bar{s}_{\mathcal{S}}$ is far from its null expectation $\bar{s}$. If $s(z)$ is one-dimensional, as in Section 9.1, we reject for extreme values of $\bar{s}_{\mathcal{S}} - \bar{s}$, either positive or negative, perhaps using permutation methods to decide what "far from" means.

The choice of $s(z)$ determines which alternatives the test will have power against. Consider two candidates,

$$s(z) = z \quad \text{or} \quad s(z) = |z|. \tag{9.29}$$

The first of these rejects if $\bar{z}_{\mathcal{S}}$ is far from $\bar{z} = \sum_1^N z_i/N$: a good choice if enrichment manifests itself as a *shift* of location in z-values, as seems to be the case with the *ras pathway* in Figure 9.1. The second candidate has power against *scale* alternatives, where the z-values in $\mathcal{S}$ are more dispersed than in the whole population.

For general use one would like to have power against both shift and scale alternatives, which suggests using a two-dimensional statistic $s(z)$. The *maxmean statistic* takes

$$s(z) = \left(s^{(1)}(z), s^{(2)}(z)\right) \qquad \begin{cases} s^{(1)}(z) = \max(z, 0) \\ s^{(2)}(z) = -\min(z, 0) \end{cases} \tag{9.30}$$

giving bivariate score function $S(\boldsymbol{z}_{\mathcal{S}}) = (\bar{s}_{\mathcal{S}}^{(1)}, \bar{s}_{\mathcal{S}}^{(2)})$. This is reduced to a one-dimensional test statistic by maximization,

$$S_{\text{maxmean}} = \max\left(\bar{s}_{\mathcal{S}}^{(1)}, \bar{s}_{\mathcal{S}}^{(2)}\right). \tag{9.31}$$

(Notice that this is not the same as $S_{\text{absmean}} = \max(\bar{z}_{\mathcal{S}}, -\bar{z}_{\mathcal{S}}) = |\bar{z}_{\mathcal{S}}|$, the two-sided version of the test based on means.)

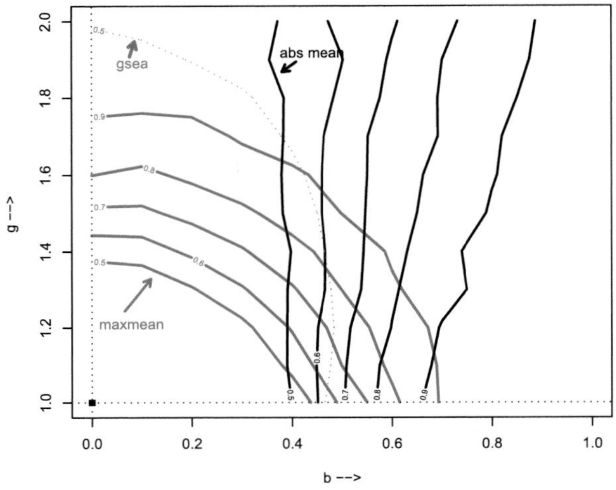

Figure 9.4 Contours of equal power for testing $(b, g) = (0, 1)$ in situation (9.32); for S_{maxmean} (9.31), GSEA (Kolmogorov–Smirnov) statistic, and $S_{\text{absmean}} = |\bar{z}_{\mathcal{S}}|$. Comparing contours for power = 0.5 shows S_{maxmean} dominating GSEA and nearly dominating S_{absmean}.

Figure 9.4 compares the power of the maxmean test with that of the GSEA (Kolmogorov–Smirnov) statistic and also with $S_{\text{absmean}} = |\bar{z}_{\mathcal{S}}|$ in an artificially simplified context: $\mathcal{S}$ consists of $m = 25$ independent normal observations,

$$z_i \overset{\text{ind}}{\sim} \mathcal{N}\left(b, g^2\right) \qquad i = 1, 2, \ldots, m = 25. \tag{9.32}$$

Contours of equal power are shown for testing

$$H_0 : (b, g) = (0, 1) \tag{9.33}$$

versus alternatives $(b \geq 0, g \geq 1)$ at level 0.95.

Better power is indicated by contours closer to the null hypothesis point $(b, g) = (0, 1)$. The maxmean test shows reasonable power in both directions. Comparing contours for power = 0.5 shows maxmean dominating

GSEA and nearly dominating the absolute mean test.[5] (The vertically oriented contours for S_{absmean} show that, as expected, it has power against shift alternatives ($b > 0$) but not scale alternatives ($s > 1$).)

Enrichment testing was carried out using S_{maxmean} on the catalog of 522 gene sets $\mathcal{S}$; $B = 1000$ restandardized values yielded p-values $p_{\mathcal{S}}^{**}$ for each set. A Benjamini–Hochberg Fdr test with control level $q = 0.1$ (4.9)–(4.10) yielded eight non-null cases, as illustrated in Figure 9.5. The top seven show shifts of their z-values, five right and two left, and would have been discovered by S_{absmean} but the eighth set shows more of a scale effect. Note that restandardization formula (9.13) was applied separately to $s^{(1)}$ and $s^{(2)}$ in (9.30) and then combined to give

$$S_{\text{maxmean}}^{**} = \max\left\{\bar{s}_{\mathcal{S}}^{(1)**}, \bar{s}_{\mathcal{S}}^{(2)**}\right\}. \tag{9.34}$$

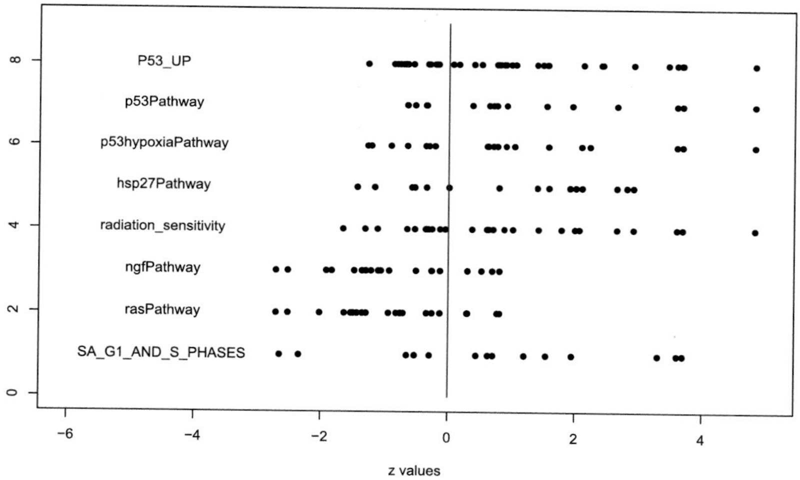

Figure 9.5 z-values for the 8 enriched gene sets, p53 data; non-null cases from catalog of 522 sets, Benjamini–Hochberg FDR control algorithm, $q = 0.10$, using p-values from maxmean statistic.

[5] The power calculations were done by simulation from model (9.32), for example comparing the distribution of S_{maxmean} for $(b, g) = (0, 1)$ with the distribution for a grid of alternatives ($b \geq 0, g \geq 1$). In this context, the GSEA statistic was essentially the Kolmogorov–Smirnov distance between the cdf of $z_1, z_2, \ldots, z_{25}$ and a standard normal cdf.

A word about computation: With 522 gene sets to consider, $B = 1000$ resamples is not excessive. The eighth largest p-value needs to be less than 0.00153 in order to be declared non-null by the BH(0.1) procedure; that is, less than two in 1000 values of $S_{\mathcal{S}}^{**}$ can exceed $S_{\mathcal{S}}$. Calculations were carried out using the `R` program `GSA`, available through the `CRAN` library. It would be convenient to pool the resamples for different gene sets. This is the usual procedure for the permutation analysis of individual genes, for example in the SAM procedure of Section 4.5. However, the gene set resampling distributions depend heavily on both the size and correlation structure of $\mathcal{S}$, making pooling dangerous.

Exercise 9.6 Verify the threshold value 0.00153 above.

The Poisson selection model (9.19)–(9.22) follows row randomization (9.5) in operating *conditionally* on the full collection $\boldsymbol{z}$ of N original z-values, with only the selection process for $\mathcal{S}$ considered random. We could instead think of $\mathcal{S}$ as fixed and the set of z-values in $\mathcal{S}$, $\boldsymbol{z}_{\mathcal{S}}$, as random, perhaps following an exponential family of densities

$$f_\beta(\boldsymbol{z}_{\mathcal{S}}) = c_\beta e^{\beta S(\boldsymbol{z}_{\mathcal{S}})} f_0(\boldsymbol{z}_{\mathcal{S}}). \tag{9.35}$$

Here $f_0(\boldsymbol{z}_{\mathcal{S}})$ is the baseline density for an unenriched set, β is an unknown parameter, c_β is a constant making f_β integrate to 1, and S is some scoring function that produces larger values under enrichment.

In this case the optimum test for null hypothesis $H_0 : \beta = 0$ rejects for $S(\boldsymbol{z}_{\mathcal{S}})$ large. By choosing S as the GSEA statistic, GSEA becomes optimum, etc., and we might very well use permutation/randomization methods to decide what "large" means. Taking $S = \sum_{\mathcal{S}} s(z_i)/m$ gets us back to our previous theory, still leaving us with the task of deciding on the appropriate form for $s(z)$.

9.3 A Correlation Model

Restandardization attempts to correct enrichment p-values for both permutation and randomization effects (9.13)–(9.14). This section[6] discusses a specialized model for correlated z-values where exact calculations are possible. We will see that restandardization fails to give exact answers, but is reasonably accurate over a range of situations. (Arguments and proofs are deferred to the section's end.)

[6] The material of this section is technical in nature and may be bypassed at first reading.

Our model assumes that the $N \times n$ data matrix $\boldsymbol{X}$ is of the form

$$\underset{N\times n}{\boldsymbol{X}} = \underset{N\times n}{\boldsymbol{U}} + \underset{N\times J}{\boldsymbol{V}}\ \underset{J\times n}{\boldsymbol{a}} \tag{9.36}$$

where $\boldsymbol{U}$ and $\boldsymbol{a}$ have mutually independent normal entries,

$$\begin{aligned} u_{ik} &\overset{\text{ind}}{\sim} \mathcal{N}(0,1) & i &= 1,2,\dots N \quad \text{and} \quad k = 1,2,\dots,n \\ a_{jk} &\overset{\text{ind}}{\sim} \mathcal{N}\left(0,\sigma_a^2\right) & j &= 1,2,\dots,J \quad \text{and} \quad k = 1,2,\dots,n; \end{aligned} \tag{9.37}$$

σ_a^2 is assumed known in what follows. J is a small integer dividing N, say with

$$H \equiv N/J \quad \text{and} \quad J_1 \equiv J - 1. \tag{9.38}$$

$\boldsymbol{V}$ is a fixed $N \times J$ matrix,

$$\underset{N\times J}{\boldsymbol{V}} = \begin{pmatrix} 1_H & -1_H/J_1 & \cdots \\ -1_H/J_1 & 1_H & \cdots \\ \vdots & \vdots & \ddots \end{pmatrix} \tag{9.39}$$

where 1_H is a vector of H 1's. (All the off-diagonal vectors in (9.39) are $-1_H/J_1$.)

Exercise 9.7 Let $\boldsymbol{I}_J$ be the $J \times J$ identity matrix and $\boldsymbol{E}_H$ the $H \times H$ matrix of 1's. Show that

$$\underset{J\times J}{\boldsymbol{V}'\boldsymbol{V}} = N\frac{J}{J_1^2}(\boldsymbol{I}_J - \boldsymbol{E}_J/J) \tag{9.40}$$

and

$$\underset{N\times N}{\boldsymbol{V}\boldsymbol{V}'} = \frac{J}{J_1}\begin{pmatrix} \boldsymbol{E}_H & -\boldsymbol{E}_H/J_1 & \cdots \\ -\boldsymbol{E}_H/J_1 & \boldsymbol{E}_H & \cdots \\ \vdots & \vdots & \ddots \end{pmatrix}. \tag{9.41}$$

We assume that n is even, and define our vector of N z-values in terms of the difference of averages of the last and first $n/2$ columns of $\boldsymbol{X}$, say $\bar{\boldsymbol{x}}_2$ and $\bar{\boldsymbol{x}}_1$,

$$\boldsymbol{z} = \sqrt{n/4}\,(\bar{\boldsymbol{x}}_2 - \bar{\boldsymbol{x}}_1)/\sigma_0 \tag{9.42}$$

with

$$\sigma_0^2 = 1 + \frac{J}{J_1}\sigma_a^2. \tag{9.43}$$

(z_i is a two-sample t-statistic for case i, with the denominator known rather than estimated.) In terms of (9.36),

$$\boldsymbol{z} = (\boldsymbol{u} + \boldsymbol{V}\boldsymbol{A})/\sigma_0 \tag{9.44}$$

where, in notation similar to (9.42), the independent vectors $\boldsymbol{u}$ and $\boldsymbol{A}$ are defined as

$$\begin{aligned} \boldsymbol{u} &= \sqrt{n/4}\left(\bar{U}_2 - \bar{U}_1\right) \sim \mathcal{N}_N(\boldsymbol{0}, \boldsymbol{I}_N), \\ \boldsymbol{A} &= \sqrt{n/4}\,(\bar{\boldsymbol{a}}_2 - \bar{\boldsymbol{a}}_1) \sim \mathcal{N}_J\left(\boldsymbol{0}, \sigma_a^2 \boldsymbol{I}_J\right); \end{aligned} \tag{9.45}$$

(9.44)–(9.45) imply that

$$\boldsymbol{z} \sim \mathcal{N}_N\left(\boldsymbol{0}, \left(\boldsymbol{I}_N + \sigma_a^2 \boldsymbol{V}\boldsymbol{V}'\right)/\sigma_0^2\right). \tag{9.46}$$

We will use $\boldsymbol{z}$ to illustrate correlation effects on the permutation and randomization calculations of Section 9.1. Formula (9.41) shows that the components z_i are $\mathcal{N}(0,1)$ with correlation between z_{i_1} and z_{i_2} depending on whether i_1 and i_2 are in the same H-dimensional block of $\boldsymbol{V}$ in (9.39),

$$\begin{pmatrix} z_{i_1} \\ z_{i_2} \end{pmatrix} \sim \mathcal{N}_2\left(\begin{pmatrix} 0 \\ 0 \end{pmatrix}, \begin{pmatrix} 1 & \rho_{i_1,i_2} \\ \rho_{i_1,i_2} & 1 \end{pmatrix}\right)$$

where

$$\rho_{i_1,i_2} = \begin{cases} \rho & \text{same block} \\ -\rho/J_1 & \text{different blocks} \end{cases} \qquad \left[\rho = \frac{\sigma_a^2}{J_1/J + \sigma_a^2}\right]. \tag{9.47}$$

The root mean square correlation α of Section 8.2 is nearly[7]

$$\alpha \doteq \frac{1}{\sqrt{J_1}}\rho \qquad \left[\sigma_a^2 = \frac{J_1}{J}\,\frac{\sqrt{J_i}\alpha}{1 - \sqrt{J_1}\alpha}\right]. \tag{9.48}$$

Given a gene set $\mathcal{S}$ of size m, let $\bar{z}_{\mathcal{S}}$ be the average z-value in $\mathcal{S}$,

$$\bar{z}_{\mathcal{S}} = \sum_{\mathcal{S}} z_i/m. \tag{9.49}$$

Our aim is to show the role of restandardization, both its benefits and limitations, in assessing the variance of $\bar{z}_{\mathcal{S}}$. To this end, let

$$m_j = \#\{\text{members of } \mathcal{S} \text{ in } j\text{th block of } V\}, \tag{9.50}$$

[7] The only approximation is setting $\rho_{ii'} = \rho$ instead of 1, so (9.48) is accurate to order $O(1/N)$.

$j = 1, 2, \ldots, J$, that is, the number of z_i values in $\mathcal{S}$ having $(j-1)H < i \le jH$ (so $\sum m_j = m$). Define

$$C = \frac{J}{J_1} \sum_{j=1}^{J} (m_j - m/J)^2 / m. \tag{9.51}$$

Lemma 9.1 *The marginal distribution of $\bar{z}_{\mathcal{S}}$ averaging over both $\boldsymbol{u}$ and $\boldsymbol{A}$ in* (9.44)–(9.45) *is*

$$\bar{z}_{\mathcal{S}} \sim \mathcal{N}\left(0, \frac{1}{m} \frac{1 + (J/J_1)\sigma_a^2 C}{1 + (J/J_1)\sigma_a^2}\right). \tag{9.52}$$

(Proof near the end of this section.)

C is a measure of correlation within $\mathcal{S}$. If all m members of $\mathcal{S}$ are in the same block, then the correlations (9.47) within $\mathcal{S}$ are big, giving $C = m$. At the other extreme, if $\mathcal{S}$ is perfectly balanced across blocks, $m_j \equiv m/J$, then most correlations are small, and $C = 0$.

Exercise 9.8 Show that if $\mathcal{S}$ is randomly selected by m draws without replacement from all N genes then

$$E\{C\} = 1. \tag{9.53}$$

C is connected with permutation calculations, since these correctly account for correlations within $\mathcal{S}$. We will see that row randomization effects are connected with

$$\hat{\sigma}_A^2 = \sum_{j=1}^{J} \left(A_j - \bar{A}\right)^2 / J_1 \qquad \left(\bar{A} = \sum_{1}^{J} A_j/J\right), \tag{9.54}$$

the sample variance of the random effects A_j in (9.44)–(9.45).

Theorem 9.2 *Conditionally on $\hat{\sigma}_A^2$, $\bar{z}_{\mathcal{S}}$ has mean 0 and variance*

$$\mathrm{var}\left\{\bar{z}_{\mathcal{S}} \middle| \hat{\sigma}_A^2\right\} = \frac{1}{m} \frac{1 + (J/J_1)\hat{\sigma}_a^2 C}{1 + (J/J_1)\sigma_a^2}. \tag{9.55}$$

(Proof near the end of this section.)

Large values of $\hat{\sigma}_A^2$ make the blocks of z-values from (9.44) more disparate, and increase the variability (9.5) of $\bar{z}_{\mathcal{S}^\dagger}$ for randomly selected subsets $\mathcal{S}^\dagger$. With $s(z) = z$ in (9.7), the component variance

$$\mathrm{sd}_s^2 = \left\|\boldsymbol{z}^2\right\| / N \tag{9.56}$$

is conditionally distributed as a scaled non-central χ^2 variate,

$$\mathrm{sd}_s^2 \left| \hat{\sigma}_A^2 \sim \chi_N^2\left(\delta^2\right) \middle/ \left(N\sigma_0^2\right) \right. \qquad \left[\delta^2 \equiv N(J/J_1)\hat{\sigma}_A^2\right] \tag{9.57}$$

as verified at the end of this section. (All the means are known to be zero in model (9.44) so we are taking $m = 0$ in (9.7).) Familiar χ^2 properties yield

$$\mathrm{sd}_s^2 \left| \hat{\sigma}_A^2 \right. = \frac{1 + (J/J_1)\hat{\sigma}_A^2}{1 + (J/J_1)\sigma_a^2} + O_p\left(N^{-\frac{1}{2}}\right). \tag{9.58}$$

This means that the factor $\hat{\sigma}_A^2$ in (9.55) is effectively recoverable from sd_s (9.56).

The unconditional distribution of $\hat{\sigma}_A^2$ in (9.54) is a scaled central χ^2,

$$\hat{\sigma}_A^2 \sim \sigma_a^2 \chi_{J_1}^2 \big/ J_1 \tag{9.59}$$

according to (9.45); $\hat{\sigma}_A^2$ has coefficient of variation $(2/J_1)^{1/2}$. Taking $J = 5$, as at the end of Section 8.2, gives CV = 0.707 so $\hat{\sigma}_A^2$ can differ considerably from σ_a^2, and similarly the conditional variance (9.55) can vary from its unconditional value in (9.52).

In any one realization of (9.44) we effectively know $\hat{\sigma}_A^2$ from (9.58), making the conditional variance (9.55) the correct one for computing the enrichment significance level of $\mathcal{S}$. Permutation calculations, on the other hand, tend to produce the unconditional variance (9.52),

$$\mathrm{var}_*\left\{\bar{z}_\mathcal{S}^*\right\} \doteq \frac{1}{m}\frac{1 + (J/J_1)\sigma_a^2 C}{1 + (J/J_1)\sigma_a^2} \tag{9.60}$$

reflecting the fact that they can "see" correlations within $\mathcal{S}$ but not the effects outside of $\mathcal{S}$ that determine $\hat{\sigma}_A^2$.

The restandardization formula (9.13) depends on $m_s = 0$, sd_s from (9.58), and

$$(m_s^*, \mathrm{sd}_s^*) \doteq (0, 1) \tag{9.61}$$

giving

$$\mathrm{var}_{**}\left\{\bar{z}_\mathcal{S}^{**}\right\} \doteq \frac{1 + (J/J_1)\hat{\sigma}_A^2}{1 + (J/J_1)\sigma_a^2}\,\mathrm{var}_*\left\{\bar{z}_\mathcal{S}^*\right\}. \tag{9.62}$$

To get from the permutation variance (9.60) to the conditional variance (9.55), we should multiply by

$$R \equiv \frac{1 + (J/J_1)\hat{\sigma}_A^2 C}{1 + (J/J_1)\sigma_a^2 C}. \tag{9.63}$$

Instead, the restandardization formula (9.62) multiplies (9.60) by

$$\hat{R} = \frac{1 + (J/J_1)\hat{\sigma}_A^2}{1 + (J/J_1)\sigma_a^2}. \tag{9.64}$$

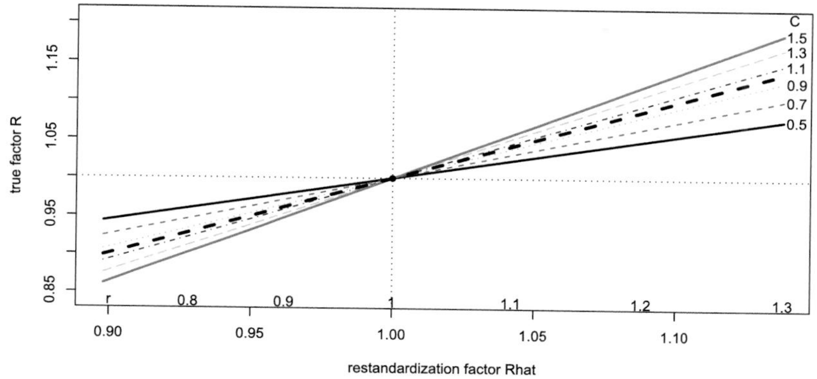

Figure 9.6 Exact variance correction factor R versus restandardization factor $\hat{R}$ (9.63)–(9.64) for values of C and $r = \hat{\sigma}_A/\sigma_a$ as labeled; $J = 5$, $\sigma_a^2 = 0.20$. $\hat{R}$ is exact for $C = 1$ (heavy dashed line).

We see that restandardization gives correct answers when $C = 1$, the average case (9.53), but is only an approximation otherwise. Figure 9.6 compares $\hat{R}$ with R for various choices of C and $r = \hat{\sigma}_A/\sigma_a$ (taking $J = 5$ and $\sigma_a^2 = 0.20$, the value from (9.48) for $\alpha = 0.10$). The restandardization results seem reasonable, though far from exact for extreme choices of C. R itself could be estimated from permutation and randomization calculations, but the special nature of our model makes this an uncertain enterprise.

Proof of Lemma 9.1 Equations (9.44) and (9.45) give

$$\sigma_0 \bar{z}_S = \bar{u}_S + \bar{v}_S \boldsymbol{A} \sim \mathcal{N}\left(0, \frac{1}{m} + \sigma_a^2 \|\bar{v}_S\|^2\right) \tag{9.65}$$

where $\bar{u}_S = \sum_S u_i/m$, and likewise $\bar{v}_S$ is the average of the rows of $\boldsymbol{V}$ indexed by $\mathcal{S}$. From (9.39) and (9.50),

$$\bar{v}_S = \frac{J}{J_1}\left(\ldots, \frac{m_j}{m} - \frac{1}{J}, \ldots\right), \tag{9.66}$$

verifying (9.52) since

$$\|\bar{v}_S\|^2 = \left(\frac{J}{J_1}\right)^2 \sum_1^5 \left(\frac{m_j}{m} - \frac{1}{J}\right)^2 = \frac{1}{m}\frac{J}{J_1}C \tag{9.67}$$

and $\sigma_0^2 = 1 + (J/J_1)\sigma_a^2$. □

Proof of Theorem 9.2 Since $\sum_j \bar{v}_{S_j} = 0$, (9.65) gives

$$\sigma_0 \bar{z}_S \big| \boldsymbol{A} \sim \mathcal{N}\left(\bar{v}_S \tilde{\boldsymbol{A}}, \frac{1}{m}\right) \tag{9.68}$$

where $\tilde{A}_j = A_j - \bar{A}$. But standard multivariate normal calculations show that

$$E\left\{\bar{v}_S \tilde{\boldsymbol{A}} \big| \hat{\sigma}_A^2\right\} = 0 \quad \text{and} \quad E\left\{\left(\bar{v}_S \tilde{\boldsymbol{A}}\right)^2 \big| \hat{\sigma}_A^2\right\} = \hat{\sigma}_A^2 \|\bar{v}_S\|^2 / J_1. \tag{9.69}$$

This combines with (9.67) to give (9.55), using $\sigma_0^2 = 1 + (J/J_1)\sigma_a^2$. □

Exercise 9.9 Verify the preceding statement.

Non-central χ^2 calculations Equation (9.44) gives $\sigma_0 \boldsymbol{z} | \boldsymbol{A} \sim \mathcal{N}_N(\boldsymbol{V}\boldsymbol{A}, I_N)$. Applying (9.40) shows that $\|\boldsymbol{V}\boldsymbol{A}\|^2 = \delta^2$, verifying (9.56)–(9.57). Distribution $\chi_N^2(\delta^2)$ has mean and variance $(\delta^2 + N, 4\delta^2 + 2N)$, showing that the variance of $\text{sd}_s^2 | \hat{\sigma}_A^2$ is $O_p(N^{-1})$ (9.58).

Permutation calculations Model (9.36) makes the elements in any one row of $\boldsymbol{X}$ i.i.d. normal variates

$$x_{ik} = u_{ik} + \sum_{j=1}^{J} v_{ij} a_{jk} \overset{\text{iid}}{\sim} \mathcal{N}\left(0, \sigma_0^2\right) \tag{9.70}$$

for $k = 1, 2, \ldots, n$. The permutation distribution of z_i (9.42) has mean and variance $(0, \hat{\sigma}_i^2/\sigma_0^2)$, where $\hat{\sigma}_i^2$, the usual unbiased variance estimate $\sum_k (x_{ik} - x_{i\cdot})^2/(n-1)$, has expectation σ_0^2. Averaging over $i = 1, 2, \ldots, N$ supports (9.61).

Let $\bar{x}_{Sk}$ be the average of x_{ik} for $i \in \mathcal{S}$. The same argument as in (9.65)–(9.67) shows that

$$\bar{x}_{Sk} \overset{\text{iid}}{\sim} \mathcal{N}\left(0, \frac{1}{m}\left[1 + \frac{J}{J_1}\sigma_a^2 C\right]\right) \quad \text{for } k = 1, 2, \ldots, N. \tag{9.71}$$

Also

$$\bar{z}_S = \sqrt{\frac{n}{4}} \frac{\bar{x}_{S(2)} - \bar{x}_{S(1)}}{\sigma_0}, \tag{9.72}$$

the notation indicating averages of $\bar{x}_{Sk}$ over $k = 1, 2, \ldots, n/2$ and over $n/2 + 1, \ldots, n$, giving approximation (9.60).

9.4 Local Averaging

Enrichment has an exotic sound in the microarray context, where the choice of gene sets seems rather mysterious, at least to those of us without the required biogenetical background. There is however one situation where gene sets are natural and enrichment a familiar tactic, though it doesn't involve genes. We will discuss the idea of *local averaging* in terms of the DTI example of Section 2.5.

The DTI data reports z-values comparing dyslexic children with normal controls at $N = 15\,443$ brain locations or *voxels*, which can be thought of as cubes roughly 1.5mm on a side. Voxel i has coordinates

$$(x_i, y_i, u_i) \tag{9.73}$$

where x_i measures units from back to front of the brain, y_i from right to left, and u_i from bottom to top. Figure 2.4 indicates the z-values on a horizontal slice of the brain about half-way from bottom to top.

Local averaging amounts to smoothing the z_i values by averaging them over nearby voxels. In the example which follows, each z_i is replaced with Z_i, the average of z_j for j in the bigger cube

$$(x_i \pm 2, y_i \pm 2, u_i \pm 2). \tag{9.74}$$

The bigger cube, call it $\mathcal{S}_i$, is five voxels on a side, containing a maximum of 125 original voxels. (Only 2% of the cubes actually had 125 voxels, the average number being 81 ± 25.)

A simple idea motivates local averaging: suppose that the voxels in $\mathcal{S}_i$ have independent normal z-values

$$z_j \overset{\text{ind}}{\sim} \mathcal{N}(\mu_j, 1) \qquad \text{for } j \in \mathcal{S}_i. \tag{9.75}$$

Then averaging results in

$$Z_i \sim \mathcal{N}(\bar{\mu}_i, 1/m_i) \tag{9.76}$$

where m_i is the number of voxels in $\mathcal{S}_i$ and $\bar{\mu}_i$ is the average of μ_j, $j \in \mathcal{S}_i$. Rescaling to variance 1 gives

$$\tilde{Z}_i = \sqrt{m_i} Z_i \sim \mathcal{N}\left(\sqrt{m_i}\bar{\mu}_i, 1\right). \tag{9.77}$$

If $\bar{\mu}_i \doteq \mu_i$, then going from $z_i \sim \mathcal{N}(\mu_i, 1)$ to $\tilde{Z}_i \dot{\sim} \mathcal{N}(\sqrt{m_i}\mu_i, 1)$ magnifies effect sizes in the non-null cubes.

Exercise 9.10 Suppose $\bar{\mu}_i = \mu_i$ and $m_i = m$ for all values of i. (a) How will averaging change g_1 in the hierarchical model (2.47), (2.49)? (b) If $m = 9$, $\pi_0 = 0.95$, and $g_1 \sim \mathcal{N}(0, 2)$ in (2.49), graph $\text{fdr}(z) = \pi_0 f_0(z)/f(z)$ for the original z_i's and the $\tilde{Z}_i$'s.

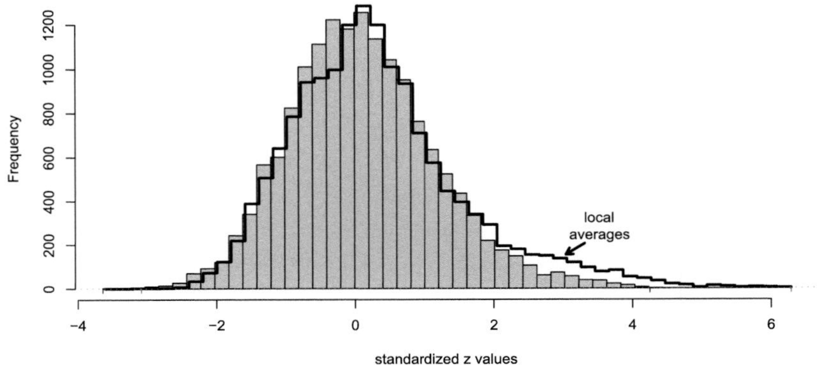

Figure 9.7a $N = 15\,443$ DTI z-values (solid histogram) compared to local averages Z_i (line histogram); standardized (9.79). Local averaging has produced a heavier right tail.

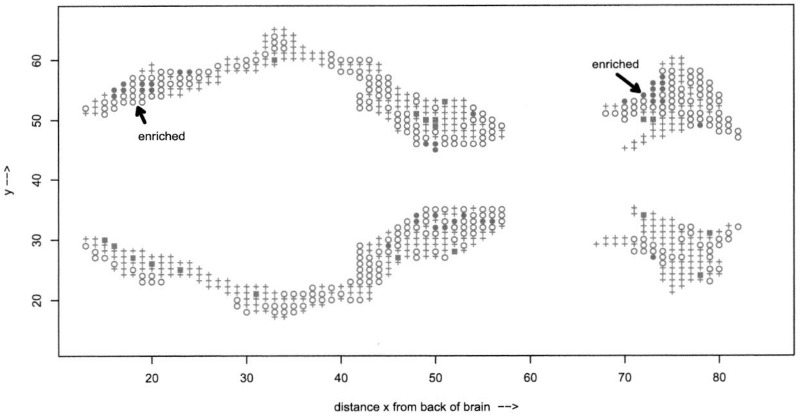

Figure 9.7b Averaged values $\tilde{Z}_i$ for DTI data (9.79); horizontal section, bottom of brain. Red, $\tilde{Z}_i > 0$; green, $\tilde{Z}_i < 0$; solid circles, $\tilde{Z}_i > 2$; solid squares, $\tilde{Z}_i < -2$. Black arrows indicate enrichment regions, $\widehat{\text{fdr}}(Z_i) \leq 0.20$.

The $N = 15\,443$ averaged values Z_i had empirical null estimates

$$\left(\hat{\delta}_0, \hat{\sigma}_0, \hat{\pi}_0\right) = (-0.15, 0.47, 0.940) \tag{9.78}$$

using the MLE method of Section 6.3. This compares with (−0.12, 1.06, 0.984) for the original z_i values pictured in Figure 5.1b. Figure 9.7a compares histograms for the standardized z-values

$$\tilde{Z}_i = (Z_i + 0.15)/0.47 \quad \text{and} \quad \tilde{z}_i = (z_i + 0.12)/1.06. \tag{9.79}$$

Averaging appears to have worked: the $\tilde{Z}$'s have a noticeably heavier right tail, though the effect is nowhere as large as suggested by (9.77). (Note that the ratio of $\hat{\sigma}_0$ values, 0.47/1.06, is closer to 1/2 than to the value 1/9 suggested by $m_i \sim 81$. Positive correlation within cubes increases the variance of Z_i in (9.76).)

A local false discovery rate analysis (6.38) based on the MLE method yielded interesting regions of difference between the dyslexic brain patterns and normal controls. Figure 9.7b indicates the $\tilde{Z}_i$ values on a horizontal slice, this time at the bottom of the brain rather than the middle slice of Figure 2.4. The black triangles indicate voxels with $\widehat{\text{fdr}}(Z_i) \leq 0.20$. These "enrichment" regions, which were not found using the original z_i's, both occur on the left side of the brain.

Local averaging is a simple form of enrichment that raises several points of comparison with the previous sections:

- The reader may wonder what has happened to restandardization (9.13)–(9.14). Our use of an empirical null in assessing $\widehat{\text{fdr}}(Z_i)$ for Figure 9.7b is in the spirit of row randomization in that it takes account of the variability of Z_j across all "gene sets" $\mathcal{S}_j$, not just within any one set of interest $\mathcal{S}_i$. Empirical null calculations can also include permutation information, as in the sixth bullet point of Section 6.5.
- There is however an important difference. Permutation calculations (9.3), (9.4) are done *separately* for each set $\mathcal{S}_i$, carrying over to restandardization (9.13)–(9.14). This takes account of set differences, e.g., different sizes m_i or correlation structures, while the $\widehat{\text{fdr}}(Z_i)$ analysis treats all sets $\mathcal{S}_i$ the same.
- On the other hand, if the experimenters had set the DTI scanner to collect data on a coarser scale, the Z_i analysis would seem natural.
- The examples of previous chapters assumed that individual cases could be treated as identical a priori. There the "gene sets" were singletons, having $m_i = 1$ and no internal correlation structure, encouraging identi-

cal treatment (though we might have Bayesian reasons for treating them differently, as in (5.21)–(5.25)).

- Suppose we did try to apply a restandardization formula to, say, the prostate data example of Section 2.1. The permutation null distribution for any one gene will be approximately $z_i^* \sim \mathcal{N}(0, 1)$, Section 6.5, making $(m_s^*, \text{sd}_s^*) \doteq (0, 1)$ in (9.13). With $s(z) = z$, we get $(m_s, \text{sd}_s) = (\bar{z}, \bar{\sigma})$, the empirical mean and standard deviation of all N z-values. Formula (9.14) then yields

$$p_i^{**} = \Phi\left(\frac{z_i - \bar{z}}{\bar{\sigma}}\right) \tag{9.80}$$

as the restandardized p-value for case i. This compares with the empirical null p-value

$$\hat{p}_i = \Phi\left(\frac{z_i - \hat{\delta}_0}{\hat{\sigma}_0}\right) \tag{9.81}$$

where $(\hat{\delta}_0, \hat{\sigma}_0)$ are the empirical null mean and standard deviation, Chapter 6.

Exercise 9.11 Verify (9.80).

The tension between row randomization and permutation brings us back to questions of "learning from the experience of others," Section 1.4 and Section 2.6. Chapter 10 deals directly with the combination and separation of cases in a false discovery rate analysis, where results like those of Figure 9.7b will be examined more skeptically.

Notes

Gene set enrichment analysis, and the GSEA algorithm, were proposed in an influential paper by Subramanian et al. (2005) where the p53 data is one of three main examples. The Bioconductor package `limma` (Smyth, 2004) offers a GSEA-like option based on $\bar{z}_S$. Other enrichment techniques appear in Pavlidis et al. (2002) and Rahnenführer et al. (2004). Most of the material in Section 9.1 and Section 9.2, including restandardization and the maxmean statistic, is taken from Efron and Tibshirani (2007).

10

Combination, Relevance, and Comparability

A tacit assumption underlies our previous examples of simultaneous inference: that all cases presented together should be analyzed together, such as all 6033 genes for the prostate data in Figure 5.2, or all 15 443 DTI voxels. This leads us down a perilous path. Omnibus combination may distort individual inferences in both directions: interesting cases may be hidden while uninteresting ones are enhanced. This chapter concerns the separation and combination of cases in a large-scale data analysis.

Figure 10.1 illustrates separation/combination difficulties in terms of the DTI data. Z-values for the 15 443 voxels have been separated into back and front halves of the brain ($x < 50$ and $x \geq 50$ in terms of Figure 2.4), 7661 voxels in back and 7782 in front. Two discrepancies strike the eye: the heavy right tail in the combined histogram of Figure 5.1b comes exclusively from the front; and the center of the back-half histogram is shifted leftwards 0.35 units relative to the front.

Separate local false discovery rate analyses, assuming a theoretical null distribution $\mathcal{N}(0, 1)$ as in Chapter 5, gave 271 voxels with $\widehat{\text{fdr}}(z_i) \leq 0.20$ for the front-half data, those having $z_i \geq 2.71$, but none for the back half. This is quite different from the combined analysis of Figure 5.2, where only the 184 voxels with $z_i \geq 3.05$ (including nine from the back) had $\widehat{\text{fdr}}(z_i) \leq 0.20$.

Scientific guidance would be most welcome at this point, but in its absence the statistical evidence of Figure 10.1 argues against a combined analysis. Section 10.2 and Section 10.3 go further, suggesting still finer separation of cases, while describing a convenient methodology for doing so. The basic methodological idea is developed in Section 10.1. Applications to enrichment are discussed in Section 10.2. Section 10.4 discusses whether false discovery rate methods maintain their inferential legitimacy in the face of separation. *Comparability*, Section 10.5, concerns the type of statistics that are appropriate for the combination of cases.

The central concern of this chapter is familiar in the family-wise error rate literature, where it takes the form of the question "What is a family?"

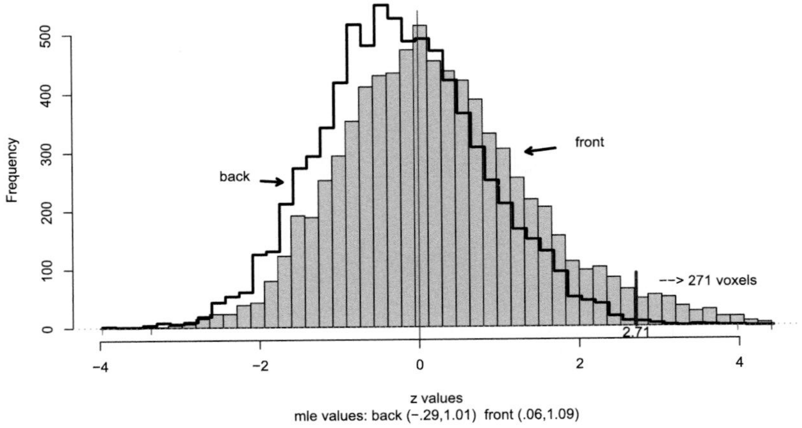

Figure 10.1 DTI data of Section 2.5 separated into back-half and front-half voxels ($x < 50$ and $x \geq 50$ in Figure 2.4). Separate analyses gave 271 voxels with $\widehat{\text{fdr}} \leq 0.20$ for front half, but none for back half. MLE empirical null estimates $(\hat{\delta}_0, \hat{\sigma}_0)$ of Section 6.3 show back-half center shifted 0.35 units to the left of front half.

Miller's introduction to the second edition of his simultaneous inference book begins in this way:

> Time has now run out. There is nowhere left for the author to go but to discuss just what constitutes a family. This is the hardest part of the entire book because it is where statistics takes leave of mathematics and must be guided by subjective judgement.[1]

We are at an advantage today because large-scale data sets allow us, sometimes, to examine directly the trade-offs between separation and combination of cases. False discovery rates, this chapter's favored methodology, help clarify the question. Basically, however, we won't be able to answer Miller's question of what is the proper family for simultaneous analysis, though evidence like that in Figure 10.1 may strongly suggest what *is not*. Miller's "subjective judgements" still dominate practice. There are some hints in what follows of a more principled approach to separation and combination, particularly in the section on *relevance*, but at this stage they remain just hints.

[1] Section 1.5 of Westfall and Young's 1993 book gives another nice discussion of basic multiple inference conundrums.

10.1 The Multi-Class Model

The two-groups model of Section 2.2 and Section 5.1 can be extended to cover the situation where the N cases are divided into distinct classes, possibly having different choices of π_0, $f_0(z)$, and $f_1(z)$ shown in Figure 2.3. Figure 10.2 illustrates the scheme: two classes A and B (e.g., the front and back halves of the brain in the DTI example) have a priori probabilities w_A and $w_B = 1 - w_A$. The two-groups model (5.1) holds separately within each class, for example with $\pi_0 = \pi_{A0}$, $f_0(z) = f_{A0}(z)$, and $f_1(z) = f_{A1}(z)$ in class A. It is important to note that the class label A or B is observed by the statistician, in contrast to the null/non-null dichotomy which must be inferred.

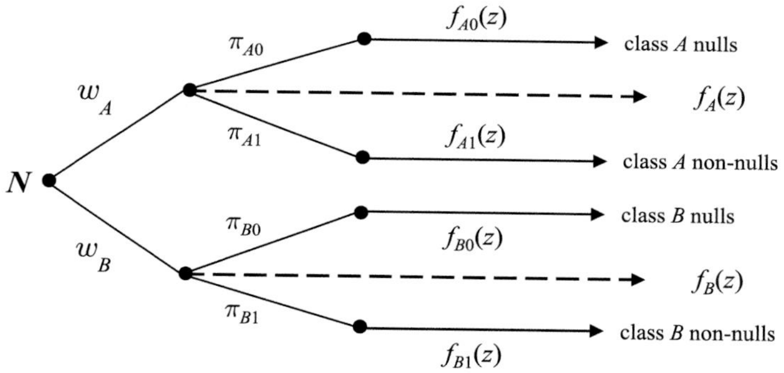

Figure 10.2 The multi-class model with two classes, A and B. Two-groups model (5.1) holds separately within each class, with possibly different parameters. Prior probabilities w_A and w_B shown for the two classes.

The definitions of Section 5.1 apply separately within classes, for instance to the class A mixture density and local false discovery rate

$$f_A(z) = \pi_{A0} f_{A0}(z) + \pi_{A1} f_{A1}(z) \quad \text{and} \quad \text{fdr}_A(z) = \pi_{A0} f_{A0}(z)/f_A(z). \tag{10.1}$$

A combined analysis, ignoring the class information in Figure 10.1, has marginal densities

$$\begin{aligned} f_0(z) &= w_A \frac{\pi_{A0}}{\pi_0} f_{A0}(z) + w_B \frac{\pi_{B0}}{\pi_0} f_{B0}(z) \\ f_1(z) &= w_A \frac{\pi_{A1}}{\pi_1} f_{A1}(z) + w_B \frac{\pi_{B1}}{\pi_1} f_{B1}(z) \end{aligned} \tag{10.2}$$

where $\pi_0 = 1 - \pi_1$ is the combined null probability,

$$\pi_0 = w_A \pi_{A0} + w_B \pi_{B0}. \tag{10.3}$$

This leads to overall marginal density and false discovery rate

$$f(z) = w_A f_A(z) + w_B f_B(z) \quad \text{and} \quad \text{fdr}(z) = \pi_0 f_0(z)/f(z) \tag{10.4}$$

just as in (5.2), (5.3). If class information were unavailable, we would have to base our inference on fdr(z), but here we have the option of separately employing $\text{fdr}_A(z)$ or $\text{fdr}_B(z)$.

Bayes theorem yields a useful relationship between the separate and combined false discovery rates.

Theorem 10.1 *Define $w_A(z)$ as the conditional probability of a case being in class A given z,*

$$w_A(z) = \Pr\{A|z\} \tag{10.5}$$

and similarly define

$$w_{A0}(z) = \Pr_0\{A|z\} \tag{10.6}$$

to be the conditional probability of class A given z for a null case. Then

$$\text{fdr}_A(z) = \text{fdr}(z) R_A(z) \qquad \textit{where } R_A(z) = \frac{w_{A0}(z)}{w_A(z)}. \tag{10.7}$$

Proof Let I be the event that a case is null, so I occurs in the two null paths of Figure 10.2 but not otherwise. We have

$$\begin{aligned} \frac{\text{fdr}_A(z)}{\text{fdr}(z)} &= \frac{\Pr\{I|A,z\}}{\Pr\{I|z\}} = \frac{\Pr\{I,A|z\}}{\Pr\{A|z\}\Pr\{I|z\}} \\ &= \frac{\Pr\{A|I,z\}}{\Pr\{A|z\}} = \frac{w_{A0}(z)}{w_A(z)}. \end{aligned} \tag{10.8}$$

□

Tail-area false discovery rates also obey the theorem. In the general notation (2.12), (2.13),

$$\text{Fdr}_A(\mathcal{Z}) = \text{Fdr}(\mathcal{Z}) R_A(\mathcal{Z}) \qquad \text{where } R_A(\mathcal{Z}) = \frac{\Pr_0\{A|z \in \mathcal{Z}\}}{\Pr\{A|z \in \mathcal{Z}\}}. \tag{10.9}$$

Exercise 10.1 Verify (10.9).

The virtue of Theorem 10.1 is that $\widehat{R_A}(z)$ is often easy to estimate, which allows $\widehat{\text{fdr}}_A(z)$ to be obtained from $\widehat{\text{fdr}}(z)$ even if class A is too small for direct estimation of fdr_A (see Section 10.2). We do not need the theorem

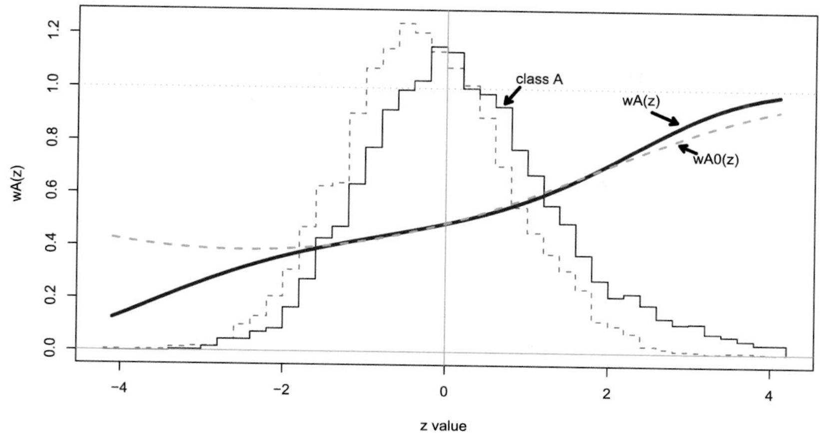

Figure 10.3 Heavy curve is logistic regression estimate of (10.5), $w_A(z) = \Pr\{A|z\}$, where class A is the front-half portion of the DTI data; dashed curve estimates $w_{A0}(z) = \Pr_0\{A|z\}$ (10.6).

for the DTI example of Figure 10.1 since class A is large, but the calculation of $R_A(z) = w_{A0}(z)/w_A(z)$ is interesting in its own right.

The heavy curve in Figure 10.3 is an estimate of $w_A(z) = \Pr\{A|z\}$ obtained by ordinary logistic regression: the class indicators

$$c_i = \begin{cases} 1 & \text{if case}_i \text{ in class } A \\ 0 & \text{if case}_i \text{ in class } B \end{cases} \tag{10.10}$$

were regressed as a cubic logistic function of z_i. As might be expected from the two histograms, $\hat{w}_A(z)$ increases smoothly with z and, in fact, a linear logistic regression would have sufficed here.

In order to estimate $w_{A0}(z) = \Pr_0\{A|z\}$, we need to make some assumptions about the null distributions in Figure 10.2. Assuming normal nulls and using the MLE method of Section 6.3 gave estimates

$$f_{A0}(z) \sim \mathcal{N}\left(\delta_{A0}, \sigma_{A0}^2\right) \quad \text{and} \quad f_{B0}(z) \sim \mathcal{N}\left(\delta_{B0}, \sigma_{B0}^2\right) \tag{10.11}$$

with

$$(\delta_{A0}, \sigma_{A0}) = (0.06, 1.09) \quad \text{and} \quad (\delta_{B0}, \sigma_{B0}) = (-0.29, 1.01). \tag{10.12}$$

The null probabilities were $\pi_{A0} = 0.969$ and $\pi_{B0} = 0.998$. The dashed curve in Figure 10.3 is $\hat{w}_{A0}(z)$ estimated, using Bayes rule, from (10.11)–(10.12).

Note: We assume $w_A = w_B = 0.50$ in Figure 10.2 since the front- and back-half sets are nearly equal in size.

Exercise 10.2 (a) Show that (10.11) gives

$$\frac{w_{A0}(z)}{w_{B0}(z)} = \frac{w_A \pi_{A0} \sigma_{B0}}{w_B \pi_{B0} \sigma_{A0}} \exp -\frac{1}{2} \left\{ \left(\frac{z - \delta_{A0}}{\sigma_{A0}} \right)^2 - \left(\frac{z - \delta_{B0}}{\sigma_{B0}} \right)^2 \right\}. \tag{10.13}$$

(b) Express $w_{A0}(z)$ in terms of $w_{A0}(z)/w_{B0}(z)$.

Looking at Figure 10.1, we might expect that $\widehat{\text{fdr}}_A(z)$ would be much lower than the combined estimate $\widehat{\text{fdr}}(z)$ for large values of z, but that is not the case. The empirical Bayes ratio

$$\hat{R}_A(z) = \hat{w}_{A0}(z)/\hat{w}_A(z) \tag{10.14}$$

is ≥ 0.94 for $z > 0$, so formula (10.7) implies only small differences. Two contradictory effects are at work: the longer right tail of the front-half distribution by itself would produce small values of $\hat{R}_A(z)$ and $\widehat{\text{fdr}}_A(z)$. However, the effect is mostly canceled by the rightward shift of the whole front-half distribution, which substantially increases the numerator of $\widehat{\text{fdr}}_A(z) = \hat{\pi}_{A0}\hat{f}_{A0}(z)/\hat{f}_A(z)$.

The close match between $\hat{w}_{A0}(z)$ and $\hat{w}_A(z)$ near $z = 0$ is no accident. Following through the definitions in Figure 10.3 and (10.1) gives, after a little rearrangement,

$$\frac{w_A(z)}{1 - w_A(z)} = \frac{w_{A0}(z)}{1 - w_{A0}(z)} \frac{1 + Q_A(z)}{1 + Q_B(z)} \tag{10.15}$$

where

$$Q_A(z) = \frac{1 - \text{fdr}_A(z)}{\text{fdr}_A(z)} \quad \text{and} \quad Q_B(z) = \frac{1 - \text{fdr}_B(z)}{\text{fdr}_B(z)}. \tag{10.16}$$

Often $\text{fdr}_A(z)$ and $\text{fdr}_B(z)$ approximately equal 1.0 near $z = 0$, reflecting a large preponderance of null cases and the fact that non-null cases tend to produce z-values farther away from zero. Then (10.15) gives

$$w_A(z) \doteq w_{A0}(z) \qquad \text{for } z \text{ near zero} \tag{10.17}$$

as seen in Figure 10.3.

Suppose we believe that $f_{A0}(z) = f_{B0}(z)$ in Figure 10.2, i.e., that the null cases are distributed identically in the two classes. (This being true in particular if we accept the usual $\mathcal{N}(0,1)$ theoretical null distribution.) Then $w_{A0}(z)$ does not depend on z, and we obtain a simplified version of Theorem 10.1.

Corollary 10.2 *If*[2]

$$f_{A0}(z) = f_{B0}(z) \tag{10.18}$$

for all z then

$$\text{fdr}_A(z) = c_0 \, \text{fdr}(z)/w_A(z) \qquad [c_0 = w_A \pi_{A0}/\pi_0]. \tag{10.19}$$

Substituting $z = 0$ in (10.19) gives $c_0 = w_A(0)\,\text{fdr}_A(0)/\,\text{fdr}(0)$ and

$$\text{fdr}_A(z) = \left(\frac{\text{fdr}_A(0)}{\text{fdr}(0)}\right) \text{fdr}(z) \cdot \frac{w_A(0)}{w_A(z)}. \tag{10.20}$$

The first factor is often near 1, as mentioned before (10.17), yielding the convenient approximation

$$\text{fdr}_A(z) \doteq \text{fdr}(z) \frac{w_A(0)}{w_A(z)}. \tag{10.21}$$

(The missing factor $\text{fdr}_A(0)/\,\text{fdr}(0)$ in (10.21) is likely to be ≤ 1 if A is an enriched class, since "enrichment" implies lower null probabilities, making approximation (10.21) conservative.) Formula (5.25) can be shown to be a special case of (10.19).

Exercise 10.3 (a) Verify that Corollary 10.2 also applies to $\text{Fdr}_A(\mathcal{Z})$ (10.9) in the form

$$\text{Fdr}_A(\mathcal{Z}) = c_0 \, \text{Fdr}(\mathcal{Z})/w_A(\mathcal{Z}) \tag{10.22}$$

where $w_A(\mathcal{Z}) = \Pr\{A|z \in \mathcal{Z}\}$ and $c_0 = w_A \pi_{A0}/\pi_0$. (b) Show that (10.17) and (10.18) together imply

$$\left.\frac{dw_A(z)}{dz}\right|_{z=0} \doteq 0. \tag{10.23}$$

Approximation (10.23) can serve as a check on assumption (10.18). For instance, the solid curve in Figure 10.3 has $d\hat{w}_A(z)/dz$ noticeably positive near $z = 0$, casting doubt on (10.18) without recourse to normal models (10.11). This kind of argument is less convincing for small "enriched" classes A where $\text{fdr}_A(0)$ may be a lot less than 1, invalidating the reasoning leading up to (10.17).

Summary Making N, the number of cases, as large as possible is the rationale for combined analysis, at least in an empirical Bayes framework. Large N makes estimates like $\overline{\text{Fdr}}(\mathcal{Z})$ better approximations to the true

[2] Notice that (10.18) is equivalent to $f_{A0}(z) = f_0(z)$, using (10.2).

Bayes value $\text{Fdr}(\mathcal{Z})$, as in Section 2.4, justifying Bayesian interpretation of our results. The danger of combination is that we may be getting an accurate estimate of the wrong quantity: if $\text{Fdr}_A(z)$ is much different than $\text{Fdr}_B(z)$, then $\overline{\text{Fdr}}(z)$ may be far from both. Theorem 10.1 and Corollary 10.2 offer the possibility of accurately estimating $\text{Fdr}_A(z)$ even when A is a small subclass of all N cases.

10.2 Small Subclasses and Enrichment

Corollary 10.2 is especially useful when A is a *small* subclass, for which direct estimates $\widehat{\text{fdr}}_A(z)$ or $\widehat{\text{Fdr}}_A(z)$ would be hopelessly inaccurate because of inadequate sample size N. We encountered small subclasses in the p53 enrichment example of Chapter 9, and enrichment will be taken up later in this section, but smallness can force itself upon our attention even in the absence of preselected enrichment classes.

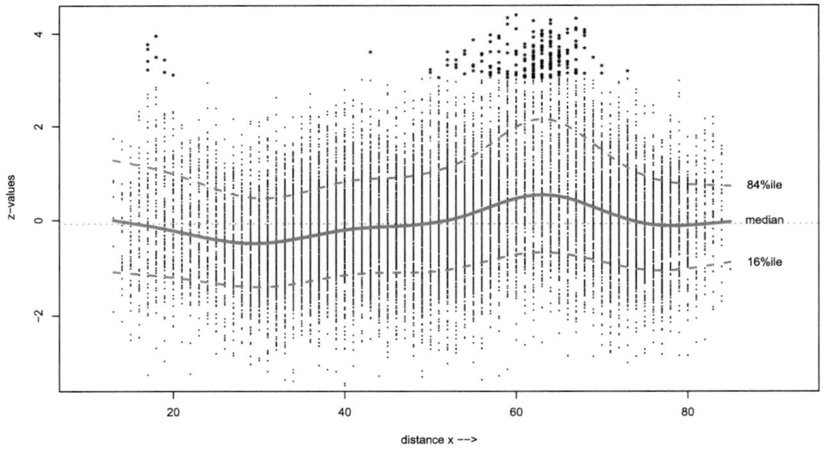

Figure 10.4 DTI z-values plotted versus distance x from back of brain; stars indicate the 184 voxels with $z_i \geq 3.05$, those having $\widehat{\text{fdr}}(z_i) \leq 0.20$ in combined analyses of all 15 443 voxels, Figure 5.2. Running percentile curves show a general upward shift of z-values around $x = 62$, the region including most of the starred voxels.

The $N = 15\,443$ DTI z-values z_i are plotted versus x_i, the voxel's distance from the back of the brain, in Figure 10.4. A clear wave is visible, cresting near $x = 62$. Most of the 184 voxels identified as non-null in the combined fdr analysis, those with $\widehat{\mathrm{fdr}}(z_i) \leq 0.20$, occurred at the top of the crest.

There is something worrisome here: the z-values around $x = 62$ are shifted upward across their entire range, not just in the upper percentiles. This might be due to a reading bias in the DTI imaging device, or a genuine difference between dyslexic and normal children for *all* brain locations near $x = 62$. In neither case would it be correct to assign some special signicance to those voxels near $x = 62$ that happen to have large z_i values.

Figure 10.4 suggests using finer subdivisions than the front/back split in Figure 10.1. As an example, let class A be the 82 voxels located at $x = 18$. These display some large z-values, attained without the benefit of riding a wave crest.[3] Their histogram is compared with that of all other voxels in Figure 10.5, along with an estimate of the factor

$$R(z) = w_A(0)/w_A(z) \tag{10.24}$$

in Corollary 10.2, formula (10.21).

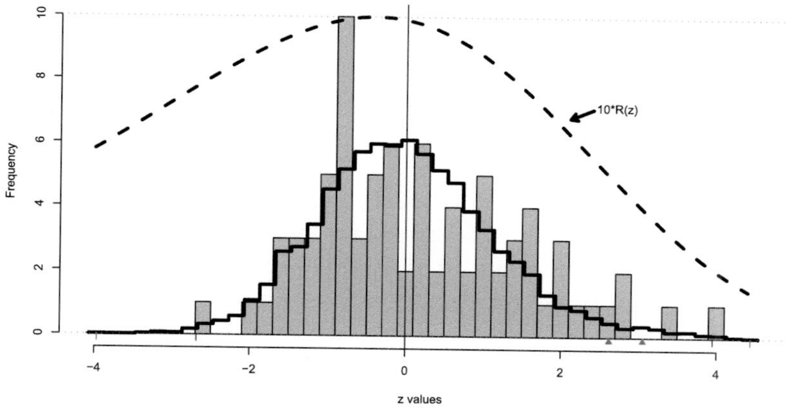

Figure 10.5 DTI z-values for the 82 voxels at $x = 18$ (solid histogram) compared with all other voxels (line histogram). Dashed curve is 10 times $\hat{R}(z) = \hat{w}_A(0)/\hat{w}_A(z)$, formula (10.21), Corollary 10.2; $\hat{R}(z)$ is less than 1 for large z, substantially reducing $\widehat{\mathrm{fdr}}_A(z)$ below $\widehat{\mathrm{fdr}}(z)$. Triangles indicate threshold values $\widehat{\mathrm{fdr}}_A(z) = 0.20$ and $\widehat{\mathrm{fdr}}(z) = 0.20$, at 2.62 and 3.05, respectively.

$\hat{R}(z)$ decreases for large values of z, making $\widehat{\mathrm{fdr}}_A(z)$ substantially smaller

[3] Figure 9.7b also suggests something interesting going on near $x = 18$.

than $\widehat{\text{fdr}}(z)$, the combined empirical null fdr estimate. The threshold value of z, that giving 0.20 as the estimated fdr, decreases from 3.05 for the combined estimate $\widehat{\text{fdr}}$ to 2.62 for $\widehat{\text{fdr}}_A$. The number of voxels having estimated fdr less than 0.20 increases from two to five. (Note that $\hat{R}(z) = \hat{w}_A(0)/\hat{w}_A(z)$ in (10.21) does not depend on how $\widehat{\text{fdr}}(z)$ is estimated. The dashed curve in Figure 10.5 applies just as well to empirical null estimation of fdr(z) instead of theoretical null, in which case the threshold values increase to 2.68 and 3.17, respectively.)

Enrichment

Returning to the enrichment calculations of Chapter 9, we can reasonably state the null hypothesis that class A is *not* enriched as $\text{fdr}_A(z) = \text{fdr}(z)$ for all z, where fdr(z) is the false discovery rate ignoring class (10.4); in other words, knowing that a case is in class A does not change our assessment of its null probability.

Assuming $f_{A0}(z) = f_0(z)$, as in Corollary 10.2, approximation (10.21) yields the

$$\textit{enrichment null hypothesis}\text{: } w_A(z) = \text{constant} \qquad (10.25)$$

which means we can use $\hat{w}_A(z)$ to test for enrichment. For instance, we might estimate $w_A(z)$ by linear logistic regression and then test (10.25) with

$$S = \hat{\beta}/\widehat{\text{se}} \qquad (10.26)$$

where $\hat{\beta}$ is the regression slope and $\widehat{\text{se}}$ its standard error.

Slope statistic (10.26) was computed for the P53_UP pathway (the top row in Figure 9.5) comprising 40 of the 10 100 p53 data z-values: a standard linear logistic model for $w_A(z)$ gave $\hat{\beta} = 0.699 \pm 0.153$ and

$$S = 0.699/0.153 = 4.58, \qquad (10.27)$$

two-sided p-value $4.7 \cdot 10^{-6}$. Permutation analysis gave $\hat{\beta} = 0.699 \pm 0.240$ and the more realistic value

$$S = 0.699/0.240 = 2.91, \qquad (10.28)$$

$p = 0.0036$.

Tests based on (10.26) can be shown to match asymptotically those based on $S = \bar{z}_A$, the average of the z-values within A. (The connection is through model (9.38).) There is no particular advantage to testing enrichment with (10.26) rather than the methods of Chapter 9. However, having found a significantly enriched class A like P53_UP in Figure 9.5, we can use Corollary 10.2 to identify individual non-null cases *within* A. Formula

(10.21), with $\hat{w}_A(z)$ obtained from the linear logistic model above, yielded nine genes in P53_UP with $\widehat{\text{fdr}}_A(z_i) \leq 0.20$, compared to just one based on $\widehat{\text{fdr}}(z_i)$.

Exercise 10.4 Assuming $f_{A0}(z) = f_0(z)$ for all z, show that the null hypothesis H_{00} : $\text{fdr}_A(z) = \text{fdr}(z)$ for all z implies H_0 : $w_A(z) = \text{constant}$, but H_0 only implies $\text{fdr}_A(z) = \text{constant} \cdot \text{fdr}(z)$. (So rejection of H_0 implies rejection of H_{00}.)

Efficiency

How efficient is the formula based on Corollary 10.2,

$$\widehat{\text{fdr}}_A(z) = c_0 \widehat{\text{fdr}}(z) \Big/ \hat{w}_A(z) \tag{10.29}$$

as an estimate of $\text{fdr}_A(z)$? The following heuristic calculations suggest substantial, and sometimes enormous, savings over direct estimation, when A is small.

Taking logarithms in (10.29) gives

$$\widehat{\text{lfdr}}_A(z) = \widehat{\text{lfdr}}(z) - \widehat{\text{l}w}_A(z) + \text{ constant} \tag{10.30}$$

with $\widehat{\text{lfdr}}_A(z) = \log(\text{fdr}_A(z))$, etc. For reasons discussed later, $\widehat{\text{lfdr}}_A(z)$ and $\widehat{\text{l}w}_A(z)$ are nearly uncorrelated, leading to a convenient approximation for the standard deviation of $\widehat{\text{lfdr}}_A(z)$,

$$\text{sd}\left\{\widehat{\text{lfdr}}_A(z)\right\} \doteq \left[\text{sd}\left\{\widehat{\text{lfdr}}(z)\right\}^2 + \text{sd}\left\{\widehat{\text{l}w}_A(z)\right\}^2\right]^{\frac{1}{2}}. \tag{10.31}$$

Of course, we expect $\widehat{\text{fdr}}_A(z)$ to be more variable than $\widehat{\text{fdr}}(z)$ since it is based on fewer cases, Nw_A rather than N. Standard "square root of sample size" considerations suggest

$$\text{sd}\left\{\widehat{\text{lfdr}}_A(z)\right\} \sim \frac{1}{\sqrt{w_A}}\,\text{sd}\left\{\widehat{\text{lfdr}}(z)\right\} \tag{10.32}$$

if $\text{fdr}_A(z)$ were estimated directly in the same manner as $\text{fdr}(z)$. The efficiency question is how does (10.31) compare with (10.32)?

Table 10.1 makes the comparison for a version of situation (5.18): independent z_i are generated as in Figure 10.2, with

$$\begin{aligned} &w_A = 0.01, \qquad \pi_{A0} = 0.50, \qquad \pi_0 = 0.95, \\ &f_{A0} = f_{B0} \sim \mathcal{N}(0,1) \quad \text{and} \quad f_{A1} = f_{B1} \sim \mathcal{N}(2.5,1), \end{aligned} \tag{10.33}$$

$N = 3000$ and 6000. So for $N = 3000$, A had 30 members, half of them of the non-null $\mathcal{N}(2.5, 1)$ variety, compared with 5% non-null overall.

Table 10.1 *Ratio* $\text{sd}\{\widehat{\text{lfdr}}_A(z)\}/\ \text{sd}\{\widehat{\text{lfdr}}(z)\}$ *using formula* (10.31) *with linear logistic estimator* $\hat{w}_A(z)$*; simulations from model* (10.33). *Bottom row is true* $\text{fdr}(z)$.

z	2	2.5	3	3.5	4	4.5
$N = 3000$	2.2	2.0	2.0	2.0	1.8	1.4
$N = 6000$	2.9	3.3	2.7	2.4	2.0	1.5
fdr(z)	.74	.46	.19	.066	.019	.006

The table shows values of $\text{sd}\{\widehat{\text{lfdr}}_A(z)\}/\ \text{sd}\{\widehat{\text{lfdr}}(z)\}$ of about 1.5 to 3.0, based on (10.31), in the range where the true fdr(z) value is less than 0.50. This compares with a ratio of 10 suggested by (10.32). Here $\widehat{\text{fdr}}(z) = \pi_0 f_0(z)/\hat{f}(z)$ was calculated assuming $\pi_0 f_0(z)$ known. Empirical null estimation of $\pi_0 f_0(z)$ sharply reduces the ratios, to near 1 for large z, because $\text{sd}\{\widehat{\text{lfdr}}(z)\}$ increases (as in Table 7.3) while $\text{sd}\{\widehat{\text{lw}}_A(z)\}$ stays the same. In practice, direct empirical null estimation is impossible for a class A of only 60 members, leaving Corollary 10.2 as the sole hope for assessing $\widehat{\text{fdr}}_A(z)$.

The standard deviation approximation (10.31) depends on $\widehat{\text{lfdr}}(z)$ and $\widehat{\text{lw}}_A(z)$ being nearly uncorrelated. Let $\boldsymbol{y}$ be the vector of discretized counts for all N z_i values, as in Section 5.2, and $\boldsymbol{y}_A$ the counts for the z-values in A. The Poisson regression estimate $\widehat{\textbf{fdr}}$ depends on $\boldsymbol{y}$ while $\widehat{\textbf{lw}}_A$ depends on the ratios $\boldsymbol{r} = (\dots y_{Ak}/y_k \dots)$ with $\boldsymbol{y}$ itself entering only in an ancillary role. In outline, the argument for (10.31) depends on the general equality

$$\text{var}\{X + Y\} = \text{var}\{X\} + \text{var}\{Y\} + 2\,\text{cov}\,\{X, E(Y|X)\} \qquad (10.34)$$

applied to $X = \widehat{\text{lfdr}}_k$ and $Y = -\widehat{\textbf{lw}}_{Ak}$. Both var$\{X\}$ and var$\{Y\}$ are $O(1/N)$, but because the expectation of $\boldsymbol{r}$ does not depend on $\boldsymbol{y}$, the covariance term in (10.31) is only $O(1/N^2)$.

Exercise 10.5 Verify (10.34).

10.3 Relevance

So far we have discussed the separation of cases into two classes A and B (or A and not-A), of possibly distinct behavior. The previous theory can be extended to include multiple classes, say one for every value of x in Figure 10.4, incorporating a notion of relevance between the classes. In the

DTI example, for instance, we might assess the relevance between voxels at x_i and x_j by some smooth kernel such as $\exp\{-|x_i - x_j|/5\}$ rather than whether they occur in the same half of the brain.

Each case is now assumed to have three components,

$$\text{case } i = (x_i, z_i, I_i) \qquad i = 1, 2, \dots, N, \tag{10.35}$$

where x_i is an observed vector of covariates, z_i is the observed z-value, and I_i is an unobservable null indicator, having I_i either 1 or 0 as case i is null or non-null. The two-groups model (5.1) is assumed to hold separately for each possible value x of the covariate, with parameters π_{x0}, $f_{x0}(z)$, and $f_{x1}(z)$, yielding

$$f_x(z) = \pi_{x0} f_{x0}(z) + (1 - \pi_{x0}) f_{x1}(z) \quad \text{and} \quad \text{fdr}_x(z) = \pi_{x0} f_{x0}(z)/f_x(z). \tag{10.36}$$

The class of cases with covariate x has a priori probability

$$w(x) = \Pr\{x_i = x\}. \tag{10.37}$$

One can picture model (10.35)–(10.37) as an extended version of Figure 10.2, now with a separate arm originating at N for each value of x.

Suppose we are interested in false discovery rates for some category A of cases, for example voxels in the hippocampus in the DTI brain study. We have available a "relevance function" $\rho_A(x)$ relating cases having covariate x to category A. Formally, we interpret $\rho_A(x)$ as the conditional probability of A given x,

$$\rho_A(x) = \Pr\{A|x\}. \tag{10.38}$$

For the example of Section 10.1, where A was the front half of the brain, $\rho_A(x)$ equaled one or zero as x was ≥ 50 or < 50 (see Figure 2.4).

Theorem 10.1 generalizes to multiple classes.

Theorem 10.3 *Under model* (10.35)–(10.37), $\text{fdr}_A(z)$, *the local false discovery rate for category A, is related to* $\text{fdr}(z)$, *the marginal false discovery rate ignoring x, by*

$$\text{fdr}_A(z) = \text{fdr}(z) R_A(z) \qquad \textit{where } R_A(z) = \frac{E_0\{\rho_A(x)|z\}}{E\{\rho_A(x)|z\}}, \tag{10.39}$$

with $E_0\{\rho_A(x)|z\}$ *the null conditional expectation of* $\rho_A(x)$ *given z,*

$$E_0\{\rho_A(x)|z\} = E\{\rho_A(x)|z, I = 1\}. \tag{10.40}$$

(Proof given below.)

As in (10.9), Theorem 10.3 extends to $\text{Fdr}_A(\mathcal{Z})$, now with $R_A(\mathcal{Z}) = E_0\{\rho_A(z)|z \in \mathcal{Z}\}/E\{\rho_A(z)|z \in \mathcal{Z}\}$.

Even if category A refers to a specific value of covariate x, say $x = x_0$, we still might wish to let $\rho_A(x)$ fall off smoothly as x moves away from x_0. This would be in the traditional spirit of regression analysis, where we are "borrowing strength" for estimating the conditional expectations in $R_A(z)$ from nearby values of x. As a side benefit, the estimated false discovery rates $\widehat{\text{fdr}}_{x_0}(z)$ would change smoothly as a function of x_0.

Exercise 10.6 Show how Theorem 10.3 reduces to Theorem 10.1 when $\rho_A(x)$ equals 1 or 0 for x equal or not equal to x_0.

There is also a general version of Corollary 10.2.

Corollary 10.4 *If the null density $f_{x0}(z)$ does not depend on x,*

$$f_{x0}(z) = f_0(z) \qquad \textit{for all } x \tag{10.41}$$

then

$$\text{fdr}_A(z) = c_0 \,\text{fdr}(z)/E\{\rho_A(x)|z\} \tag{10.42}$$

where

$$c_0 = E\{\rho_A(x)|I = 1\} = \int \rho_A(x)w(x)\pi_{x0}\, dx/\pi_0. \tag{10.43}$$

Exercise 10.7 Use (10.39)–(10.40) to verify Corollary 10.4.

One can imagine a grand theory of large-scale testing in which every case (10.35) is related to every other by similarities in both their covariates and z-values. The relevance function theory of this section is only a small step in that direction. Its main advantage lies in bringing regression theory, our most flexible "relevance" methodology, to bear on empirical Bayes testing problems.

Proof of Theorem 10.3 From (10.37) and (10.40), the conditional distribution of x given category A is

$$w(x|A) = \frac{w(x)\rho_A(x)}{\rho_A} \qquad \text{where } \rho_A = \int w(x)\rho_A(x)\, dx = \Pr\{A\}. \tag{10.44}$$

The null probability given A is

$$\pi_{A0} = \int w(x|A)\pi_{x0}\, dx \tag{10.45}$$

with corresponding null density

$$f_{A0}(z) = \int w(x|A)\pi_{x0} f_{x0}(z)\, dx / \pi_{A0} \tag{10.46}$$

and marginal density

$$f_A(z) = \int w(x|A) f_x(z)\, dx. \tag{10.47}$$

Together, (10.44)–(10.46) yield

$$\text{fdr}_A(z) = \frac{\pi_{A0} f_{A0}(z)}{f_A(z)} = \frac{\int \rho_A(x) w(x) \pi_{x0} f_{x0}(z)\, dx}{\int \rho_A(x) w(x) f_x(z)\, dx}. \tag{10.48}$$

The overall marginal quantities are

$$f(z) = \int w(x) f_x(z)\, dx \qquad f_0(z) = \int w(x)\pi_{x0} f_{x0}(z)\, dx / \pi_0 \tag{10.49}$$

with $\pi_0 = \int w(x)\pi_{x0}\, dx$ and

$$\text{fdr}(z) = \frac{\int w(x)\pi_{x0} f_{x0}(z)\, dx}{\int w(x) f_x(z)\, dx}. \tag{10.50}$$

Dividing (10.48) by (10.50) gives

$$\begin{aligned} \frac{\text{fdr}_A(z)}{\text{fdr}(z)} &= \left[\frac{\int \rho_A(x) w(x) \pi_{x0} f_{x0}(z)\, dx}{\int w(x)\pi_{x0} f_{x0}(z)\, dx} \right] \bigg/ \left[\frac{\int \rho_A(x) w(x) f_x(z)\, dx}{\int w(x) f_x(z)\, dx} \right] \\ &= E_0\{\rho_A(x)|z\} / E\{\rho_A(x)|z\}, \end{aligned} \tag{10.51}$$

using Bayes theorem separately on the numerator and denominator of the top equation, which verifies (10.39). □

10.4 Are Separate Analyses Legitimate?

The example of Section 10.1 split the DTI data into two halves, performing separate false discovery rate analyses on each. Is this a legitimate tactic, or do separate analyses compromise inferential properties such as false discovery rate estimates? The answer will turn out to depend on the two faces of empirical Bayes methodology: separation is fine from a Bayesian viewpoint, but some caution is called for in the frequentist domain.

The Bonferroni FWER control method of Chapter 3 suggests why caution might be necessary. Applied to N simultaneous hypothesis tests, it rejects the null for those cases having p-values p_i sufficiently small to account for multiplicity,

$$p_i \leq \alpha/N \tag{10.52}$$

where α is the significance level. If we separate the cases into two classes of size $N/2$, rejecting for $p_i \leq \alpha/(N/2)$, we effectively double α. Some adjustment of Bonferroni's method is necessary after separation. Changing α to $\alpha/2$ corrects things here, but more complicated situations, like that suggested by Figure 10.4, require careful thought. False discovery rate methods are more forgiving, often (but not always) requiring no adjustment for separate analysis.

This is certainly true from a Bayesian point of view. Let (X, Z, I) be random variables defined as in (10.35), with X indexing possible separation strata (e.g., x in Figure 10.4), Z for z-values, and I equaling 1 or 0 as a random case is null or non-null. Suppose that for each value $X = x$ we have a set $\mathcal{Z}_x$ with conditional Bayesian false discovery rate (2.13) equaling a target value q,

$$\mathrm{Fdr}_x\{\mathcal{Z}_x\} = \Pr\{I = 1|X = x \text{ and } Z \in \mathcal{Z}_x\} = q. \tag{10.53}$$

Then the composite region

$$\mathcal{Z} = \bigcup_x \mathcal{Z}_x \tag{10.54}$$

has unconditional Bayesian Fdr equal to q,

$$\mathrm{Fdr}(\mathcal{Z}) = \Pr\{I = 1|Z \in \mathcal{Z}\} = q. \tag{10.55}$$

So separate Fdr level-q control rules can be combined without fear of increasing the overall Fdr level, which is to say that separation is a legitimate tactic here. This might seem obvious, but it is not true for FWER control rules.

Exercise 10.8 Prove (10.55).

Frequentist Fdr properties require more care. It is convenient to work within the Poisson-independence assumptions of Section 2.4: that the number of cases N has Poisson distribution $\mathrm{Poi}(\eta)$ and that the z_i are independent. We suppose there are J possible values x of X, and for each x we have a decision rule that decides "non-null" if

$$X = x \quad \text{and} \quad Z \in \mathcal{Z}_x \tag{10.56}$$

(*not* requiring condition (10.53)). The unbiased false discovery rate estimate (2.45) for stratum x is

$$\widetilde{\text{Fdr}}_x = e_{x0}/(N_x + 1) \tag{10.57}$$

where N_x is the number of cases (x_i, z_i) satisfying (10.56) and e_{x0} is the null hypothesis expectation of N_x,

$$e_{x0} = \eta w(x) F_{x0}(\mathcal{Z}_x), \tag{10.58}$$

$F_{x0}(\mathcal{Z}_x) = \Pr\{z \in \mathcal{Z}_x | X = x \text{ and } I = 1\}$.

The modified estimate $\widetilde{\text{Fdr}}_x$ is conditionally unbiased for the expected false discovery proportion in stratum x,

$$E\left\{\widetilde{\text{Fdr}}_x \middle| X = x\right\} = E\{\text{Fdp}_x | X = x\} \tag{10.59}$$

as in Lemma 2.4. The combined estimate, ignoring x,

$$\widetilde{\text{Fdr}}_{\text{comb}} = \frac{e_{\text{tot},0}}{N_{\text{tot}} + 1} \qquad \left[e_{\text{tot},0} = \sum_x e_{x0},\ N_{\text{tot}} = \sum_x N_x\right] \tag{10.60}$$

is unbiased for the overall expected false discovery proportion.

Simple algebra relates $\widetilde{\text{Fdr}}_{\text{comb}}$ to a weighted average of the $\widetilde{\text{Fdr}}_x$.

Theorem 10.5 *Let*[4]

$$\tilde{w}(x) = \frac{N_x + 1}{N_{\text{tot}} + J} \quad \text{and} \quad \widetilde{\text{Fdr}}_{(\cdot)} = \sum_x \tilde{w}(x) \widetilde{\text{Fdr}}_x \tag{10.61}$$

where J is the number of strata x. Then

$$\widetilde{\text{Fdr}}_{\text{comb}} = \frac{N_{\text{tot}} + J}{N_{\text{tot}} + 1} \widetilde{\text{Fdr}}_{(\cdot)}. \tag{10.62}$$

The factor

$$\text{Pen} \equiv (N_{\text{tot}} + J)/(N_{\text{tot}} + 1) \tag{10.63}$$

can be thought of as a penalty for separating Fdr estimation into J classes. If all the separate unbiased estimates $\widetilde{\text{Fdr}}_x$ equal q, for example, the combined unbiased estimate $\widetilde{\text{Fdr}}_{\text{comb}}$ takes the greater value Pen$\cdot q$.

The largest penalty occurs for "single event" situations, say with $N_1 = 1$ and $N_2 = N_3 = \cdots = N_J = 0$, where Pen $= (J + 1)/2$. In this case a separate small value of $\widetilde{\text{Fdr}}_1$ must be multiplied by a factor roughly proportional to

[4] The weights $\tilde{w}(x)$ are a regularized version of the empirical weights $\hat{w}(x) = N_x/N_{\text{tot}}$ and can be motivated by a vague Dirichlet prior distribution on the stratum probabilities.

the number J of strata, reminiscent of the Bonferroni bound and its adjusted p-values (3.15).

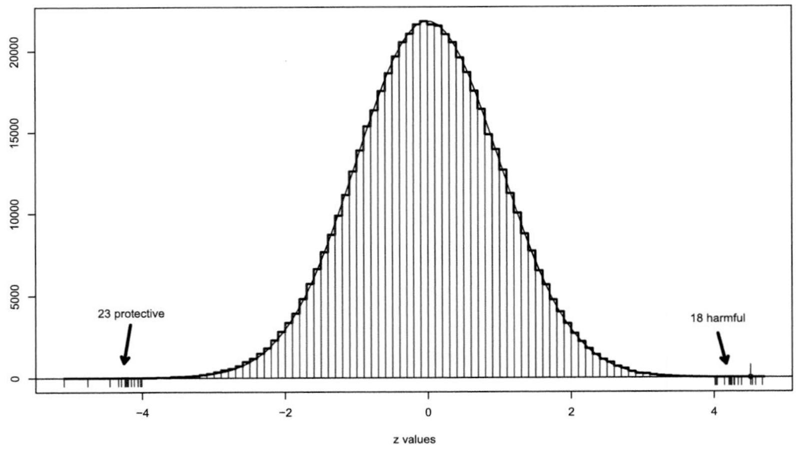

Figure 10.6 Histogram of z-values at $N = 547\,458$ chromosome sites, snp study, comparing 505 cardiovascular patients with 515 controls. Hash marks indicate 41 z-values having $|z_i| \geq 4.0$. Smooth curve is proportional to $\mathcal{N}(0, 1)$ density function, almost perfectly matching the histogram.

Figure 10.6 reports on a snp (*single nucleotide polymorphism*) study examining $N = 547\,458$ chromosome sites on $n = 1020$ subjects, 515 controls and 505 cardiovascular patients. The response variable y_{ij} at site i for person j was the number of observed polymorphisms (i.e., atypical base pairs): $y_{ij} = 0, 1,$ or 2. For each site, a linear model was fit predicting y_{ij}, $j = 1, 2, \ldots, 1020$, from disease status (healthy or sick) along with gender and race covariates. A z-value z_i was then computed as in (2.5) from the t-value

$$t_i = \hat{\beta}_i / \widehat{\mathrm{se}}_i \tag{10.64}$$

where $\hat{\beta}_i$ and $\widehat{\mathrm{se}}_i$ were the usual regression coefficient and its standard error for disease status.

The histogram of the 547 458 z-values is almost perfectly matched by a $\mathcal{N}(0, 1)$ density. In this example the theoretical null (2.6) seems trustworthy, and the null probability π_0 (2.7) must be nearly 1. Only 41 of the snps give z-values exceeding 4 in absolute value, 18 on the right and 23 on the left. Table 10.2 shows the 41 z-values, their chromosome, and the number of sample sites per chromosome. For instance, chromosome 9 with $N(9) = 26\,874$ sites had only one z-value making the list, at $z_i = 4.51$.

Table 10.2 *The 41 snps with* $|z_i| \geq 4$, *snp study, Figure 10.6, showing chromosome, number of measured snps per chromosome, z-value, unbiased Fdr estimate* $\widetilde{\text{Fdr}}_x$ (10.57), *penalized version* $\widetilde{\text{Fdr}}_x^+$ (10.66), *combined q-value* $\hat{q}_{\text{comb}}$ (10.67), *and separate-chromosome q-value* $\hat{q}_{\text{sep}}$, *as in* (10.68).

Chrom x	# snps	z-value	$\widetilde{\text{Fdr}}_x$	$\widetilde{\text{Fdr}}_x^+$	$\hat{q}_{\text{comb}}$	$\hat{q}_{\text{sep}}$
1	42 075	−4.21	.53	.80	.57	.27
1	42 075	4.38	.53	.80	.72	.50
1	42 075	−4.22	.53	.80	.62	.34
1	42 075	−4.06	.53	.80	.88	1.00
2	45 432	−4.16	.72	1.00	.65	.72
2	45 432	−4.02	.72	1.00	.90	1.00
2	45 432	−4.46	.72	1.00	.64	.12
5	34 649	−4.21	1.00	1.00	.60	.88
6	36 689	4.25	.58	.87	.74	.39
6	36 689	4.14	.58	.87	.67	1.00
6	36 689	−4.34	.58	.87	.79	.17
7	30 170	4.28	.48	.72	.74	.19
7	30 170	−4.23	.48	.72	.64	.35
7	30 170	−4.06	.48	.72	.87	1.00
8	31 880	4.33	.67	1.00	.73	.48
8	31 880	4.53	.67	1.00	.63	.09
9	**26 874**	**4.51**	**.85**	**1.00**	**.59**	**.17**
10	29 242	−4.01	.93	1.00	.84	1.00
11	27 272	4.22	.43	.65	.63	.22
11	27 272	−4.77	.43	.65	.51	.05
11	27 272	−4.02	.43	.65	.84	.79
12	27 143	4.01	.34	.52	.83	1.00
12	27 143	4.58	.34	.52	.64	.04
12	27 143	−4.11	.34	.52	.74	.54
12	27 143	−5.10	.34	.52	.19	.00
13	20 914	−4.01	.33	.50	.81	1.00
13	20 914	−4.20	.33	.50	.56	.19
13	20 914	−4.03	.33	.50	.91	.58
15	16 625	4.03	.26	.39	.93	.93
15	16 625	−4.46	.26	.39	.57	.05
15	16 625	−4.23	.26	.39	.66	.19
17	14 341	−4.25	.30	.45	.78	.15
17	14 341	−4.24	.30	.45	.70	.32
18	16 897	4.24	.27	.40	.68	.19
18	16 897	4.02	.27	.40	.87	.98
18	16 897	4.67	.27	.40	.54	.02
19	9501	4.04	.20	.30	.90	.51
19	9501	4.21	.20	.30	.59	.12
20	14 269	4.28	.30	.45	.77	.27
20	14 269	−4.30	.30	.45	.79	.12
21	8251	4.02	.26	.39	.88	.48

As an example of Theorem 10.5, suppose we let $x = 1, 2, \ldots, 22$ index chromosomes (chromosome 23 was excluded from the study) and take

$$\mathcal{Z}_x = \{|z| \geq 4\} \qquad \text{for } x = 1, 2, \ldots, 22. \tag{10.65}$$

In this case $N_{\text{tot}} = 41$, $J = 22$, and $\text{Pen} = (41 + 22)/(41 + 1) = 1.50$. Columns 4 and 5 of Table 10.2 show $\widetilde{\text{Fdr}}_x$ and the penalized value

$$\widetilde{\text{Fdr}}_x^{+} = \text{Pen} \cdot \widetilde{\text{Fdr}}_x. \tag{10.66}$$

The idea behind $\widetilde{\text{Fdr}}_x^{+}$ is to adjust the separate estimate $\widetilde{\text{Fdr}}_x$ so that the adjusted average equals the combined estimates $\widetilde{\text{Fdr}}_{\text{comb}}$ (10.62). None of the separate estimates are strikingly small, though chromosome 19 gets down to $\widetilde{\text{Fdr}}_{19} = 0.20$, adjusted to $\widetilde{\text{Fdr}}_{19}^{+} = 0.30$.

Exercise 10.9 Verify $\widetilde{\text{Fdr}}_{19} = 0.20$.

The sixth column of Table 10.2 concerns application of the Benjamini–Hochberg FDR control algorithm to the combined set of all 547 458 snps. Tabled is $\hat{q}_{\text{comb},i}$, the minimum value of q for which BH(q) would declare z_i non-null. Ordering the z_i values according to absoute value, $|z_{(1)}| > |z_{(2)}| > \ldots$, the q-value corresponding to the ith ordered case is

$$\hat{q}_{\text{comb},(i)} = 2 \cdot N \cdot \Phi\left(|z_{(i)}|\right)/i. \tag{10.67}$$

Only the most extreme case, $z_i = -5.10$, has even a moderately interesting result, with $\hat{q}_{\text{comb}} = 0.19$. (Note that (10.67) is based on two-sided p-values.)

Exercise 10.10 Justify (10.67) in terms of Corollary 4.2 (4.30).

By contrast, q-values calculated separately for the different chromosomes yield several interestingly small values $\hat{q}_{\text{sep},i}$, the last column in Table 10.2: $z_i = -4.51$ on chromosome 9, for instance, has

$$\hat{q}_{\text{sep}} = 2 \cdot 26874 \cdot \Phi(-4.51)/1 = 0.17. \tag{10.68}$$

The histogram of the 26 874 chromosome 9 z-values looks just like Figure 10.2 except with 4.51 as the only outlier (indicated by the dot and bar at the far right of Figure 10.6), giving a strong visual impression of a significant discovery. Sixteen of the 41 snps have $\hat{q}_{\text{sep}} \leq 0.20$, with six of them ≤ 0.10.

Can these much more optimistic results be trusted? Probably not, at least not from the point of view of Theorem 10.5. Changing 4 in (10.65) to 4.5 increases the penalty factor to $\text{Pen} = 5.2$, weakening any interest in the chromosome 9 outcome. We do not have a neat formula like (10.66) for adjusting the q-values, but if we have been cherry-picking results from all 22 chromosomes then some sort of adjustment must be called for.

The snp study raises several points of concern relating to false discovery rate analysis, especially as carried out at the extremes of a data set:

- In actuality, the data for chromosome 9 was analyzed first, at which point 4.51 looked like a clear discovery. The problem here, as with all frequentist methods, lies in defining an appropriate frame of reference. Whether mathematical statistics can improve on what Miller lamented as "subjective judgement" remains an open question, though perhaps the relevance considerations of the previous section offer some guidance.
- The variability of $\overline{\text{Fdr}}(\mathcal{Z})$ or $\widetilde{\text{Fdr}}(\mathcal{Z})$ as an estimate of the Bayes probability $\text{Fdr}(\mathcal{Z})$ is enormous at the extremes of the z-values, where the crucial quantity $e_+(\mathcal{Z})$ in (2.39) may be 1 or less. This undercuts faith in the Bayesian interpretation of our Fdr estimates (which, in (10.53)–(10.55), supported the legitimacy of separate analyses). Fdr and FWER methods coalesce at the extremes, both requiring frequentist caution in their use.
- These concerns vanish as we move in from the extremes of the z-value histogram. The separation penalty factor Pen (10.63) decreases to 1 when N_{tot} is much larger than J, while the implication of large $e_+(\mathcal{Z})$ supports Bayesian interpretation of our Fdr estimates.
- Local fdr methods are of no help in the extreme tails, where semi-parametric fitting methods for $f(z)$, Section 5.2, cannot be trusted.
- There is a certain "angels on the head of a pin" quality to the Fdr results in Table 10.2. They require belief in calculations such as

$$\Pr_0\{z_i \geq 4\} = \Phi(-4) = 3.2 \cdot 10^{-5}, \tag{10.69}$$

  well beyond the range of accuracy of the normal-theory approximations for statistics (2.5), (10.64). Such beliefs are necessary for any multiple testing procedure applied to the snp data.
- If all $N = 547\,458$ snps were null, we would expect 34.7 z-values exceeding 4 in absolute value (according to the two-sided version of (10.69)). The "higher criticism" test (4.42) rejects this hypothesis at the 0.05 level for $N_{\text{tot}} \geq 45$ (4.43), well beyond the observed value 41.
- Searching for enriched classes of snps is a promising tactic. However, in this case grouping snps from the same gene and applying the slope test (10.26) yielded no new discoveries.

10.5 Comparability

Empirical Bayes methods involve each case learning from the experience of others: other baseball players, other genes, other policemen, other snps. To make this believable, the "others" have to be similar in nature to the case at hand, or at least not obviously dissimilar. We wouldn't try to learn about prostate cancer genes from a leukemia study, even if the z-value histograms resembled each other. *Comparability*, our subject here, refers to questions of fair comparison in large-scale studies. It relates to the relevance calculations of Section 10.3, but with more emphasis on purely statistical corrections and less on the underlying covariate connections.

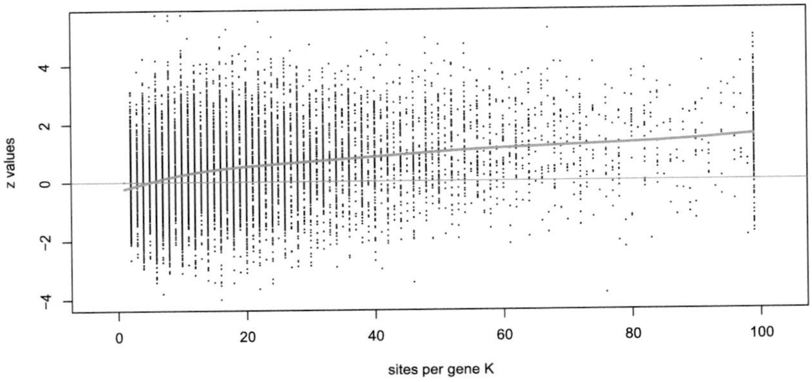

Figure 10.7a z-values for chi-square data, Section 6.1B, plotted versus number K of binding sites for each gene. Smoothing spline regression (heavy curve) shows that sites with large K tend toward larger z-values.

Figure 10.7a and Figure 10.7b illustrate two examples. The first refers to the chi-square data of Figure 6.1b. Here the $N = 16\,882$ z-values are plotted versus K, the number of binding sites within each gene. We see that the z-value algorithm of Section 6.1B tended to produce larger values for larger K. Recentering the z-values by subtraction of the regression curve produces a fairer comparison. (Repeating the $\widehat{\text{fdr}}$ analysis of Figure 6.1b after recentering, the seven z-values on the extreme right no longer had $\widehat{\text{fdr}} < 0.2$.)

A more egregious example appears in Figure 10.7b. Thirteen brain tumor patients were measured on two microarrays each, one from the tumor center, the other from its periphery. One-sample t-statistics with 12 degrees

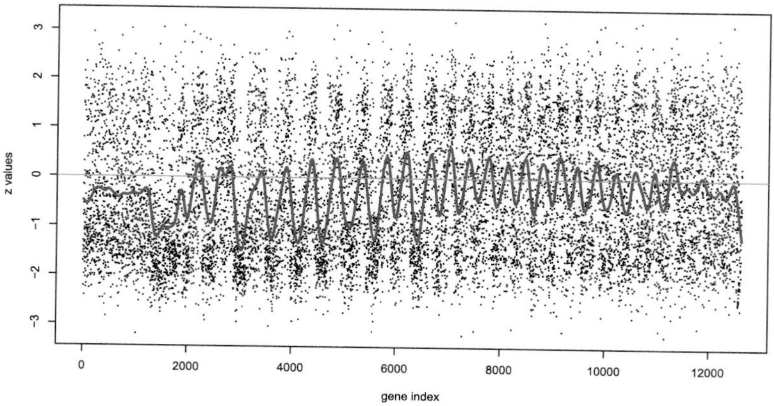

Figure 10.7b z-values for $N = 12\,625$ genes, $n = 13$ patients, brain tumor study, plotted versus their order of being recorded. Smoothing spline (heavy curve) reveals a strong periodic disturbance.

of freedom, based on center/periphery differences, were calculated for each of $N = 12\,625$ genes and converted to z-values as in (2.5). The z_i values are plotted versus i, the order in which they were read. A substantial periodic disturbance is evident, probably caused by defects in the microarray reading mechanism.[5]

Subtracting off the oscillating curve is a minimal first step toward improved comparability across genes. It may not be enough. In the DTI example of Figure 10.4, the *variability* of the z-values changes along with their running median. Application of Corollary 10.2, as in Figure 10.5, offers a more forceful equilibration of the z-values.

Questions of comparability can arise more subtly. The early microarray literature focused on "fold-change" rather than t-statistics in two-sample comparisons. If the expression-level entries x_{ij} are logarithms of the original measurements (as with the prostate data), this amounts to analyzing the numerators of the t-statistic (2.2),

$$d_i = \bar{x}_i(2) - \bar{x}_i(1) \tag{10.70}$$

rather than $t_i = d_i/s_i$.

A recent experimental comparison of microarray platforms used *spike-in*

[5] The period matches the width of the rectangular microarray chip from which they were read.

samples[6] to test agreement among different commercial technologies. The good news was that the different microarray types agreed in locating non-null genes. Less good was the authors' recommendation of fold-change rather than t-statistics for testing purposes, on the grounds that the d_i gave more stable results.

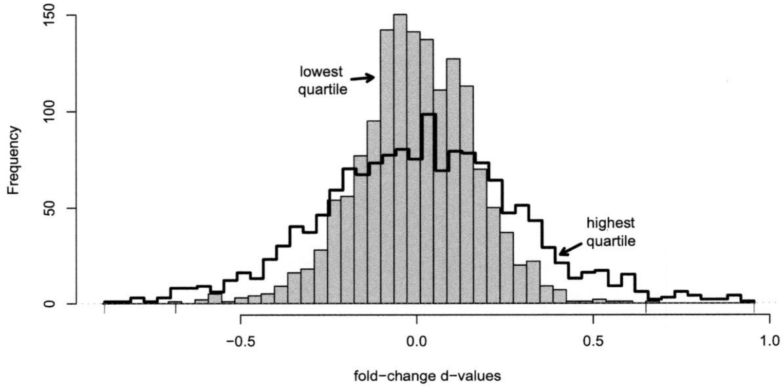

Figure 10.8 Fold-change statistics d_i (10.70) for the prostate data of Section 2.1. The genes have been divided into quartiles on the basis of s_i, the estimated standard error. The d_i histogram for the highest s_i quartile is much more dispersed than that for the lowest quartile.

This last statement almost has to be true: $t_i = d_i/s_i$ is usually noisier than d_i because of the randomness in the denominator. Stability is less important than comparability though, as illustrated in Figure 10.8. There the prostate data of Section 2.1 has been reanalyzed in terms of fold-changes d_i. The $N = 6033$ genes have been divided into quartiles according to the size of s_i, the estimated standard error. Histograms show d_i distributions in the lowest and highest quartiles. Not surprisingly, the histogram is much wider for the highest quartile.

Exercise 10.11 Why isn't this surprising? *Hint*: Consider the situation $x_{ij} \overset{\text{ind}}{\sim} \mathcal{N}(0, \sigma_i^2)$.

A false discovery rate analysis based on the d_i values would preferentially select genes i with high variability in their expression levels x_{ij}, which

[6] Artificially constructed biological samples where it is known which genes are null or non-null.

is probably irrelevant to their null/non-null status. The obvious way to correct for this is to divide d_i by s_i, getting us back to t-statistics t_i. (The corresponding figure for the t_i values shows no difference between the two histograms.)

The popular SAM algorithm described in Section 4.5 advocates use of a compromise test statistic,

$$u_i = \frac{d_i}{a_0 + s_i}. \tag{10.71}$$

Choosing $a_0 = 0$ makes $u_i = t_i$, while letting a_0 approach infinity makes u_i equivalent to d_i for testing purposes. As a_0 increases, stability improves (especially for genes that have very small values of s_i) but at the cost of decreased comparability. With a_0 equal the median of the N s_i values, one of the SAM choices, the equivalent of Figure 10.8 still shows a large discrepancy between the two histograms.

Double standardization (8.2) improves comparability across cases. It makes fold-changes d_i equivalent to t-statistics. Full double standardization isn't a necessity, but if the rows of X are of vastly different scales, non-comparability may be a concern, particularly for more complicated situations than two-sample testing.

Empirical Bayes methods require the pooling of information across sets of, hopefully, similar cases. This chapter's topic, combination and separation, relevance, and comparability, concerned when and how the pooling should be done. The discussion raised more questions than it settled, but they are good questions that deserve some thought before embarking on any large-scale data analysis.

Notes

Most of the material in Section 10.1 through Section 10.4 originated in Efron (2008b). Genovese et al. (2006) consider more qualitative situations than those of Section 10.1, where A or B might be classes of greater or less a priori null probability. Their "weighted BH" rule transforms z_i into values z_{Ai} or z_{Bi} depending on the class and then carries out a combined Fdr analysis, rather than keeping z_i the same but using different Fdr_A or Fdr_B analyses. Ferkingstad et al. (2008) explore the dependence of Fdr(z) on x by means of explicit parametric models.

Q values, Section 10.4, were introduced by Storey (2002). Efron and Morris (1971, 1972) used relevance functions for James–Stein estimation where, as in Section 10.3, the question was one of combining data from more or less similar situations. The snp study, ongoing, comes from the Stanford laboratory of T. Quertermous. Figure 10.7b is based on an unpublished study from the Stanford School of Medicine.

11

Prediction and Effect Size Estimation

A prediction problem begins with the observation of n independent vectors, the "training set",

$$(\boldsymbol{x}_j, y_j) \qquad j = 1, 2, \ldots, n, \tag{11.1}$$

where $\boldsymbol{x}_j$ is an N-vector of predictors and y_j a real-valued response. Using the training set, the goal is to construct an effective prediction rule $r(\boldsymbol{x})$: having observed a new vector $\boldsymbol{x}$ but not y, it is hoped that $r(\boldsymbol{x})$ will accurately predict y. An insurance company, for instance, might collect predictors (age, gender, smoking habits) and be interested in predicting imminent heart attacks.

Classical prediction methods depend on Fisher's linear discriminant function. Here the response variable is dichotomous, $y = 1$ or 2 (perhaps representing "healthy" or "sick") and $\boldsymbol{x}_j$ has a multivariate normal distribution whose mean vector depends on y,

$$y = \begin{cases} 1 & \boldsymbol{x} \sim \mathcal{N}_N(\boldsymbol{\delta}_1, \boldsymbol{\Sigma}) \\ 2 & \boldsymbol{x} \sim \mathcal{N}_N(\boldsymbol{\delta}_2, \boldsymbol{\Sigma}). \end{cases} \tag{11.2}$$

In the Bayesian situation, where y has prior probabilities[1]

$$p_1 = \Pr\{y = 1\} \quad \text{and} \quad p_2 = \Pr\{y = 2\} \tag{11.3}$$

an application of Bayes rule gives log posterior odds ratio

$$\log\left(\frac{\Pr\{y = 2|\boldsymbol{x}\}}{\Pr\{y = 1|\boldsymbol{x}\}}\right) = \beta_0 + \boldsymbol{\beta}'\boldsymbol{x} \tag{11.4}$$

with

$$\begin{aligned} \boldsymbol{\beta}' &= (\boldsymbol{\delta}_2 - \boldsymbol{\delta}_1)' \, \boldsymbol{\Sigma}^{-1} \\ \beta_0 &= \log(p_2/p_1) + \left(\boldsymbol{\delta}_2'\boldsymbol{\Sigma}^{-1}\boldsymbol{\delta}_2 - \boldsymbol{\delta}_1'\boldsymbol{\Sigma}^{-1}\boldsymbol{\delta}_1\right)\big/2. \end{aligned} \tag{11.5}$$

[1] Notice that this is different than the two-groups model (2.7), where the distinction is between the unobserved null vs non-null dichotomy, rather than healthy vs sick, which is observable in the training set.

Exercise 11.1 Verify (11.4)–(11.5).

The optimum Bayes rule predicts $y = 1$ or 2 as $\beta_0 + \boldsymbol{\beta}'\boldsymbol{x}$, *Fisher's linear discriminant function*, is less than or greater than 0; (11.4) is a nice linear function of $\boldsymbol{x}$, neatly dividing the space of possible $\boldsymbol{x}$ vectors into prediction regions separated by the hyperplane $\beta_0 + \boldsymbol{\beta}'\boldsymbol{x} = 0$. In practice, β_0 and $\boldsymbol{\beta}$ are estimated from the training set using the usual unbiased estimates of $\boldsymbol{\delta}_1, \boldsymbol{\delta}_2$, and $\boldsymbol{\Sigma}$ in model (11.2), with $\log(p_2/p_1)$ perhaps estimated by $\log(n_2/n_1)$, the observed numbers in the two categories.

All of this works well in the classic situation where the number of independent observations n is much bigger than N, the predictor dimension. This is just what we *don't* have in large-scale applications. Suppose we wish to use the prostate data set of Section 2.1 to develop a prediction rule for prostate cancer: given $\boldsymbol{x}$, a microarray expression vector for the $N = 6033$ genes, we want a rule $r(\boldsymbol{x})$ that predicts whether the man who provided $\boldsymbol{x}$ will develop prostate cancer. Now $N = 6033, n = 102$, and the usual unbiased estimates of (11.4) will probably be useless, unbiasedness being an unaffordable luxury in high dimensions.

A variety of computer-intensive prediction algorithms has been developed recently that trade bias for stability and accuracy: boosting, bagging, support vector machines, LARS, Lasso, and ridge regression being some of the prominent candidates. This chapter will discuss a comparatively simple empirical Bayes approach to large-scale prediction. Besides adding to the store of useful methodology, this will give us a chance to put prediction problems into a false discovery rate context.

In the prostate example it seems obvious that a good prediction rule should depend on genes that express themselves much differently in sick and healthy subjects, perhaps those falling beyond the small triangles in Figure 5.2. Fdr analysis aims to identify such genes. In Section 11.4 we will go further, and try to estimate *effect size*, a quantitative assessment of just how non-null a gene may be.

We begin in Section 11.1 with a simple model that leads to an empirical Bayes algorithm for prediction and, eventually, to effect size estimation. We restrict attention to the *dichotomous case* where each y_j equals 1 or 2, calling the two states "healthy" and "sick" for the sake of definiteness, and refer to the components of $\boldsymbol{x}_j$ as "gene expression" or just "genes", though of course the theory has nothing particularly to do with genomics.

11.1 A Simple Model

Motivation for our empirical Bayes prediction rules starts from a drastically idealized model for the predictor vectors $\boldsymbol{x}_j$ in (11.1). We assume that the components x_i of a typical such vector $\boldsymbol{x} = (x_1, x_2, \ldots, x_i, \ldots, x_N)'$ are independently normal, with location and scale parameters μ_i and σ_i, and with possibly different expectations in the two subject categories,

$$\frac{x_i - \mu_i}{\sigma_i} \overset{\text{ind}}{\sim} \mathcal{N}\left(\pm\frac{\delta_i}{2c_0}, 1\right) \quad \begin{cases} \text{“}-\text{”} & \text{healthy class } (y=1) \\ \text{“}+\text{”} & \text{sick class } (y=2). \end{cases} \tag{11.6}$$

The constant c_0 equals

$$c_0 = (n_1 n_2 / n)^{1/2} \tag{11.7}$$

where n_1 and n_2 are the number of healthy and sick subjects in the training set, $n = n_1 + n_2$. (Dividing by c_0 in (11.6) will make δ_i the effect size in what follows.) Null genes have $\delta_i = 0$, indicating no difference between the two categories; non-null cases, particularly those with large values of $|\delta_i|$, are promising ingredients for effective prediction.

Let

$$u_i = (x_i - \mu_i)/\sigma_i \qquad i = 1, 2, \ldots, N, \tag{11.8}$$

be the standardized version of x_i in (11.7) so that $\boldsymbol{u}$ has two possible N-dimensional normal distributions,

$$\boldsymbol{u} \sim \mathcal{N}_N\left(\pm\boldsymbol{\delta}/(2c_0), \boldsymbol{I}\right), \tag{11.9}$$

“$-$” or “$+$” as $y = 1$ or 2, with $\boldsymbol{\delta} = (\delta_1, \delta_2, \ldots, \delta_N)'$ the vector of effect sizes and $\boldsymbol{I}$ the $N \times N$ identity matrix.

The ideal prediction rule depends on the weighted sum

$$S = \sum_{i=1}^{N} \delta_i u_i \sim \mathcal{N}\left(\pm\|\boldsymbol{\delta}\|^2/2c_0, \|\boldsymbol{\delta}\|^2\right) \tag{11.10}$$

with “$-$” and “$+$” applying as in (11.6). If the two categories in (11.3) have equal prior probabilities,

$$p_1 = p_2 = 0.50 \tag{11.11}$$

we predict

$$\begin{aligned} &\text{healthy } (y=1) \text{ if } S < 0 \\ &\text{sick } (y=2) \text{ if } S > 0. \end{aligned} \tag{11.12}$$

Prediction error rates of the first and second kinds, i.e., confusing healthy with sick or vice versa, both equal

$$\alpha \equiv \Phi(-\|\boldsymbol{\delta}\|/2c_0). \tag{11.13}$$

We need $\|\boldsymbol{\delta}\| = (\sum \delta_i^2)^{1/2}$, the length of the effect size vector, to be large for successful prediction.

Exercise 11.2 Use Fisher's linear discriminant function to verify (11.10)–(11.13).

The ideal rule (11.12) is unavailable even if we believe model (11.6): in practice we need to estimate the parameters

$$(\mu_i, \sigma_i, \delta_i) \qquad i = 1, 2, \ldots, N, \tag{11.14}$$

entering into S (11.10), more than 18 000 of them for the prostate data. This is where the training data (11.1) comes in,

$$\underset{N\times n}{X} = (\boldsymbol{x}_1, \boldsymbol{x}_2, \ldots, \boldsymbol{x}_n) \quad \text{and} \quad \boldsymbol{y} = (y_1, y_2, \ldots, y_n)'. \tag{11.15}$$

By reordering the subjects, we can take the first n_1 entries $y_j = 1$ and the last n_2 entries $y_j = 2$, that is, healthy subjects first. Let $\bar{x}_{i1}$ and SS_{i1} be the mean and within-group sum of squares for gene i measurements in the healthy subjects,

$$\bar{x}_{i1} = \sum_{j=1}^{n_1} x_{ij}/n_1 \quad \text{and} \quad SS_{i1} = \sum_{j=1}^{n_1} \left(x_{ij} - \bar{x}_{i1}\right)^2 \tag{11.16}$$

and similarly $\bar{x}_{i2}$ and SS_{i2} for the n_2 sick subjects. Then

$$\hat{\mu}_i = \frac{\bar{x}_{i1} + \bar{x}_{i2}}{2} \quad \text{and} \quad \hat{\sigma}_i^2 = \frac{SS_{i1} + SS_{i2}}{n-2} \tag{11.17}$$

are unbiased estimates of μ_i and σ_i^2.

The crucial prediction parameters are the effect sizes δ_i. This is where unbiasedness fails us. Assume for the moment that σ_i is known, in which case

$$\bar{\delta}_i \equiv c_0 \frac{\bar{x}_{i2} - \bar{x}_{i1}}{\sigma_i} \sim \mathcal{N}(\delta_i, 1) \tag{11.18}$$

unbiasedly estimates δ_i. Substituting $\hat{\sigma}_i$ for σ_i gives the two-sample t-statistic t_i (2.2) that we can transform to z_i as in (2.5) with approximate distribution

$$z_i \mathrel{\dot\sim} \mathcal{N}(\delta_i, 1), \tag{11.19}$$

which becomes increasingly accurate for large values of n. For now we will treat it as exact, though the prediction program `Ebay` introduced in Section 11.2 makes a small change to accommodate the effects of transformation (2.5).

This looks promising. For prediction purposes, however, we are hoping to find extreme values of δ_i, either positive or negative, while *selection bias* makes the extreme z_i values overinflated estimates: $z_{610} = 5.29$ is the most extreme z-value for the prostate data, but it is likely that δ_{610} is less than 5.29; z_{610} has won a "farthest from the origin" contest among $N = 6033$ contenders, partly no doubt from having δ_{610} large, but also from having a positive random error (or else it probably would not have won the contest), which is what selection bias means.

The `pamr` algorithm[2] counteracts selection bias by shrinking the estimates z_i toward zero according to what is called a *soft thresholding rule*,

$$\hat{\delta}_i = \text{sign}\,(z_i) \cdot (|z_i| - \lambda)_+ \tag{11.20}$$

where x_+ equals $\max(x, 0)$. That is, each z_i is shrunk toward zero by some fixed amount λ, under the restriction that shrinking never goes past zero. A range of possible shrinkage parameters λ is tried, and for each one a statistic like (11.10) is formed from the estimates $(\hat{\mu}_i, \hat{\sigma}_i, \hat{\delta}_i)$,

$$\hat{S}_\lambda(\boldsymbol{x}) = \sum_{i=1}^{N} \hat{\delta}_i \hat{u}_i \qquad [\hat{u}_i = (x_i - \hat{\mu}_i)/\hat{\sigma}_i] \tag{11.21}$$

leading to the rule $r_\lambda(\boldsymbol{x})$ that predicts $y = 1$ or 2 as $\hat{S}_\lambda(\boldsymbol{x})$ is less than or greater than zero (11.12). A cross-validation method, described below, is then employed to estimate α_λ, the error rate for $r_\lambda(\boldsymbol{x})$ (as well as providing separate estimates for errors of the first and second kinds).

Table 11.1 shows `pamr` output for the prostate data. As λ increased from 0, the estimated error rate $\hat{\alpha}_\lambda$ decreased, reaching a minimum of 0.08 at $\lambda = 2.16$ and then increasing again. This suggests that rule $r_{2.16}(\boldsymbol{x})$ would make only 8% errors in predicting prostate cancer from microarray measurements.

Notice that the prediction statistic $\hat{S}_\lambda(\boldsymbol{x})$ involves only those genes with $|z_i| > \lambda$, since $\hat{\delta}_i$ (11.20) is zero for $|z_i| \leq \lambda$. At $\lambda = 2.16$, only 377 of the 6033 genes are involved in the prediction. Less is more as far as prediction is concerned, rules based on fewer predictions being generally easier to implement and more reproducible.

[2] *Prediction analysis for microarrays*, also known as the "nearest shrunken centroid" algorithm, is available from the R library `CRAN`.

Table 11.1 *Cross-validated error rates* $\hat{\alpha}_\lambda$, `pamr` *prediction rules applied to prostate cancer data, for increasing choices of shrinkage parameter* λ (11.20)–(11.21). *Taking* $\lambda = 2.16$ *gave* $\hat{\alpha}_\lambda = 0.08$, *using a prediction rule based on 377 genes.*

Shrinkage λ	# Genes	$\hat{\alpha}_\lambda$
.00	6033	.34
.54	3763	.33
1.08	1931	.23
1.62	866	.12
2.16	**377**	**.08**
2.70	172	.10
3.24	80	.16
3.78	35	.30
4.32	4	.41
4.86	1	.48
5.29	0	.52

Cross-validation

Ten-fold cross-validation, used in `pamr`, randomly splits the n subjects into ten "folds" with roughly proportional numbers of "healthy" and "sick" in each fold. The prediction algorithm is refit ten times with the cases of each fold withheld from the training set in turn, the cross-validated error rate $\hat{\alpha}_{\text{CV}}$ being the average error rate on the withheld cases. A typical fold contained ten subjects for the prostate data, five healthy and five sick, who were then predicted by the rule constructed from the data of the other 92 subjects; `pamr` actually repeats ten-fold cross-validation several times, averaging the cross-validated error rates over all the repetitions.

Apparent error, how accurately a rule predicts its own training set, is usually too optimistic. Cross-validation is a simple way to estimate, almost unbiasedly, how well the rule will do on a new set of data. It is important to remember that $\hat{\alpha}_{\text{CV}}$ is *not* an estimate of error for a specific rule, for instance $\hat{r}_{2.16}(\boldsymbol{x})$ in Table 11.1. Rather, it is the expected error rate for rules selected according to the same recipe. Cross-validation is not perfect and can give misleadingly optimistic estimates in situations like (6.36), but it is almost always preferable to the apparent error rate.

Examining Table 11.1, it seems we should use $\lambda = 2.16$ for our prediction rule. However, there is a subtle danger lurking here: because we have looked at all the data to select the "best" λ, this choice is not itself cross-validated, and the corresponding rate 0.08 may be optimistic.

A small simulation study was run with $N = 1000$, $n_1 = n_2 = 10$, and all $x_{ij} \overset{\text{ind}}{\sim} \mathcal{N}(0, 1)$. In this situation, all δ_i in (11.6) are zero and $\alpha = 0.50$ at (11.13); but the minimum cross-validated `pamr` error rates in 100 simulations of this set-up had median 0.30 with standard deviation ± 0.16. This is an extreme example. Usually the over-optimism is less severe, particularly when good prediction is possible.

11.2 Bayes and Empirical Bayes Prediction Rules

We now assume a Bayes prior distribution $g(\delta)$ for the effect size parameters δ_i in model (11.6). Reducing the data to z-values (11.19), we have, as in (2.47),

$$\delta \sim g(\cdot) \quad \text{and} \quad z|\delta \sim \mathcal{N}(\delta, 1) \tag{11.22}$$

for a typical (δ, z) pair. For convenience we will treat $g(\delta)$ as a density function but allow it to have discrete atoms, perhaps with an atom π_0 at $\delta = 0$ as in the two-groups model (5.1), though that is not assumed here.

The *normal hierarchical model* (11.22) is a favorite in the empirical Bayes literature because it leads to particularly simple Bayesian estimates.

Theorem 11.1 *Let $f(z)$ be the marginal density of z in model* (11.22),

$$f(z) = \int_{-\infty}^{\infty} \varphi(z - \delta) g(\delta)\, d\delta \qquad \left[\varphi(z) = e^{-z^2/2} \big/ \sqrt{2\pi}\right]. \tag{11.23}$$

Then the posterior density of δ given z is

$$g(\delta|z) = e^{z\delta - \psi(z)} \left(e^{-\delta^2/2} g(\delta)\right) \tag{11.24}$$

where

$$\psi(z) = \log\left(f(z)/\varphi(z)\right). \tag{11.25}$$

Proof According to Bayes theorem,

$$g(\delta|z) = \varphi(z - \delta) g(\delta) / f(z). \tag{11.26}$$

□

Exercise 11.3 Complete the proof.

Formula (11.24) is an exponential family of densities for δ with canonical, or natural, parameter z. The standard exponential family properties reviewed in Appendix A then yield the following result.

Corollary 11.2 *Under model* (11.22), *δ has posterior mean and variance*

$$E\{\delta|z\} = \psi'(z) \quad \text{and} \quad \text{var}\{\delta|z\} = \psi''(z) \tag{11.27}$$

where ψ' and ψ'' indicate the first and second derivatives of $\psi(z)$. Going further, the kth cumulant of δ given z equals $d^k\psi(z)/dz^k$. Letting $l(z) =$ $\log(f(z))$, *we can rewrite* (11.27) *as*

$$E\{\delta|z\} = z + l'(z) \quad \text{and} \quad \text{var}\{\delta|z\} = 1 + l''(z), \tag{11.28}$$

now with primes indicating derivatives of $l(z)$.

The great advantage of Corollary 11.2 is that the moments of δ given z are obtained directly from the marginal density of $f(z)$ without requiring explicit calculation of the prior $g(\delta)$, thus avoiding the usual difficulties of deconvolution. This is essential for empirical Bayes applications, where now we need only estimate $f(z)$, not $g(\delta)$. The `R` algorithm[3] `Ebay` uses Poisson regression on histogram counts (5.11)–(5.15) to obtain a smooth estimate $\hat{l}(z)$ of $l(z) = \log(f(z))$, and then estimates of $E\{\delta|z\}$ and $\text{var}\{\delta|z\}$,

$$\boldsymbol{z} \longrightarrow \hat{l}(\boldsymbol{z}) \longrightarrow \hat{l}'(z) \longrightarrow \hat{E}\{\delta|z\} = z + l'(z) \tag{11.29}$$

and $\widehat{\text{var}}\{\delta|z\} = 1 + l''(z)$.

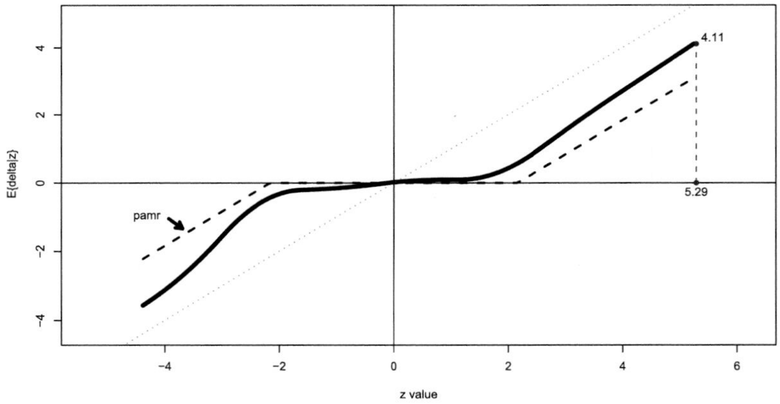

Figure 11.1 Estimated effect size $\hat{E}\{\delta|z\}$, from application of `Ebay` algorithm to prostate cancer z-values. Dashed curve is soft thresholding rule (11.20) using `pamr` value $\lambda = 2.16$.

The heavy curve in Figure 11.1 shows $\hat{E}\{\delta|z\}$ obtained by application

[3] See Appendix B.

of Ebay to the prostate cancer z-values. We see that $\hat{E}\{\delta|z\}$ is nearly zero for z in $(-2, 2)$, then grows almost linearly for $|z| > 2$. Gene 610, with $z_{610} = 5.29$, is estimated to have effect size 4.11. $\hat{E}\{\delta|z\}$ resembles a soft thresholding rule (11.20), but with λ noticeably smaller than the pamr choice 2.16, the latter being a more vigorous shrinker, particularly for negative values of z.

Beginning with a vector $\boldsymbol{z}$ of z-values obtained from t-statistics as in (2.5), the Ebay algorithm produces prediction rules in the following steps:

1 A target error rate α_0 is selected (default $\alpha_0 = 0.025$).

2 Poisson regression is used to generate effect size estimates

$$\hat{\delta}_i = \hat{E}\{\delta|z_i\} \tag{11.30}$$

as at (11.29), with an adjustment to account for Student-t effects described below at (11.33)–(11.34).

3 Letting $\hat{\boldsymbol{\delta}}_m$ be the vector of m largest $\hat{\delta}_i$ values in absolute value, m is selected to be the smallest integer such that the nominal error rate $\hat{\alpha}_m = \Phi(-\|\hat{\boldsymbol{\delta}}_m\|/2c_0)$ (11.13) is less than α_0, i.e., the minimum m making

$$\|\hat{\boldsymbol{\delta}}_m\| \geq 2c_0\Phi^{-1}(1 - \alpha_0). \tag{11.31}$$

4 The resulting empirical Bayes prediction rule is based on the sign of

$$\hat{S} = \sum_m \hat{\delta}_i \hat{u}_i \tag{11.32}$$

as in (11.12), (11.21), the summation being over the m selected cases.

5 Repeated ten-fold cross-validation is used to provide an unbiased estimate of the rule's prediction error, as described near the end of Section 11.1.

Table 11.2 displays Ebay's output for the prostate data. At the first step, gene 610 with $z_{610} = 5.29$ and $\hat{\delta}_{610} = 4.11$ was selected, giving $\hat{\alpha} = \Phi(-4.11/2c_0) = 0.342$ ($c_0 = 5.05$ (11.7)). The algorithm stopped at step 55, at which point $\hat{\alpha}_m$ first went below the target value $\alpha_0 = 0.025$.

Cross-validation estimates of prediction error were based on 20 independent replications of ten-fold CV, $\hat{\alpha}_{\text{CV}} = 0.090$ being the overall average error (with standard error 0.003, obtained in the usual way for an average). Averaging the CV prediction errors separately over the 50 healthy and 52 sick subjects indicated much poorer accuracy in predicting the latter. Note that here $\hat{\alpha}_{\text{CV}}$ is honest in the sense that the form of the empirical Bayes rule is completely determined, as opposed to our pamr example where the best value $\lambda = 2.16$ was selected *after* examining the cross-validation results.

Table 11.2 Ebay *output for prostate cancer prediction rule* (11.32). *The rule employs data from the 55 genes having the largest values of* $|\hat{\delta}_i|$, *the effect size estimates. Also shown is variance estimates* $\widehat{\text{var}}$ *from* (11.27), *theoretical prediction error estimates* $\hat{\alpha}$, *and* $\hat{\alpha}_{\text{cor}}$, *error estimates including correlations between the genes* (11.38).

Step	Index	z	$\hat{\delta}$	$\widehat{\text{var}}$	$\hat{\alpha}$	$\hat{\alpha}_{\text{cor}}$
1	610	5.29	4.11	.87	.342	.342
2	1720	4.83	3.65	.89	.293	.289
3	364	−4.42	−3.57	.92	.258	.258
4	3940	−4.33	−3.52	.92	.231	.231
5	4546	−4.29	−3.47	.93	.208	.223
6	4331	−4.14	−3.30	.97	.190	.197
7	332	4.47	3.24	.91	.175	.189
8	914	4.40	3.16	.92	.162	.175
⋮	⋮	⋮	⋮	⋮	⋮	⋮
45	4154	−3.38	−2.23	1.18	.032	.054
46	2	3.57	2.22	.97	.031	.055
47	2370	3.56	2.20	.97	.031	.053
48	3282	3.56	2.20	.97	.030	.052
49	3505	−3.33	−2.15	1.19	.029	.050
50	905	3.51	2.15	.97	.028	.051
51	4040	−3.33	−2.14	1.19	.027	.052
52	3269	−3.32	−2.12	1.19	.027	.050
53	805	−3.32	−2.12	1.19	.026	.048
54	4552	3.47	2.09	.97	.025	.048
55	721	3.46	2.09	.97	.025	.044

There are many reasons why $\hat{\alpha}_{\text{CV}}$ might exceed the ideal theoretical rate α_0: $(\hat{\mu}_i, \hat{\sigma}_i)$ (11.17) does not equal (μ_i, σ_i); the measurements x_i in (11.6) are not normally distributed; they are not independent; the empirical Bayes estimates $\hat{\delta}_i$ differ from the true Bayes estimates $\delta_i = E\{\delta|z_i\}$ (11.27).

This last point causes trouble at the extreme ends of the z scale, where $|\hat{\delta}(z)|$ is largest but where we have the least amount of data for estimating $E\{\delta|z\}$. An option in Ebay allows for truncation of the $\hat{\delta}$ estimation procedure at some number k_{trunc} of observations in from the extremes. With $k_{\text{trunc}} = 10$, used in the second row of Table 11.3, $\hat{\delta}_i$ for the ten largest z_i values is set equal to $\max\{\hat{\delta}_i : i \leq N - 10\}$, and similarly at the negative end of the z scale. This was moderately effective in reducing the cross-validated prediction errors in Table 11.3.

Table 11.3 *Cross-validation estimates of prediction error for prostate cancer data (standard errors in parentheses). Prediction errors are much larger in the sick category. Truncating the prediction rule reduces prediction errors.*

	$\hat{\alpha}_{\text{CV}}$	Healthy	Sick
no truncation	**.090** (.003)	**.044** (.005)	**.128** (.005)
$k_{\text{trunc}} = 10$	**.073** (.004)	**.032** (.005)	**.107** (.006)

Student-t effects

Section 7.4 showed that the distribution of $z_i = \Phi^{-1}(F_\nu(t_i))$ of a non-central t-statistic with ν degrees of freedom is well-approximated by a modified version of (11.22),

$$z \sim \mathcal{N}\left(\delta, \sigma^2(\delta)\right) \qquad \left[\sigma^2(\delta) \le 1\right] \tag{11.33}$$

where the function $\sigma^2(\delta)$ depends on ν (see Figure 7.6). Rather than formulas (11.29), Ebay uses

$$\hat{E}\{\delta|z\} = z + \sigma^2(z)\hat{l}'(z) \quad \text{and} \quad \widehat{\text{var}}\{\delta|z\} = \sigma^2(z)\left[1 + \sigma^2(z)\hat{l}''(z)\right]. \tag{11.34}$$

This produces slightly larger $\hat{\delta}_i$ estimates since $\sigma^2(z_i) \le 1$.

Exercise 11.4 (a) Suppose $z \sim \mathcal{N}(\delta, \sigma_0^2)$ for some fixed variance σ_0^2. Show that

$$E\{\delta|z\} = z + \sigma_0^2 l'(z) \quad \text{and} \quad \text{var}\{\delta|z\} = \sigma_0^2\left[1 + \sigma_0^2 l''(z)\right]. \tag{11.35}$$

(b) Suggest an improvement on (11.34).

Correlation corrections

The assumption of gene-wise independence in our basic model (11.6) is likely to be untrue, perhaps spectacularly untrue, in applications. It is not difficult to make corrections for correlation structure. Suppose that the vector $\boldsymbol{u}$ of standardized predictors in (11.8) has covariance matrix $\boldsymbol{\Sigma}$ rather than $\boldsymbol{I}$, but we continue to use prediction rule (11.10), (11.12). Then the error probability α (11.13) becomes

$$\alpha = \Phi(-\Delta_0 \cdot \gamma) \qquad \text{where} \begin{cases} \Delta_0 = \|\boldsymbol{\delta}\|/2c_0 \\ \gamma = (\boldsymbol{\delta}'\boldsymbol{\delta}/\boldsymbol{\delta}'\boldsymbol{\Sigma}\boldsymbol{\delta})^{1/2}. \end{cases} \tag{11.36}$$

Here Δ_0 is the independence value while γ is a correction factor, usually less than 1, that increases the error rate α.

Exercise 11.5 Verify (11.36).

The standardized variates u_i have variance 1 so $\boldsymbol{\Sigma}$ is $\boldsymbol{u}$'s correlation matrix. At the mth step of the Ebay algorithm we can compute the obvious estimate of γ,

$$\hat{\gamma}_m = \left(\hat{\boldsymbol{\delta}}'_m \hat{\boldsymbol{\delta}}_m \Big/ \hat{\boldsymbol{\delta}}'_m \hat{\boldsymbol{\Sigma}}_m \hat{\boldsymbol{\delta}}_m\right)^{\frac{1}{2}} \tag{11.37}$$

where $\hat{\boldsymbol{\Sigma}}_m$ is the $m \times m$ sample correlation matrix for the first-selected m genes. The last column of Table 11.2 gives the correlation-corrected error estimate

$$\hat{\alpha}_{\text{cor}} = \Phi\left(-\frac{||\hat{\boldsymbol{\delta}}_m||}{2c_0}\hat{\gamma}_m\right). \tag{11.38}$$

These are nearly twice the size of the uncorrected values.

Table 11.4 Ebay *output for the Michigan lung cancer study. Correlation error estimates $\hat{\alpha}_{\text{cor}}$ are much more pessimistic, as confirmed by cross-validation.*

Step	Index	z-value	$\hat{\delta}$	$\hat{\alpha}$	$\hat{\alpha}_{\text{cor}}$
1	3144	4.62	3.683	.3290	.329
2	2446	4.17	3.104	.2813	.307
3	4873	4.17	3.104	.2455	.256
4	1234	3.90	2.686	.2234	.225
5	621	3.77	2.458	.2072	.213
6	676	3.70	2.323	.1942	.228
7	2155	3.69	2.313	.1824	.230
8	3103	3.60	2.140	.1731	.236
9	1715	3.58	2.103	.1647	.240
10	452	3.54	2.028	.1574	.243
⋮	⋮	⋮	⋮	⋮	⋮
193	3055	2.47	.499	.0519	.359
194	1655	−2.21	−.497	.0518	.359
195	2455	2.47	.496	.0517	.359
196	3916	2.47	.496	.0516	.359
197	4764	2.47	.495	.0515	.359
198	1022	−2.20	−.492	.0514	.359
199	1787	−2.19	−.490	.0513	.360
200	901	−2.18	−.486	.0512	.360

Table 11.4 shows Ebay output for the Michigan lung cancer study, a microarray experiment with $N = 5217$ genes and $n = 86$ subjects, $n_1 = 62$ "good outcomes" and $n_2 = 24$ "poor outcomes". Correlation problems are much more severe here. Ebay stopped after $m = 200$ steps (the default stopping rule) without $\hat{\alpha}$ reaching the target value $\alpha_0 = 0.025$. The correlation-corrected errors $\hat{\alpha}_{\text{cor}}$ are much more pessimistic, increasing after the first six steps, eventually to $\hat{\alpha}_{\text{cor}} = 0.36$. A cross-validation error rate of $\hat{\alpha}_{\text{CV}} = 0.37$ confirmed the pessimism.

We could employ more elaborate recipes for selecting an empirical Bayes prediction rule, for example "Use the rule corresponding to the minimum value of $\hat{\alpha}_{\text{cor}}$ in the first 200 steps." (This would select the rule based on the first five genes in Table 11.4.) If Ebay were programmed to follow such recipes automatically, which it isn't, we could then get honest cross-validation error rates $\hat{\alpha}_{\text{CV}}$ just as before.

11.3 Prediction and Local False Discovery Rates

The local false discovery rate $\text{fdr}(z)$ (5.2) is

$$\text{fdr}(z) = \pi_0 \varphi(z)/f(z) \tag{11.39}$$

under the theoretical null distribution $f_0(z) = \varphi(z)$ (2.6). Comparing (11.39) with (11.25) and (11.27) gives the following.

Corollary 11.3 *Under the normal hierarchical model* (11.22), *δ has posterior mean and variance*

$$E\{\delta|z\} = -\frac{d}{dz}\log \text{fdr}(z) \quad \textit{and} \quad \text{var}\{\delta|z\} = -\frac{d^2}{dz^2}\log \text{fdr}(z) \tag{11.40}$$

when $\text{fdr}(z)$ *is calculated using the theoretical null* (11.39).

There is something a little surprising about this result. It seemingly makes sense that genes with low false discovery rates should be the ones utilized in prediction rules, but (11.40) shows that this is not exactly true: large values of $\tilde{\delta}_i = E\{\delta_i|z_i\}$, the Bayes effect size estimate, depend on the rate of change of $\log \text{fdr}(z_i)$, not on $\text{fdr}(z_i)$ itself. Small values of $\text{fdr}(z_i)$ usually correspond to large values of $|\tilde{\delta}_i|$, but this doesn't have to be the case.

Let $A(x)$ be the area under the curve $E\{\delta|z\}$ between 0 and x,

$$A(x) = \int_0^x E\{\delta|z\}\,dz. \tag{11.41}$$

The first equation in (11.40) then gives the following.

Corollary 11.4 *Under* (11.22), (11.39),

$$\frac{\text{fdr}(x)}{\text{fdr}(0)} = e^{-A(x)}. \tag{11.42}$$

(This remains true for $x < 0$ with the definition $A(x) = -\int_x^0 E\{\delta|z\}dz$.)

Looking at $\hat{E}\{\delta|z\}$ for the prostate data, Figure 11.1, we can see that $A(x)$ is nearly zero for $x < 2$ but then increases roughly quadratically with x, driving down fdr(x) according to (11.42).

Suppose that the appropriate null hypothesis distribution is $\mathcal{N}(\delta_0, \sigma_0^2)$ as in (6.9), rather than $\mathcal{N}(0, 1)$. We can change the normal hierarchical model (11.22) to

$$\delta \sim g(\cdot) \quad \text{and} \quad \left.\frac{z - \delta_0}{\sigma_0}\right|\delta \sim \mathcal{N}(\delta, 1) \tag{11.43}$$

so that δ still represents effect size on a $\mathcal{N}(0, 1)$ scale. The case $\delta = 0$ corresponds to null density $f_0(z) = \varphi((z - \delta_0)/\sigma_0)/\sigma_0$, giving fdr$(z) = \pi_0 f_0(z)/f(z)$, where the marginal density $f(z)$ is

$$f(z) = \int_{-\infty}^{\infty} \frac{1}{\sigma_0}\varphi\left(\frac{z - \delta_0 - \delta}{\sigma}\right) g(\delta)\, d\delta. \tag{11.44}$$

Differentiating log fdr(z) gives

$$-\frac{d}{dz}\log \text{fdr}(z) = \frac{z - \delta_0}{\sigma_0^2} + l'(z) \qquad \left[l'(z) = \frac{d}{dz}\log f(z)\right] \tag{11.45}$$

while an exponential family calculation similar to that for Theorem 11.1 yields a transformed version of (11.28),

$$E\{\delta|z\} = \frac{z - \delta_0}{\sigma_0} + \sigma_0 l'(z) \quad \text{and} \quad \text{var}\{\delta|z\} = 1 + \sigma_0^2 l''(z). \tag{11.46}$$

So $d \log \text{fdr}(z)/dz = E\{\delta|z\}/\sigma_0$, and the transformed version of Corollary 11.4 is

$$\frac{\text{fdr}(x)}{\text{fdr}(\delta_0)} = e^{-A_0(x)} \qquad \left[A_0(x) = \int_{\delta_0}^{x} E\{\delta|z\}\, dz \Big/ \sigma_0\right]. \tag{11.47}$$

Exercise 11.6 Verify (11.45) and (11.46). Why does (11.46) differ from (11.35)?

Figure 11.2 compares $E\{\delta|z\}$ from models (11.22) and (11.43) for the prostate and leukemia data. Here we have chosen (δ_0, σ_0) in (11.43) to be the MLE empirical null estimates of Section 6.3,

$$\textit{prostate}\ (\delta_0, \sigma_0) = (0.003, 1.06) \qquad \textit{leukemia} = (0.094, 1.68). \tag{11.48}$$

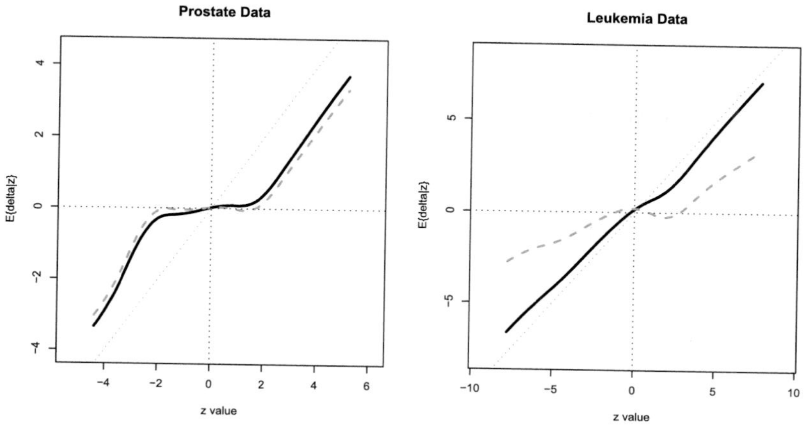

Figure 11.2 $E\{\delta|z\}$ from the normal hierarchical model (11.22), solid curves; and from the transformed version (11.43), dashed curves, with (δ_0, σ_0) the MLE empirical null values (11.48). Changing from the theoretical to empirical null flattens $E\{\delta|z\}$ near $z = 0$.

Note that the slight slope of $\hat{E}\{\delta|z\}$ between -2 and 2 seen in Figure 11.1 for the prostate data is flattened by changing from $(\delta_0, \sigma_0) = (0, 1)$ to the empirical null values. The flattening effect is much more dramatic for the leukemia data. Area $A_0(x)$ is much smaller than $A(x)$ and (11.47) is bigger than (11.42), demonstrating in geometric terms why it is more difficult in this case to get small values of fdr with the empirical null.

Large-scale hypothesis testing usually begins with the assumption that most of the effects are zero, or at least small, as in (2.7)–(2.8), which suggests that $E\{\delta|z\}$ should be near zero and flat for central values of z. The main point of Figure 11.2 is diagnostic: $\hat{E}\{\delta|z\} = z + \hat{l}'(z)$ passing through $z = 0$ at a sharp angle is a warning against casual use of the theoretical null.

Figure 11.3 concerns another, more vivid, geometric representation of the relationship between false discovery rate and effect size. Let $e(z)$ equal $E\{\delta|z\}$ in the original hierarchical model (11.22),

$$e(z) = z + l'(z) \tag{11.49}$$

(11.29), and define (δ_0, σ_0) from $l(z) = \log(f(z))$ according to

$$\delta_0 : l'(\delta_0) = 0 \quad \text{and} \quad \sigma_0 = 1/(-l''(\delta_0))^{1/2}; \tag{11.50}$$

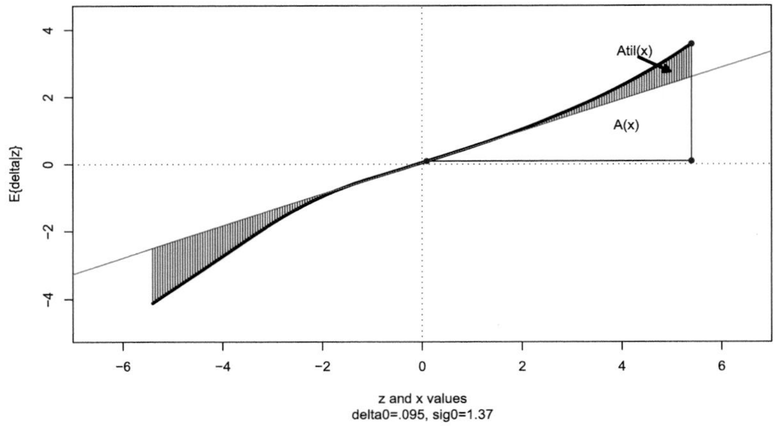

Figure 11.3 Corollary 11.5 illustrated for the police data of Figure 6.1c; hatched area $\tilde{A}(x)$ determines the local false discovery rate based on empirical null (11.53). It is much smaller than $A(x)$, the area of the outlined region (including $\tilde{A}(x)$) that determines the theoretical null fdr (11.42). (Heavy curve is $E\{\delta|z\}$ calculated according to the original hierarchical model (11.22).)

(δ_0, σ_0) are the central matching estimates (6.11). The tangent line to $e(z)$ at $z = \delta_0$ is

$$e_{\tan}(z) = e(\delta_0) + e(\delta_0)'(z - \delta_0) \qquad [e(z)' = de(z)/dz]. \tag{11.51}$$

Corollary 11.5 *Let $\tilde{A}(x)$ be the area between $e(z)$ and $e_{\tan}(z)$,*

$$\tilde{A}(x) = \int_{\delta_0}^{x} [e(z) - e_{\tan}(z)]\, dz. \tag{11.52}$$

Then

$$\frac{\widetilde{\text{fdr}}(x)}{\widetilde{\text{fdr}}(\delta_0)} = e^{-\tilde{A}(x)} \tag{11.53}$$

where $\widetilde{\text{fdr}}(z)$ is the local false discovery rate $\pi_0 f_0(z)/f(z)$ with $f_0(z)$ the central matching empirical null distribution.[4]
(Proof at the end of this section.)

[4] The `locfdr` central matching algorithm actually applies calculations (11.50) to a local quadratic fit to $\log \hat{f}(z)$. It gave $(\hat{\delta}_0, \hat{\sigma}_0) = (0.04, 1.45)$ for the police data, rather than $(0.09, 1.37)$ from (11.50).

Exercise 11.7 Show that for (δ_0, σ_0), the central matching values (11.50), the transformed version of $E\{\delta|z\}$ in (11.46) satisfies

$$\left.\frac{dE\{\delta|z\}}{dz}\right|_{z=\delta_0} = 0 \tag{11.54}$$

(so that $E\{\delta|z\}$ is perfectly flattened near $z = 0$).

One thing is noticeably absent from Corollary 11.4 and Corollary 11.5: π_0, the prior null probability in $\text{fdr}(z) = \pi_0 f_0(z)/f(z)$. Its place is taken by $\text{fdr}(0)$ or $\widetilde{\text{fdr}}(\delta_0)$ in (11.42), (11.53). For the two-groups model (5.2),

$$\text{fdr}(0) = 1/(1 + \text{OR}(0)) \qquad [\text{OR}(z) = (\pi_1 f_1(z))/(\pi_0 f_0(z))], \tag{11.55}$$

$\text{OR}(z)$ being the posterior odds ratio for *non-null* given z. Under the *zero assumption* (4.44) we employed to estimate π_0, $\text{OR}(0) = 0$ and $\text{fdr}(0) = 1$. It is quite reasonable to interpret the corollaries as saying

$$\text{fdr}(x) = e^{-A(x)} \quad \text{and} \quad \widetilde{\text{fdr}}(x) = e^{-\tilde{A}(x)}, \tag{11.56}$$

thereby finessing the problem of estimating π_0.

$A(x)$ is easier to estimate than $\tilde{A}(x)$ in Figure 11.3. The latter requires estimating the lower tangent line, adding variability to the assessment of a smaller quantity. This results in less accurate fdr estimates when based on empirical null distributions, as seen in Table 7.3. Of course there is really no choice in the police data example, where the theoretical null is unrealistic.

Proof of Corollary 11.5 From $e(z) = z + l'(z)$ (11.49) and $e(z)' = 1 + l''(z)$, (11.50) gives

$$e_{\tan}(z) = \delta_0 + \left(1 - 1/\sigma_0^2\right)(z - \delta_0) \tag{11.57}$$

and

$$\tilde{e}(z) \equiv e(z) - e_{\tan}(z) = \frac{z - \delta_0}{\sigma_0^2} + l'(z). \tag{11.58}$$

But this equals $-d \log \text{fdr}(z)/dz$ (11.45), yielding (11.53). □

11.4 Effect Size Estimation

The traditional purpose of simultaneous significance testing has been to identify non-null cases, usually assumed to be a small subset of all N possibilities in large-scale situations. Here we take up a more ambitious goal: to assess effect sizes for those so identified, that is, to estimate how far away non-null cases lie from the null hypothesis. Not all non-null cases are created equal: those with large effect sizes δ_i or $-\delta_i$ are the most useful for prediction, and most likely to play key roles in an ongoing process of scientific discovery.

We now combine the normal hierarchical model (11.22) with the two-groups model (5.1) by assuming that the prior density $g(\delta)$ has an atom of probability π_0 at $\delta = 0$, leaving probability $\pi_1 = 1 - \pi_0$ for the non-zero values,

$$g(\delta) = \pi_0 \Delta_0(\delta) + \pi_1 g_1(\delta) \quad \text{and} \quad z|\delta \sim \mathcal{N}(\delta, 1). \tag{11.59}$$

Here $\Delta_0(\cdot)$ represents a delta function at 0 while $g_1(\cdot)$ is the prior density for non-null δ values; (11.59) gives a two-groups model (5.1) with $f_0(z) = \varphi(z)$, the $\mathcal{N}(0, 1)$ density, and

$$f_1(z) = \int_{-\infty}^{\infty} \varphi(z - \delta) g_1(\delta)\, d\delta. \tag{11.60}$$

We also define J to be the indicator of the null status of a case,

$$J = \begin{cases} 0 & \text{null } (\delta = 0) \\ 1 & \text{non-null } (\delta \neq 0) \end{cases} \tag{11.61}$$

(the reverse of the indicator I used previously).

By definition,

$$\text{fdr}(z) = \Pr\{J = 0|z\}, \tag{11.62}$$

the conditional probability of nullity given z. For any value j,

$$\begin{aligned} E\{\delta^j|z\} &= E\{\delta^j|z, J = 0\}\, \text{fdr}(z) + E\{\delta^j|z, J = 1\}\, (1 - \text{fdr}(z)) \\ &= E\{\delta^j|z, J = 1\}\, (1 - \text{fdr}(z)) \end{aligned} \tag{11.63}$$

since $\delta = 0$ if $J = 0$.

For convenient notation, let

$$E_1(z) = E\{\delta|z, J = 1\} \quad \text{and} \quad \text{var}_1(z) = \text{var}\{\delta|z, J = 1\}, \tag{11.64}$$

the conditional mean and variance for a non-null case.

Exercise 11.8 Use (11.63) to show that

$$E_1(z) = \frac{E\{\delta|z\}}{1 - \text{fdr}(z)} \quad \text{and} \quad \text{var}_1(z) = \frac{\text{var}\{\delta|z\}}{1 - \text{fdr}(z)} - \text{fdr}(z) E_1(z)^2. \tag{11.65}$$

Combined with (11.28), this yields

$$E_1(z) = \frac{z + l'(z)}{1 - \text{fdr}(z)} \quad \text{and} \quad \text{var}_1(z) = \frac{1 + l''(z)}{1 - \text{fdr}(z)} - \text{fdr}(z) E_1(z)^2. \tag{11.66}$$

Plugging in estimates of $l(z)$ and $\mathrm{fdr}(z)$ then gives $\hat{E}_1(z)$ and $\widehat{\mathrm{var}}_1(z)$. Figure 11.4 shows the resulting 90% empirical Bayes posterior bands for the right tail of the prostate data,

$$\hat{E}_1(z) \pm 1.645\, \widehat{\mathrm{var}}_1(z)^{1/2}. \tag{11.67}$$

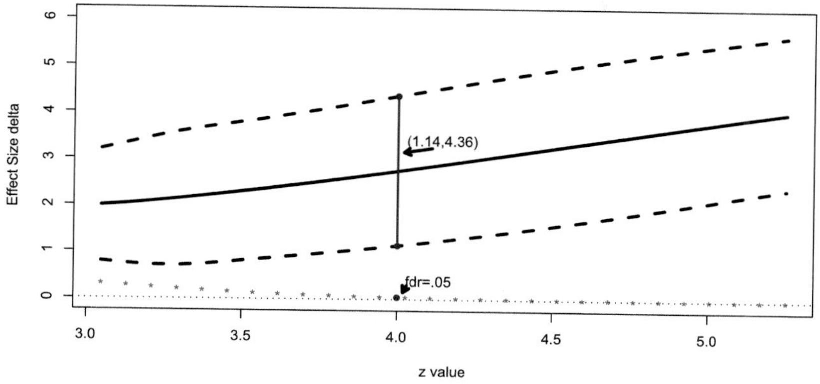

Figure 11.4 Estimated 90% empirical Bayes posterior band for a non-null effect (11.67), prostate data, and estimated local false discovery rate $\widehat{\mathrm{fdr}}(z)$. At $z = 4$, $\widehat{\mathrm{fdr}} = 0.05$, and the 90% interval is (1.14, 4.36).

At $z = 4$ we have $\widehat{\mathrm{fdr}}(z) = 0.05$, $\hat{E}_1(z) = 2.75$, and $\widehat{\mathrm{var}}_1(z)^{1/2} = 0.98$, so interval (11.67) is (1.14, 4.36). The literal interpretation is that, given $z = 4$, there is a 5% chance that $\delta = 0$ and, if not, a 90% chance that δ lies in (1.14, 4.36). In other words, δ is positive with high probability and, if so, the effect size is probably substantial.

From a traditional point of view, it seems strange that our assessment of δ is split into two parts: a point probability at zero and an interval that does not contain zero. But this is a natural consequence of the "mixed" prior $g(\delta)$ in (11.59) which combines an atom at zero with a possibly diffuse, non-zero component. This reinforces the point made in previous chapters that large-scale significance testing is more than just a collection of individual tests done at the same time. Estimation and prediction are also involved, as indirect evidence of an empirical Bayes nature, "learning from the experience of others," makes itself felt.

A version of Theorem 11.1 and Corollary 11.3 is available that applies directly to the non-null posterior distribution $g_1(\delta) = g(\delta|z, J = 1)$.

Theorem 11.6 *Under model* (11.59)*,* $g_1(\delta|z)$ *is an exponential family,*

$$g_1(\delta) = e^{\delta z - \psi_1(z)} \left[e^{-\delta^2/2} g_1(\delta) \right] \tag{11.68}$$

where

$$\psi_1(z) = \log \left\{ \frac{1 - \text{fdr}(z)}{\text{fdr}(z)} \middle/ \frac{\pi_1}{\pi_0} \right\}. \tag{11.69}$$

Proof Bayes rule says that $g_1(\delta|z) = \varphi(z - \delta) g_1(\delta)/f_1(z)$, yielding

$$g_1(\delta|z) = e^{\delta z - \log\{f_1(z)/\varphi(z)\}} \left[e^{-\delta^2/2} g_1(\delta) \right]. \tag{11.70}$$

□

Exercise 11.9 Complete the proof by using $\text{fdr}(z) = \pi_0 \varphi(z)/f(z)$ to show that

$$\frac{f_1(z)}{\varphi(z)} = \frac{\pi_0}{\pi_1} \frac{1 - \text{fdr}(z)}{\text{fdr}(z)}. \tag{11.71}$$

Differentiating $\psi_1(z)$ yields non-null posterior cumulants of δ given z, the jth cumulant being

$$-\frac{d^j}{dz^j} \log \left\{ \frac{\text{fdr}(z)}{1 - \text{fdr}(z)} \right\}; \tag{11.72}$$

$j = 1$ and 2 are the non-null conditional mean and variance (as compared with (11.40)), agreeing of course with expressions (11.65).

False coverage rate control

A more classically oriented frequentist approach to effect size estimation aims to control FCR, the false coverage rate. We are given a method of discovering non-null cases, for example the BH(q) algorithm of Chapter 4, and also of assigning confidence intervals for the effect sizes δ_i of the discovered cases. The method's FCR is the *expected* proportion of discovered cases in which δ_i is not in its assigned confidence interval.

FCR control is a laudable criterion, aimed against the bad practice of assigning the usual confidence intervals to those cases selected as non-null — say 5.29 ± 1.645 for δ_{610} in Table 11.2 — which completely ignores selection bias.

The BY(q) FCR control algorithm[5] works as follows: let $\text{CI}_z(p)$ be the usual two-sided normal confidence interval of coverage probability $1 - p$,

$$\text{CI}_z(p) = z \pm \Phi^{-1}\left(1 - \frac{p}{2}\right) \tag{11.73}$$

[5] Benjamini–Yekutieli; see Notes at this chapter's end.

(so $p = 0.10$ corresponds to $z \pm 1.645$). Perform the BH(q) FDR control algorithm of Section 4.2 on $p_1, p_2, \dots, p_N$ with $p_i = 2\Phi(-|z_i|)$, declaring, say, R of the z_i non-null. To each corresponding δ_i assign the confidence interval

$$\mathrm{CI}_{z_i}\left(\frac{Rq}{N}\right) = z_i \pm \Phi^{-1}\left(1 - \frac{Rq}{2N}\right). \tag{11.74}$$

Then, assuming the z_i are *independent* normal unbiased estimates of their effect sizes δ_i,

$$z_i \overset{\text{ind}}{\sim} \mathcal{N}(\delta_i, 1) \qquad i = 1, 2, \dots, N, \tag{11.75}$$

the false coverage rate will be no greater than q.

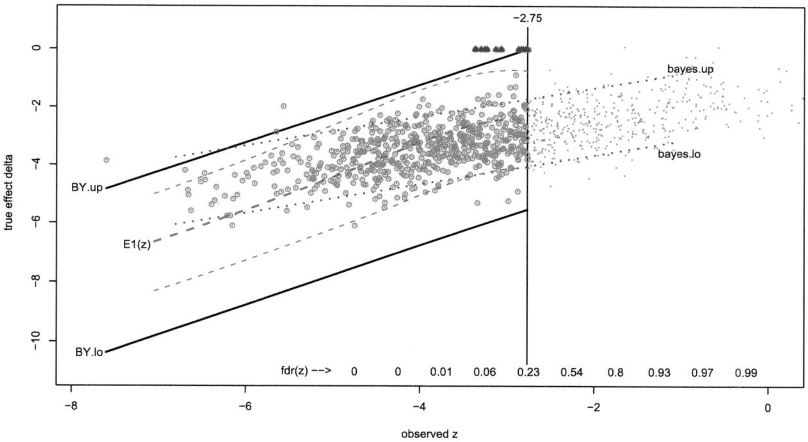

Figure 11.5 False coverage rate example: $N = 10\,000$ pairs (z_i, δ_i) from model (11.75)–(11.76). Green points, the 1000 non-null cases, $\delta_i \sim \mathcal{N}(-3, 1)$; circled points, those 563 declared non-null by BH(0.10) algorithm. *Heavy black lines*, BY(0.10) FCR control confidence limits (11.74); *blue dotted lines*, Bayes 90% limit for non-null cases (11.80); *red dashed lines*, $\hat{E}_1(z)$ (11.65) and nominal 90% empirical Bayes limits (11.67). Bottom numbers list fdr(z) for model (11.75)–(11.76). Dark triangles are 14 null cases ($\delta_i = 0$) having $z_i < -2.75$; 26 more, not pictured, had $z_i > 2.75$.

Figure 11.5 shows an application of the BY(q) algorithm, $q = 0.10$, to a simulated data set in which $N = 10\,000$ independent (δ_i, z_i) pairs were generated as in (11.59), with

$$g(\delta) = 0.90\Delta_0(\delta) + 0.10\varphi_{-3,1}(\delta). \tag{11.76}$$

That is, 90% of the effects were zero and 10% $\mathcal{N}(-3, 1)$. The BH(0.10) rule declared $R = 603$ cases non-null, those having $|z_i| \geq 2.75$. Using (11.74), the BY(q) algorithm assigned confidence intervals

$$z_i \pm 2.75 \tag{11.77}$$

to the 603 cases declared non-null.

Exercise 11.10 Why is it *not* a coincidence that the BH cutoff $|z_i| \geq 2.75$ is nearly the same as the half-width 2.75 in (11.77)?

Because this is a simulation, we can see how well the BY algorithm performed. Figure 11.5 plots (z_i, μ_i) points for the 1000 *actual* non-null cases, those having $\delta_i \sim \mathcal{N}(-3, 1)$; 563 of these were *declared* non-null, those with $z_i \leq -2.75$. There were 40 false discoveries, those having $\delta_i = 0$, 14 with $z_i \leq -2.75$ and 26 with $z_i \geq 2.75$ (not pictured).

The first thing to notice is that the FCR property is satisfied: only 43 of the $R = 603$ confidence intervals fail to cover δ_i, those for the 40 false discoveries and three of 563 true discoveries, giving $43/603 = 0.07$, substantially less than $q = 0.10$. However, the second thing is that the intervals (11.77) are frighteningly wide — about 67% longer than the individual 90% intervals $z_i \pm 1.645$ — and poorly centered, particularly at the left where all the δ_i fall in the upper halves of their intervals, as seen from the heavy lines for the BY confidence limits.

An interesting comparison is with Bayes rule applied to (11.75)–(11.76). This yields

$$\Pr\{\delta = 0|z_i\} \equiv \text{fdr}(z_i) = \frac{0.90\varphi(z)}{0.90\varphi(z) + 0.10\varphi_{-3,\sqrt{2}}(z)} \tag{11.78}$$

$(\varphi_{\delta,\sigma}(z) = \varphi((z-\delta)/\sigma)/\sigma)$ and non-null posterior density

$$g_1(\delta_i|z_i) \equiv g(\delta_i|z_i, J_i = 1) \sim \mathcal{N}\left(\frac{z_i - 3}{2}, \frac{1}{2}\right). \tag{11.79}$$

That is, δ_i is null with probability fdr(z_i) and $\mathcal{N}((z_i - 3)/2, 1/2)$ with probability $1 - \text{fdr}(z_i)$. The blue dotted lines indicate the 90% posterior intervals given that δ_i is non-null,

$$(z_i - 3)/2 \pm 1.645/\sqrt{2}, \tag{11.80}$$

now a factor $\sqrt{2}$ shorter than $z_i \pm 1.645$.

The Bayes intervals and the frequentist BY(q) intervals are pursuing the same goal, to include the effect sizes δ_i with 90% certainty. At $z = -2.75$, the BY(q) assessment is $\Pr\{\delta \in (-5.50, 0)\} = 0.90$, while Bayes rule states that $\delta = 0$ with probability $\text{fdr}(-2.75) = 0.23$, and otherwise $\delta \in (-4.04, -1.71)$ with probability 0.90. This kind of bifurcated assessment, as in Figure 11.4, is inherent to model (11.59). A principal cause of BY(q)'s oversized intervals comes from using a connected set to describe a disconnected situation.

As described in Chapter 4, the BH(q) FDR control algorithm enjoyed both frequentist and Bayesian support, and both philosophies played roles in the subsequent chapters' empirical Bayes development. The same cannot be said for FCR control and the BY(q) algorithm. Frequentist methods like BY(q) enjoy the considerable charm of exact error control, without requiring prior distributions, but the discrepancy evident in Figure 11.5 is disconcerting.

The red dashed lines in Figure 11.5 are the 90% empirical Bayes intervals (11.67). (Actually using $\widehat{\text{var}}\{\delta|z\} = 1 + \hat{l}''(z)$ (11.28), rather than $\widehat{\text{var}}_1(z)$. These are nearly the same for $\widehat{\text{fdr}}(z)$ small, while the former is a more stable estimator.) These are closer to the Bayes intervals, and $\widehat{\text{fdr}}(z)$ is close to the true $\text{fdr}(z)$, e.g., $\text{fdr}(-2.75) = 0.27$, but the results are far from perfect.

There is a hierarchy of estimation difficulties connected with the methods of this chapter: $\text{fdr}(z)$ is the easiest to estimate since it only requires assessing the marginal density $f(z)$ or its logarithm $l(z)$; $E\{\delta|z\} = z + l'(z)$ requires estimating the *slope* of $l(z)$, while $\text{var}\{\delta|z\} = 1 + l''(z)$ requires its curvature. The Poisson regression estimates of Section 5.2 are quite dependable for $\text{fdr}(z)$, adequate for $E\{\delta|z\}$, and borderline undependable for $\text{var}\{\delta|z\}$. Empirical Bayes assessment of effect size is a worthy goal, and we can hope for the development of a more stable methodology.

11.5 The Missing Species Problem

We conclude with perhaps the first empirical Bayes success story, the solution of the "missing species problem." The original motivating data set is shown in Table 11.5. Alexander Corbet, a naturalist, spent two years in the early 1940s trapping butterflies in Malaysia (then Malaya). Let

$$n_j = \#\{\text{species trapped exactly } j \text{ times in two years}\}. \tag{11.81}$$

The table shows $n_1 = 118$, this being the number of species so rare that Corbet captured just one each; $n_2 = 74$ species were trapped twice each, n_3 three times each, etc.

Table 11.5 *Corbet's Malaysian butterfly data; n_j is the number of butterfly species captured exactly j times in two years of trapping:* 118 *species were trapped just once each,* 74 *two times each, etc.*

j	1	2	3	4	5	6	7	8	9	10	11	12	...
n_j	118	74	44	24	29	22	20	19	20	15	12	14	...

Corbet then asked a very interesting prediction question: if he spent an additional one year trapping — half as long as he had already been there — how many *new* species could he expect to capture? One reason the question is interesting is that it seems impossible to answer. It refers to n_0, the number of Malaysian butterfly species *not* previously captured. And of course n_0 is missing from Table 11.5.

Fortunately, Corbet asked the right person: R. A. Fisher, the founding father of modern statistical theory. We will return later to Fisher's solution, which involves a parametric empirical Bayes argument, but first we present a non-parametric solution developed in the 1950s. Let $\mathcal{S}$ be the set of all Malaysian butterfly species and let $S = \#\mathcal{S}$ be their number ($S \geq 435$ for Corbet's problem, the total of Table 11.5). The key assumption is that each species s is trapped according to a Poisson process with intensity parameter λ_s,

$$\lambda_s = \text{Expected number of species } s \text{ captured per unit time.} \tag{11.82}$$

The time unit is two years in Corbet's example.

Letting x_s be the number of times species s is trapped in the original one unit of observation time, x_s has Poisson probability density function

$$p_s(j) = \Pr\{x_s = j\} = e^{-\lambda_s}\lambda_s^j/j! \tag{11.83}$$

for $j = 0, 1, 2, \ldots$. The expectation η_j of n_j is then given by

$$\eta_j = E\{n_j\} = \sum_s e^{-\lambda_s}\lambda_s^j/j! = S\int_0^\infty e^{-\lambda}\lambda^j/j!\,dG(\lambda) \tag{11.84}$$

where $G(\lambda)$ is the (unobservable) empirical cdf of the λ_s values,

$$G(\lambda) = \#\{\lambda_s \leq \lambda\}/S. \tag{11.85}$$

The essence of Poisson model (11.82) is that species s is trapped independently in non-overlapping time intervals. In particular, for a new trapping period of length t, the number of times species s will be captured follows a Poi($\lambda_s t$) distribution, independent of its original capture number $x_s \sim$ Poi(λ_s). Therefore species s has probability

$$q_s(t) = e^{-\lambda_s}\left(1 - e^{-\lambda_s t}\right) \tag{11.86}$$

of *not being seen* during the original period and *being seen* in the new trapping period.

Adding up over species, $\mu(t)$, the expected number of new species seen in succeeding time t is

$$\begin{aligned}\mu(t) &= \sum_s q_s(t) = S \int_0^\infty e^{-\lambda}\left(1 - e^{-\lambda t}\right) dG(\lambda) \\ &= S \int_0^\infty e^{-\lambda}\left[\lambda t - (\lambda t)^2/2! + (\lambda t)^3/3! + \dots\right] dG(\lambda),\end{aligned} \tag{11.87}$$

the last line following by Taylor expansion of $1 - \exp(-\lambda t)$. Comparing (11.87) with (11.84) yields an intriguing formula,

$$\mu(t) = \eta_1 t - \eta_2 t^2 + \eta_3 t^3 \dots . \tag{11.88}$$

Each η_j is unbiasedly estimated by its corresponding count n_j (11.81), giving an unbiased estimate of $\mu(t)$,

$$\hat{\mu}(t) = n_1 t - n_2 t^2 + n_3 t^3 \dots, \tag{11.89}$$

not involving the zero count n_0.

Now we can answer Corbet's question: for a proposed additional trapping period of one year, i.e., $t = 1/2$ as long as the original period, our prediction for the number of new species captured is[6]

$$\hat{\mu}(t) = 118\left(\frac{1}{2}\right) - 74\left(\frac{1}{4}\right) + 44\left(\frac{1}{8}\right) - 24\left(\frac{1}{16}\right) \dots = 45.2. \tag{11.90}$$

How accurate is $\hat{\mu}(t)$? Just as for the empirical Bayes estimates of Chapter 2, the unbiasedness of $\hat{\mu}(t)$ does not depend on independence among the species-capture values x_s, but its accuracy does. A useful bound on the variance of $\hat{\mu}(t)$, developed next, is available assuming independence,

$$x_s \overset{\text{ind}}{\sim} \text{Poi}(\lambda_s) \qquad \text{for } s \in \mathcal{S}. \tag{11.91}$$

Let $\boldsymbol{p}_s = (p_s(0), p_s(1), p_s(2), \dots)'$ be the vector[7] of Poisson probabilities (11.83) and $\boldsymbol{I}_s$ the indicator vector

$$\boldsymbol{I}_s = (0, 0, \dots, 0, 1, 0, \dots)' \tag{11.92}$$

[6] Perhaps a discouraging figure; it is not clear if Corbet continued his stay.

[7] Theoretically, $\boldsymbol{p}_s$ and $\boldsymbol{I}_s$ have an infinite number of components, but that does not affect the calculations that follow.

with 1 in the jth place if $x_s = j$. A standard multinomial calculation shows that $\boldsymbol{I}_s$ has expectation $\boldsymbol{p}_s$ and covariance matrix

$$\text{cov}(\boldsymbol{I}_s) = \text{diag}(\boldsymbol{p}_s) - \boldsymbol{p}_s\boldsymbol{p}_s', \tag{11.93}$$

"diag" denoting a diagonal matrix with the indicated entries (so $\text{var}(p_s(j)) = p_s(j)(1 - p_s(j))$, the usual binomial formula).

The vectors $\boldsymbol{n} = (n_0, n_1, n_2, \dots)'$ and $\boldsymbol{\eta} = (\eta_0, \eta_1, \eta_2, \dots)'$ are given by

$$\boldsymbol{n} = \sum_s \boldsymbol{I}_s \quad \text{and} \quad \boldsymbol{\eta} = \sum_s \boldsymbol{p}_s. \tag{11.94}$$

Assuming independence,

$$\text{cov}(\boldsymbol{n}) = \sum_s \text{cov}(\boldsymbol{I}_s) = \text{diag}(\boldsymbol{\eta}) - \sum_s \boldsymbol{p}_s\boldsymbol{p}_s'. \tag{11.95}$$

Defining

$$\boldsymbol{u}(t) = \left(0, t, -t^2, t^3, \dots\right)', \tag{11.96}$$

(11.89) gives

$$\hat{\mu}(t) = \sum_j u(t)_j n_j = \boldsymbol{u}(t)'\boldsymbol{n} \tag{11.97}$$

and

$$\begin{aligned} \text{var}\left(\hat{\mu}(t)\right) &= \boldsymbol{u}(t)' \,\text{cov}(\boldsymbol{n})\boldsymbol{u}(t) \\ &= \boldsymbol{u}(t)' \,\text{diag}(\boldsymbol{\eta})\boldsymbol{u}(t) - \sum_s \left(\boldsymbol{u}(t)'\boldsymbol{p}_s\right)^2. \end{aligned} \tag{11.98}$$

We see that an upper bound on $\text{var}(\hat{\mu}(t))$ is

$$\text{var}\left(\hat{\mu}(t)\right) \le \boldsymbol{u}(t)' \,\text{diag}(\boldsymbol{\eta})\boldsymbol{u}(t) = \sum_{j\ge1} \eta_j t^{2j}. \tag{11.99}$$

Substituting n_j for η_j provides an approximation for the standard deviation of $\hat{\mu}(t)$,

$$\widehat{\text{sd}}\left\{\hat{\mu}(t)\right\} = \sum_{j\ge1} n_j t^{2j}. \tag{11.100}$$

This yields $\widehat{\text{sd}} = 9.3$ for Corbet's data, so the coefficient of variation of estimate 45.2 is about 20%. The term we are ignoring in (11.98), $\sum_s(\boldsymbol{u}(t)'\boldsymbol{p}_s)^2$, is usually small. Without it, (11.99) gives $\text{var}(\hat{\mu}(t))$ under the Poisson model $n_j \overset{\text{ind}}{\sim} \text{Poi}(\eta_j)$, not unreasonable in its own right.

Exercise 11.11 Show that the ignored term equals

$$\sum_s (\boldsymbol{u}(t)'\boldsymbol{p}_s)^2 = S \int_0^\infty q_\lambda(t)^2\, dG(\lambda) \qquad (11.101)$$
$$\left[q_\lambda(t) = e^{-\lambda}\left(1 - e^{-\lambda t}\right)\right].$$

For an additional trapping period of two years, i.e., $t = 1$, (11.89), (11.100) predict 75 ± 20 new species. The series in (11.89) diverges for $t > 1$, invalidating the predictor $\hat{\mu}(t)$. Fisher's original approach to the missing species problem avoids this limitation at the expense of assuming that the intensity cdf (11.85) is obtained from a gamma density function for $\lambda \geq 0$,

$$g(\lambda) = c(\alpha,\beta)\lambda^{\alpha-1}e^{-\lambda/\beta} \qquad \left[c(\alpha,\beta) = (\beta^\alpha\Gamma(\alpha))^{-1}\right]. \qquad (11.102)$$

The previous expressions now take on specific forms; for instance,

$$\eta_j = S\frac{\Gamma(\alpha + j)}{j!\Gamma(\alpha)} \qquad (11.103)$$

for (11.84). Some algebra yields Fisher's prediction formula:

Exercise 11.12 Show that the gamma assumption (11.102) results in

$$\mu(t) = \eta_1\left\{1 - 1/(1 + \gamma t)^\alpha\right\}/(\gamma\alpha) \qquad [\gamma = \beta/(1 + \beta)]. \qquad (11.104)$$

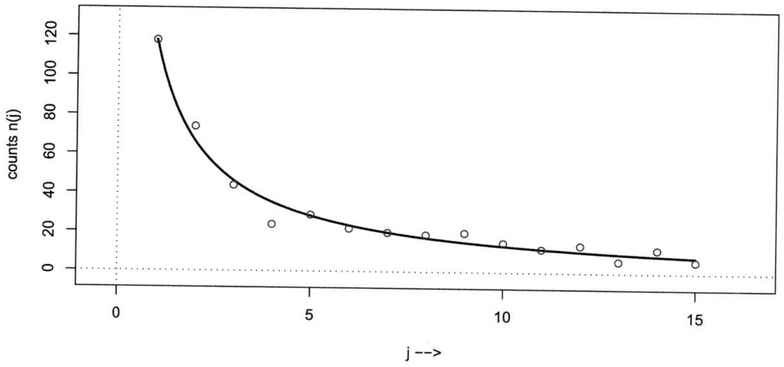

Figure 11.6 Solid curve is Fisher's negative binomial estimates $\hat{\eta}_j$ (11.105), compared to counts n_j from Corbet's Table 11.5; $\hat{\eta}_1 = 118$, $\hat{\alpha} = 0.166$, $\hat{\beta} = 27.06$.

Formula (11.103) can be re-expressed in *decapitated negative binomial* form

$$\eta_j = \eta_1 \frac{\Gamma(\alpha + j)}{j!\Gamma(\alpha + 1)} \gamma^{j-1}, \tag{11.105}$$

decapitation referring to the elimination of η_0. Maximum likelihood estimation in the model $n_j \overset{\text{ind}}{\sim} \text{Poi}(\eta_j)$ gave MLE values $\hat{\alpha} = 0.166$ and $\hat{\beta} = 27.06$. Figure 11.6 compares Fisher's estimate $\hat{\eta}_j$ from (11.105) (taking $\hat{\eta}_1 = n_1$) with n_j from Table 11.5. The fit is not wonderful, but the results are similar to those obtained non-parametrically: $\hat{\mu}(0.5) = 46.6$, $\hat{\mu}(1) = 78.1$.

We can go as far into the future as we wish now — for example, $\hat{\mu}(2) = 120.4$ expected new species for four additional trapping years — but the predictions become increasingly speculative. The crucial function

$$q_\lambda(t) = e^{-\lambda}\left(1 - e^{-\lambda t}\right) \tag{11.106}$$

which determines the probability that a species of intensity λ will be a new capture (11.86), concentrates ever nearer zero as t increases. It is easy to show that $q_\lambda(t)$ is maximized at

$$\lambda_{\max} = (\log(t + 1))/t, \tag{11.107}$$

$\lambda_{\max} = 0.37$ for $t = 2$. The number of new captures depends ever more on the behavior of $g(\lambda)$ near $\lambda = 0$, where it cannot be well-estimated, and there is no particular reason to trust Fisher's negative binomial assumption (11.102).

Table 11.6 *Shakespeare word count data;* 14 376 *distinct words appeared just once each in the canon,* 4343 *twice each,* 305 *eleven times each, etc.*

	1	2	3	4	5	6	7	8	9	10
0+	14 376	4343	2292	1463	1043	837	638	519	430	364
10+	305	259	242	223	187	181	179	130	127	128
20+	104	105	99	112	93	74	83	76	72	63
30+	73	47	56	59	53	45	34	49	45	52
40+	49	41	30	35	37	21	41	30	28	19
50+	25	19	28	27	31	19	19	22	23	14
60+	30	19	21	18	15	10	15	14	11	16
70+	13	12	10	16	18	11	8	15	12	7
80+	13	12	11	8	10	11	7	12	9	8
90+	4	7	6	7	10	10	15	7	7	5

The fascinating formula (11.89) has applications outside the biological realm. Table 11.6 gives a statistical view of Shakespeare. Spevack (1968), a "canon" or authoritative volume of all Shakespeare's plays and poems, comprised 884 647 total words, among which were 31 534 distinct words. Of these, 14 376 were so rare they showed up just once each, 4343 twice each, etc. The table gives n_j for $j = 1, 2, \ldots, 100$, $n_1 = 14\,376, n_2 = 4343, \ldots, n_{100} = 5$.

Suppose one found proportion t of a "new" canon, that is, previously unknown Shakespeare works of $884\,647 \cdot t$ total words. How many previously unused words would be seen? The predicted number is given by $\hat{\mu}(t)$ (11.89). For $t = 1$, that is, for a whole new canon of Shakespeare, the prediction[8] is

$$\hat{\mu}(1) = 11,430 \pm 178. \tag{11.108}$$

Fisher's model gives $\hat{\mu}(1) = 11\,483$, nearly the same result using MLE estimates $\hat{\alpha} = -0.3954, \hat{\beta} = 104.26$.

A variation on missing species is the "missing mass problem." Define

$$\nu_0 = \sum_0 \lambda_s \Big/ \sum \lambda_s \tag{11.109}$$

where $\sum \lambda_s$ is the sum over all species of the Poisson intensity parameters (11.82), while $\sum_0 \lambda_s$ is the corresponding sum over species *not* seen in the original observation period; ν_0 is the missing proportion of species, weighted according to their natural prevalence.

The expected value of the numerator in (11.109) is

$$\sum_s \lambda_s e^{-\lambda_s} = S \int_0^\infty \lambda e^{-\lambda}\, dG(\lambda) = \eta_1 \tag{11.110}$$

while that for the denominator is

$$\sum_s \lambda_s = \sum_s E\{x_s\} = E\left\{\sum_s x_s\right\} = E\{T\} \tag{11.111}$$

where T is the total number of individuals observed, $T = \sum_j j n_j$. The obvious missing mass estimator is then

$$\hat{\nu}_0 = n_1/T. \tag{11.112}$$

For the Shakespeare word counts,

$$\hat{\nu}_0 = 14\,376/884\,647 = 0.016, \tag{11.113}$$

[8] Checks have been run, with reassuring results, by removing one of the plays, recomputing the now reduced counts n_j, and seeing how well (11.89) predicts the "new" words in the removed material.

suggesting that we have seen all but 1.6% of Shakespeare's vocabulary, when weighted by his usage frequency. (But not nearly that much in a missing species sense: there is at least proportion $11\,430/31\,534 = 0.36$ *distinct* missing words (11.108), and other methods raise this to more than 1.00.)

The early empirical Bayes literature is filled with other fascinating prediction results concerning the hierarchical Poisson model

$$\lambda \sim g(\cdot) \quad \text{and} \quad x|\lambda \sim \text{Poi}(\lambda). \tag{11.114}$$

For example, (11.84) gives, almost immediately, Robbins' prediction formula

$$E\{\lambda|x = j\} = (j+1)\eta_{j+1}/\eta_j \tag{11.115}$$

and the useful estimate $\hat{E}\{\lambda|x = j\} = (j+1)n_{j+1}/n_j$, an analog of the normal-theory result (11.28).

Exercise 11.13 Verify (11.115).

The missing species formula, like the James–Stein results of Chapter 1, had a near-magical quality when it first appeared in the 1950s and 60s. But, viewed in the context of this book, these are recognizable, if flamboyant, relatives of our more workaday empirical Bayes methods for estimation, testing, and prediction.

My own belief is that empirical Bayes theory is still in its adolescence. The use of indirect evidence — what I called learning from the experience of others — was not much needed for most of 20th century statistics, and actively disparaged in important areas such as drug testing and clinical trials. A total reliance on direct evidence is an unaffordable luxury in large-scale applications. I expect, or at least hope, that 21st century statisticians will perfect an empirical Bayes theory as complete and satisfying for large-scale problems as linear regression and ANOVA are within their ambit.

Notes

The `pamr` prediction procedure was developed, under the name "nearest shrunken centroids," in Tibshirani et al. (2002). Tibshirani and Tibshirani (2009) discuss a bias-correction for the kind of minimum-seeking cross-validation suggested in Table 11.1. A full review of modern prediction methods is available in Hastie et al.'s very readable 2009 text on "statistical learning." The `Ebay` empirical Bayes prediction algorithm appears in

Efron (2009b). Subramanian et al. (2005) discuss the Michigan lung cancer study.

The normal hierarchical model (11.22) goes far back into the empirical Bayes literature, and played important roles in the profound investigations of multidimensional normal estimation of Brown (1971) and Stein (1981). Robbins (1956) includes the formula for $E\{\delta|z\}$ in (11.40), prediction formula (11.115), and much more.

Benjamini and Yekutieli (2005) pioneered large-scale effect size estimation, proposing the BY(q) algorithm. A version of Figure 11.5 appears in Efron (2008a).

The missing species problem has a distinguished pedigree, with key contributions by Good (1953), Good and Toulmin (1956) (crediting Alan Turing of Turing machine fame), as well as Robbins (1968). Fisher's original work with Corbet appears in Fisher et al. (1943). The Shakespeare data and analysis is from Efron and Thisted (1976), where linear programming methods are invoked to provide a lower bound of 35 000 on Shakespeare's unseen vocabulary. A later paper, Thisted and Efron (1987), applies versions of formula (11.89) to the question of authenticity of a newly discovered poem attributed to Shakespeare.

Hierarchical models such as (2.47) and (11.114) are available for all one-parameter exponential families, as discussed in Muralidharan (2009).

Appendix A

Exponential Families

Exponential families are the main connecting idea of classical statistics. Here we will provide a very brief sketch of the theory as it applies to the previous material, particularly that in Chapter 5. Advanced texts such as Lehmann and Romano (2005b) discuss exponential families, and their applications to testing and estimation, in more detail.

A one-parameter exponential family $\mathcal{F}$ is a family of probability densities on a subset $\mathcal{X}$ of the real line,

$$f_\eta(x) = e^{\eta x - \psi(\eta)} f_0(x) \qquad (x \in \mathcal{X}) \tag{A.1}$$

where η is the *natural* or *canonical* parameter, x is the *sufficient statistic*, $f_0(x)$ is the *carrier*, $\mathcal{X}$ is the *sample space*, and $\psi(\eta)$ is the *cumulant generating function*, or more simply the *normalizer*, so called because $\exp\{-\psi(\eta)\}$ is the positive constant required to make $f_\eta(x)$ integrate to 1,

$$e^{\psi(\eta)} = \int_{\mathcal{X}} e^{\eta x} f_0(x)\, dx. \tag{A.2}$$

The family $\mathcal{F}$ consists of all exponential "tilts" of $f_0(x)$, $e^{\eta x} f_0(x)$, normalized to integrate to 1. Exponential tilting enjoys a crucial statistical property: if $x_1, x_2, \ldots, x_N$ is an i.i.d. sample from $f_\eta(x)$, then $\bar{x} = \sum x_i/N$ is a sufficient statistic for estimating η, with densities

$$f_\eta^{\bar{X}}(\bar{x}) = e^{N[\eta\bar{x} - \psi(\eta)]} f_0^{\bar{x}}(\bar{x}), \tag{A.3}$$

again a one-parameter exponential family. This greatly simplifies inference theory for η, the original motivation for definition (A.1) in the 1930s. Perhaps surprisingly, it then turned out that most familiar statistical families — normal binomial, Poisson, gamma — were in fact of form (A.1).

As a first example, $x \sim \mathcal{N}(\mu, 1)$ has density functions

$$f(x) = \frac{1}{\sqrt{2\pi}} e^{-\frac{1}{2}(x-\mu)^2} = e^{\mu x - \mu^2/2} \cdot \varphi(x) \tag{A.4}$$

for μ any value in $(-\infty, \infty)$, where $\varphi(x)$ is the standard normal density $\exp\{-x^2/2\}/\sqrt{2\pi}$. This is of form (A.1), having sufficient statistic x, natural parameter $\eta = \mu$, $\mathcal{X}$ equaling the whole real line, $\psi(\eta) = \eta^2/2$, and carrier $\varphi(x)$. (So the entire normal family (A.4) can be obtained from exponential tilts of $\varphi(x)$.) Repeated sampling $x_1, x_2, \dots, x_n \overset{\text{iid}}{\sim} \mathcal{N}(\mu, 1)$ as in (A.3) produces essentially the same family, but now with sufficient statistic $\bar{x}$.

The sample space for the Poisson family is the non-negative integers $\mathcal{X} = \{0, 1, 2, \dots\}$. Its (discrete) density function can be written in its usual form as

$$f(x) = e^{-\mu}\mu^x/x! = e^{(\log\mu)x-\mu}/x! \tag{A.5}$$

for any $\mu > 0$. This is again of type (A.1), with

$$\eta = \log\mu, \qquad \psi(\eta) = e^\eta, \tag{A.6}$$

and $f_0(x) = 1/x!$. (It is only necessary for $f_0(x)$ to be positive, not to integrate to 1.)

The cumulant generating function is so named because differentiating it gives the cumulants of x. In particular, the mean and variance are the first two derivatives with respect to η, the usual notation being

$$\mu \equiv E_\eta\{x\} = \partial\psi(\eta)/\partial\eta \tag{A.7}$$

and

$$V \equiv \text{var}_\eta\{x\} = \partial^2\psi(\eta)/\partial\eta^2. \tag{A.8}$$

For the Poisson family, where $\psi(\eta) = \exp(\eta)$, we get

$$\mu = e^\eta \quad \text{and} \quad V = e^\eta, \tag{A.9}$$

that is, the variance equals the expectation, and in fact all the higher cumulants equal μ. Note that formulas (A.7) and (A.8) do not involve $f_0(x)$ or any integration, which, if $\psi(\eta)$ is available, makes them wonderfully convenient for calculating moments.

The *deviance function* $D(\eta_1, \eta_2)$ is a measure of distance between two members of an exponential family,

$$D(\eta_1, \eta_2) = 2 \cdot E_{\eta_1}\left\{\log\left(\frac{f_{\eta_1}(x)}{f_{\eta_2}(x)}\right)\right\} = 2\int_{\mathcal{X}} \log\left(\frac{f_{\eta_1}(x)}{f_{\eta_2}(x)}\right) f_{\eta_1}(x)\,dx. \tag{A.10}$$

It is always non-negative, but usually not symmetric, $D(\eta_1, \eta_2) \neq D(\eta_2, \eta_1)$. ($D(\eta_1, \eta_2)/2$ is also known as the *Kullback–Leibler distance*, another related term being *mutual information*.) Deviance generalizes the concept of squared error distance $(\eta_1 - \eta_2)^2$ or $(\mu_1 - \mu_2)^2$, this being $D(\eta_1, \eta_2)$ in the normal family (A.4). For the Poisson family,

$$D(\eta_1, \eta_2) = 2\mu_1 \left[\log\left(\frac{\mu_1}{\mu_2}\right) - \left(1 - \frac{\mu_2}{\mu_1}\right) \right]. \tag{A.11}$$

In a regression situation such as (5.15)–(5.17), minimizing the sum of deviances between the observations and a parameterized regression function is equivalent to maximum likelihood estimation. This amounts to least squares fitting in a normal regression problem, but "least deviance" is more efficient in a Poisson regression context like (5.17).

A.1 Multiparameter Exponential Families

By changing η and x in (A.1) to K-dimensional vectors $\boldsymbol{\eta}$ and $\boldsymbol{x}$ we obtain a K-parameter exponential family of densities

$$f_{\boldsymbol{\eta}}(\boldsymbol{x}) = e^{\boldsymbol{\eta}'\boldsymbol{x} - \psi(\boldsymbol{\eta})} f_0(\boldsymbol{x}). \tag{A.12}$$

Now the factor $\exp\{\sum_1^K \eta_k x_k\}$ can tilt the carrier $f_0(\boldsymbol{x})$ in K possible directions. The sufficient statistic $\boldsymbol{x}$ is usually some K-dimensional function of a more complicated set of data. Formula (A.12) provides a convenient way for the statistician to model complex data sets flexibly, by focusing on K summary statistics x_k, for example powers of z in (5.10).

Once again, most of the familar multiparameter families — normal, Dirichlet, multinomial — are of form (A.12). The multinomial family was seen to play an important role in Section 5.2, where count y_k was the number of the N z-values falling into the kth bin, for $k = 1, 2, \ldots, K$. If the z_i are independent, vector $\boldsymbol{y} = (y_1, y_2, \ldots, y_K)'$ follows a multinomial distribution on K categories, sample size N,

$$\boldsymbol{y} \sim \text{mult}_K(N, \boldsymbol{\pi}) \tag{A.13}$$

with

$$\boldsymbol{\pi} = (\pi_1, \pi_2, \ldots, \pi_K)' \tag{A.14}$$

the vector of bin probabilities, $\pi_k = \Pr\{z \in \mathcal{Z}_k\}$ in notation (5.11).

The multinomial density function

$$f(\boldsymbol{y}) = c(\boldsymbol{y}) \prod_{k=1}^K \pi_k^{y_k} \qquad \left[c(\boldsymbol{y}) = N! \Big/ \prod_1^K y_k! \right] \tag{A.15}$$

is a $(K - 1)$-dimensional exponential family, one degree of freedom being lost to the constraint $\sum_1^K \pi_k = 1$. A convenient parameterization lets $\boldsymbol{\eta} = (\eta_1, \eta_2, \ldots, \eta_K)'$ be any vector in K-dimensional Euclidean space $\mathcal{R}^K$, and defines

$$\pi_k = e^{\eta_k} \bigg/ \sum_1^K e^{\eta_j} \qquad \text{for } k = 1, 2, \ldots, K. \tag{A.16}$$

Then $\log(\pi_k) = \eta_k - \log(\sum e^{\eta_j})$, and (A.15) takes the form of (A.12),

$$f_{\boldsymbol{\eta}}(\boldsymbol{y}) = e^{\boldsymbol{\eta}'\boldsymbol{y} - \psi(\boldsymbol{\eta})} c(\boldsymbol{y}) \qquad \left[\psi(\boldsymbol{\eta}) = \log \sum_1^K e^{\eta_j}\right]. \tag{A.17}$$

This looks like a K-parameter family, but every scalar multiple $m \cdot \boldsymbol{\eta}$ produces the same vector $\boldsymbol{\pi}$ in (A.16) and (A.13), removing one degree of freedom.

Moment properties (A.7)–(A.8) extend directly to multivariate exponential families,

$$\boldsymbol{\mu} = E_{\boldsymbol{\eta}}\{\boldsymbol{x}\} = (\cdots \partial\psi/\partial\eta_k \cdots)' \tag{A.18}$$

and

$$V = \text{cov}_{\boldsymbol{\eta}}\{\boldsymbol{x}\} = \left(\cdots \partial^2\psi/\partial\eta_j\partial\eta_k \cdots\right), \tag{A.19}$$

the last notation indicating a $K \times K$ matrix of second derivatives. As in (A.10), the deviance function is

$$D(\boldsymbol{\eta}_1, \boldsymbol{\eta}_2) = 2E_{\boldsymbol{\eta}_1}\left\{\log\left(f_{\boldsymbol{\eta}_1}(\boldsymbol{x})/f_{\boldsymbol{\eta}_2}(\boldsymbol{x})\right)\right\}. \tag{A.20}$$

Exponential families and maximum likelihood estimation are closely related, both historically and mathematically. The "score function" equals

$$\frac{\partial \log f_{\boldsymbol{\eta}}(\boldsymbol{x})}{\partial \boldsymbol{\eta}} = \left(\cdots \frac{\partial \log f_{\boldsymbol{\eta}}(\boldsymbol{x})}{\partial \eta_k} \cdots\right)' = \boldsymbol{x} - \boldsymbol{\mu} \tag{A.21}$$

from (A.12) and (A.18), so the MLE $\hat{\boldsymbol{\eta}}$ is the value of $\boldsymbol{\eta}$ making the expectation vector $\boldsymbol{\mu}$ equal the observed vector $\boldsymbol{x}$. The second derivative matrix is

$$\frac{\partial^2 \log f_{\boldsymbol{\eta}}(\boldsymbol{x})}{\partial \boldsymbol{\eta}^2} = \left(\cdots \frac{\partial^2 \log f_{\boldsymbol{\eta}}(\boldsymbol{x})}{\partial \eta_j \partial \eta_k} \cdots\right) = -V \tag{A.22}$$

according to (A.18)–(A.19). This implies that $\log f_{\boldsymbol{\eta}}(\boldsymbol{x})$ is a concave function of $\boldsymbol{\eta}$ (since the covariance matrix V is positive semidefinite for every choice of $\boldsymbol{\eta}$), usually making it straightforward to calculate iteratively the MLE $\hat{\boldsymbol{\eta}}$.

These days multivariate exponential families play their biggest role in *generalized linear models*. A GLM supposes that the K-vector of natural parameters $\boldsymbol{\eta}$ in (A.12) is a known linear function of a J-vector $\boldsymbol{\beta}$ ($J \leq K$), say

$$\boldsymbol{\eta} = \underset{K\times J}{M}\,\boldsymbol{\beta}. \tag{A.23}$$

This construction extends the notion of ordinary linear models to non-normal exponential families. It reduces the K-parameter family (A.12) to a J-parameter exponential family, with natural parameter vector $\boldsymbol{\beta}$ and sufficient statistic $M'\boldsymbol{x}$. The MLE equations for $\boldsymbol{\beta}$ take the form

$$M'\left(\boldsymbol{x} - \boldsymbol{\mu}(\boldsymbol{\beta})\right) = \mathbf{0} \tag{A.24}$$

where $\boldsymbol{\mu}(\boldsymbol{\beta})$ is the expectation vector in (A.12) corresponding to $\boldsymbol{\eta} = M\boldsymbol{\beta}$. McCullagh and Nelder (1989) is the standard reference for generalized linear models.

A.2 Lindsey's Method

Model (5.10) is a J-parameter exponential family with natural parameter $\boldsymbol{\beta} = (\beta_1, \beta_2, \dots, \beta_J)'$, sufficient statistic $\boldsymbol{x} = (z, z^2, \dots, z^J)'$, and normalizer $\psi(\boldsymbol{\beta})$ equaling $-\beta_0$. If $z_1, z_2, \dots, z_N$ is an i.i.d. sample from $f(z)$ then, as in (A.3), we still have a J-parameter exponential family, now with sufficient statistic $\bar{\boldsymbol{x}} = (\bar{x}_1, \bar{x}_2, \dots, \bar{x}_J)'$,

$$\bar{x}_j = \sum_{i=1}^{N} z_i^j \Big/ N, \tag{A.25}$$

and natural parameter vector $N\boldsymbol{\beta}$.

Solving for the MLE $\hat{\boldsymbol{\beta}}$ requires special programming, essentially because there is no closed form for $\psi(\boldsymbol{\beta})$. Lindsey's method (Lindsey, 1974; Efron and Tibshirani, 1996, Sect. 2) is a way of using standard Poisson GLM software to find $\hat{\boldsymbol{\beta}}$. The N z-values are binned into K counts $\boldsymbol{y} = (y_1, y_2, \dots, y_K)'$ as in (5.11)–(5.12) so that the density $f_{\boldsymbol{\beta}}(z)$ is reduced to a multinomial probability vector $\boldsymbol{\pi}(\boldsymbol{\beta})$ (A.14), with $\pi_k(\boldsymbol{\beta}) = d \cdot f_{\boldsymbol{\beta}}(x_k)$ in notation (5.14).

A special relationship between multinomial and Poisson distributions is then invoked: assuming that the y_k's are independent Poisson observations,

$$y_k \overset{\text{ind}}{\sim} \text{Poi}(\nu_k), \quad \nu_k = N\pi_k(\boldsymbol{\beta}) \qquad \text{for } k = 1, 2, \dots, K, \tag{A.26}$$

the MLE $\hat{\boldsymbol{\beta}}$ in (A.26) equals that in the multinomial model of (A.13),

$\boldsymbol{y} \sim \text{mult}_K(N, \boldsymbol{\pi}(\boldsymbol{\beta}))$. Family (5.10), for example, becomes a Poisson GLM (A.23) in this formulation, with $\eta_k = \log(\nu_k)$ and M having kth row

$$\left(x_k, x_k^2, \ldots, x_k^J\right)', \tag{A.27}$$

x_k being the kth bin midpoint (5.13).

In computing language R, the single call

$$\textsf{glm}(\textbf{y} \sim \textsf{M, family=Poisson}) \tag{A.28}$$

produces $\hat{\boldsymbol{f}}$, the MLE for density $f(z)$ in (5.10), evaluated at $z = x_k$ for $k = 1, 2, \ldots, K$. These are the heavy curves plotted, for example, in Figure 5.1 and Figure 6.1. Calling $\hat{\boldsymbol{f}}$ the MLE assumes that the z_i are independent, which is usually untrue; $\hat{\boldsymbol{f}}$ is better thought of as a smooth parametric fit to the histogram heights y_k, minimizing the total Poisson deviance (5.17) as a fitting criterion.

Appendix B

Data Sets and Programs

Most of the data sets and R programs featured in the book are available at the following web site:

`http://statistics.stanford.edu/~brad`

Follow the link to Data Sets and Programs.

Data Sets

- **Kidney data** (Figure 1.2): 157×2 matrix kidneydata giving ages and kidney scores.
- **Prostate data** (Figure 2.1): 6033×102 matrix prostatedata, columns labeled 1 and 2 for Control and Treatment; also prostz, the 6033 z-values in Figure 2.1 and Figure 4.6.
- **DTI data** (Figure 2.4): $15\,443 \times 4$ matrix DTIdata; first three columns give (x,y,z) brain coordinates, with z-values in fourth column.
- **Leukemia data** (Figure 6.1a): 7128×72 matrix leukdata, columns labeled 0 and 1 for AML and ALL; also leukz, the 7128 z-values shown in Figure 4.6 and Figure 6.1a.
- **Chi-square data** (Figure 6.1b): $321\,010 \times 3$ matrix chisquaredata, column 1 = gene, columns 2 and 3 conditions 1 and 2, as in Table 6.1 (which is actually gene 23 except that one of the entries for condition 2 has been changed from 1 to 0); also chisqz, z-values (6.2) calculated for the 16 882 of the 18 399 genes having at least three sites.
- **Police data** (Figure 6.1c): 2749-vector policez of z-values.
- **HIV data** (Figure 6.1d): 7680×8 matrix hivdata, columns labeled 0 and 1 for healthy controls and HIV patients; also hivz, vector of 7680 z-values obtained from two-sample t-tests on logged and standardized version of hivdata.
- **Cardio data** (Figure 8.1): $20\,426 \times 63$ matrix cardiodata, first 44 columns

healthy controls, last 19 cardiovascular patients; the columns are standardized but the matrix is not doubly standardized.

- **P53 data** (Figure 9.1): 10 100 × 50 matrix p53data, first 33 columns mutated cell lines, last 17 unmutated; also p53z, vector of 10 100 z-values.
- **Brain data** (Figure 10.7b): vector of 12 625 t-values braintval; also (x,y) coordinates for smoothing spline brainsm.
- **Michigan data** (Table 11.4): 5217 × 86 matrix michigandata, first 24 columns bad outcomes, last 62 good outcomes.
- **Shakespeare data** (Table 11.6): 10 × 10 matrix shakedata.

Programs

- `locfdr`: Produces estimates of the local false discovery rate fdr(z) (5.2), both assuming the theoretical null hypothesis as at (5.5) and using an empirical null estimate as in Section 6.2 and Section 6.3. Only a vector $\boldsymbol{z}$ of z-values need be entered, but the help file details a list of possible special adjustments. Particularly valuable is the histogram plot superimposing the estimated null and mixture densities $\hat{f}_0(z)$ and $\hat{f}(z)$, which should always be inspected for goodness of fit.
- `simz`: Generates an $N \times n$ matrix X of correlated z-values $z_{ij} \sim \mathcal{N}(0, 1)$, with root mean square correlation (8.15) approximately equaling a target value α; see the paragraph on simulating correlated z-values in Section 8.2. Comments at the beginning of `simz` show its use. Non-null versions of X can be obtained by adding appropriate constants to selected entries, e.g., the top 5% of rows in the last $n/2$ columns.
- `alpha`: Inputs an $N \times n$ matrix X and outputs estimate (8.18) of the root mean square correlation α. It is best to first remove potential systematic effects from X. For instance, with the prostate data of Section 2.1, one might subtract gene-wise means separately from the control and cancer groups. (Or, more simply, separately apply `alpha` to the 50 control and 52 cancer columns of X.)
- `Ebay`: Produces the empirical Bayes estimated effect size $\hat{E}\{\delta|z\}$ as in Figure 11.1. All that need be provided is the "training data," the $N \times n$ predictor matrix X and the n-vector Y of zero/one responses, but the help file lists a range of user modifications, including the folds call for cross-validation.

References

Allison, David B., Gadbury, Gary L., Heo, Moonseong, Fernández, José R., Lee, Cheol-Koo, Prolla, Tomas A., and Weindruch, Richard. 2002. A mixture model approach for the analysis of microarray gene expression data. *Comput. Statist. Data Anal.*, **39**(1), 1–20.

Ashley, Euan A., Ferrara, Rossella, King, Jennifer Y., Vailaya, Aditya, Kuchinsky, Allan, He, Xuanmin, Byers, Blake, Gerckens, Ulrich, Oblin, Stefan, Tsalenko, Anya, Soito, Angela, Spin, Joshua M., Tabibiazar, Raymond, Connolly, Andrew J., Simpson, John B., Grube, Eberhard, and Quertermous, Thomas. 2006. Network analysis of human in-stent restenosis. *Circulation*, **114**(24), 2644–2654.

Aubert, J., Bar-Hen, A., Daudin, J.J., and Robin, S. 2004. Determination of the differentially expressed genes in microarray experiments using local FDR. *BMC Bioinformatics*, **5**(September).

Benjamini, Yoav, and Hochberg, Yosef. 1995. Controlling the false discovery rate: A practical and powerful approach to multiple testing. *J. Roy. Statist. Soc. Ser. B*, **57**(1), 289–300.

Benjamini, Yoav, and Yekutieli, Daniel. 2001. The control of the false discovery rate in multiple testing under dependency. *Ann. Statist.*, **29**(4), 1165–1188.

Benjamini, Yoav, and Yekutieli, Daniel. 2005. False discovery rate-adjusted multiple confidence intervals for selected parameters. *J. Amer. Statist. Assoc.*, **100**(469), 71–93. With comments and a rejoinder by the authors.

Broberg, Per. 2004. A new estimate of the proportion unchanged genes in a microarray experiment. *Genome Biology*, **5**(5), P10.

Brown, L.D. 1971. Admissible estimators, recurrent diffusions, and insoluble boundary value problems. *Ann. Math. Statist.*, **42**, 855–903.

Clarke, Sandy, and Hall, Peter. 2009. Robustness of multiple testing procedures against dependence. *Ann. Statist.*, **37**(1), 332–358.

Desai, K., Deller, J., and McCormick, J. 2010. The distribution of number of false discoveries for highly correlated null hypotheses. *Ann. Appl. Statist.* submitted, under review.

Donoho, David, and Jin, Jiashun. 2009. Higher criticism thresholding: Optimal feature selection when useful features are rare and weak. *Proc. Natl. Acad. Sci. USA*, **105**, 14 790–14 795.

Dudoit, Sandrine, and van der Laan, Mark J. 2008. *Multiple Testing Procedures with Applications to Genomics*. Springer Series in Statistics. New York: Springer.

Dudoit, Sandrine, Shaffer, Juliet Popper, and Boldrick, Jennifer C. 2003. Multiple hypothesis testing in microarray experiments. *Statist. Sci.*, **18**(1), 71–103.

Dudoit, Sandrine, van der Laan, Mark J., and Pollard, Katherine S. 2004. Multiple testing. I. Single-step procedures for control of general type I error rates. *Stat. Appl. Genet. Mol. Biol.*, **3**, Art. 13, 71 pp. (electronic).

Efron, Bradley. 1969. Student's t-test under symmetry conditions. *J. Amer. Statist. Assoc.*, **64**, 1278–1302.

Efron, Bradley. 1987. Better bootstrap confidence intervals. *J. Amer. Statist. Assoc.*, **82**(397), 171–200. With comments and a rejoinder by the author.

Efron, Bradley. 1996. Empirical Bayes methods for combining likelihoods. *J. Amer. Statist. Assoc.*, **91**(434), 538–565. With discussion and a reply by the author.

Efron, Bradley. 2003. Robbins, empirical Bayes and microarrays. *Ann. Statist.*, **31**(2), 366–378.

Efron, Bradley. 2004. Large-scale simultaneous hypothesis testing: The choice of a null hypothesis. *J. Amer. Statist. Assoc.*, **99**(465), 96–104.

Efron, Bradley. 2007a. Correlation and large-scale simultaneous significance testing. *J. Amer. Statist. Assoc.*, **102**(477), 93–103.

Efron, Bradley. 2007b. Size, power and false discovery rates. *Ann. Statist.*, **35**(4), 1351–1377.

Efron, Bradley. 2008a. Microarrays, empirical Bayes and the two-groups model. *Statist. Sci.*, **23**(1), 1–22.

Efron, Bradley. 2008b. Simultaneous inference: When should hypothesis testing problems be combined? *Ann. Appl. Statist.*, **2**(1), 197–223.

Efron, Bradley. 2009a. Are a set of microarrays independent of each other? *Ann. Appl. Statist.*, **3**(September), 922–942.

Efron, Bradley. 2009b. Empirical Bayes estimates for large-scale prediction problems. *J. Amer. Statist. Assoc.*, **104**(September), 1015–1028.

Efron, Bradley. 2010. Correlated z-values and the accuracy of large-scale statistical estimates. *J. Amer. Statist. Assoc.* to appear.

Efron, Bradley, and Gous, Alan. 2001. Scales of evidence for model selection: Fisher versus Jeffreys. Pages 208–256 of: *Model Selection*. IMS Lecture Notes Monogr. Ser., vol. 38. Beachwood, OH: Inst. Math. Statist. With discussion by R. E. Kass, G. S. Datta, and P. Lahiri, and a rejoinder by the authors.

Efron, Bradley, and Morris, Carl. 1971. Limiting the risk of Bayes and empirical Bayes estimators. I. The Bayes case. *J. Amer. Statist. Assoc.*, **66**, 807–815.

Efron, Bradley, and Morris, Carl. 1972. Limiting the risk of Bayes and empirical Bayes estimators. II. The empirical Bayes case. *J. Amer. Statist. Assoc.*, **67**, 130–139.

Efron, Bradley, and Morris, Carl. 1973. Stein's estimation rule and its competitors – An empirical Bayes approach. *J. Amer. Statist. Assoc.*, **68**, 117–130.

Efron, Bradley, and Thisted, Ronald. 1976. Estimating the number of unseen species: How many words did Shakespeare know? *Biometrika*, **63**(3), 435–447.

Efron, Bradley, and Tibshirani, Robert. 1993. *An Introduction to the Bootstrap*. Monographs on Statistics and Applied Probability, vol. 57. New York: Chapman and Hall.

Efron, Bradley, and Tibshirani, Robert. 1996. Using specially designed exponential families for density estimation. *Ann. Statist.*, **24**(6), 2431–2461.

Efron, Bradley, and Tibshirani, Robert. 2007. On testing the significance of sets of genes. *Ann. Appl. Statist.*, **1**(1), 107–129.

Efron, Bradley, Tibshirani, Robert, Storey, John D., and Tusher, Virginia. 2001. Empirical Bayes analysis of a microarray experiment. *J. Amer. Statist. Assoc.*, **96**(456), 1151–1160.

Ferkingstad, Egil, Frigessi, Arnoldo, Rue, Håvard, Thorleifsson, Gudmar, and Kong, Augustine. 2008. Unsupervised empirical Bayesian multiple testing with external covariates. *Ann. Appl. Statist.*, **2**(2), 714–735.

Fisher, R.A. 1935. *The Design of Experiments*. Edinburgh: Oliver and Boyd.

Fisher, R.A., Corbet, A.S., and Williams, C.B. 1943. The relation between the number of species and the number of individuals in a random sample of an animal population. *J. Anim. Ecol.*, **12**, 42–58.

Genovese, Christopher, and Wasserman, Larry. 2002. Operating characteristics and extensions of the false discovery rate procedure. *J. Roy. Statist. Soc. Ser. B*, **64**(3), 499–517.

Genovese, Christopher, and Wasserman, Larry. 2004. A stochastic process approach to false discovery control. *Ann. Statist.*, **32**(3), 1035–1061.

Genovese, Christopher R., Roeder, Kathryn, and Wasserman, Larry. 2006. False discovery control with p-value weighting. *Biometrika*, **93**(3), 509–524.

Golub, T.R., Slonim, D.K., Tamayo, P., Huard, C., Gaasenbeek, M., Mesirov, J.P., Coller, H., Loh, M.L., Downing, J.R., Caligiuri, M.A., Bloomfield, C.D., and Lander, E.S. 1999. Molecular classification of cancer: Class discovery and class prediction by gene expression monitoring. *Science*, **286**(5439), 531–537.

Good, I.J. 1953. The population frequencies of species and the estimation of population parameters. *Biometrika*, **40**, 237–264.

Good, I.J., and Toulmin, G.H. 1956. The number of new species, and the increase in population coverage, when a sample is increased. *Biometrika*, **43**, 45–63.

Gottardo, Raphael, Raftery, Adrian E., Yeung, Ka Yee, and Bumgarner, Roger E. 2006. Bayesian robust inference for differential gene expression in microarrays with multiple samples. *Biometrics*, **62**(1), 10–18, 313.

Hall, Peter. 1992. *The Bootstrap and Edgeworth Expansion*. Springer Series in Statistics. New York: Springer-Verlag.

Hastie, Trevor, Tibshirani, Robert, and Friedman, Jerome. 2009. *The Elements of Statistical Learning*. 2nd edn. Springer Series in Statistics. New York: Springer.

Hedenfalk, I., Duggan, D., Chen, Y.D., Radmacher, M., Bittner, M., Simon, R., Meltzer, P., Gusterson, B., Esteller, M., Kallioniemi, O.P., Wilfond, B., Borg, A., Trent, J., Raffeld, M., Yakhini, Z., Ben-Dor, A., Dougherty, E., Kononen, J., Bubendorf, L., Fehrle, W., Pittaluga, S., Gruvberger, S., Loman, N., Johannsoson, O., Olsson, H., and Sauter, G. 2001. Gene-expression profiles in hereditary breast cancer. *N. Engl. J. Med.*, **344**(8), 539–548.

Heller, G., and Qing, J. 2003. *A mixture model approach for finding informative genes in microarray studies*. Unpublished.

Hochberg, Yosef. 1988. A sharper Bonferroni procedure for multiple tests of significance. *Biometrika*, **75**(4), 800–802.

Hoeffding, Wassily. 1952. The large-sample power of tests based on permutations of observations. *Ann. Math. Statist.*, **23**, 169–192.

Holm, Sture. 1979. A simple sequentially rejective multiple test procedure. *Scand. J. Statist.*, **6**(2), 65–70.

Hommel, G. 1988. A stagewise rejective multiple test procedure based on a modified Bonferroni test. *Biometrika*, **75**(2), 383–386.

James, W., and Stein, Charles. 1961. Estimation with quadratic loss. Pages 361–379 of: *Proc. 4th Berkeley Sympos. Math. Statist. and Prob., Vol. I.* Berkeley, Calif.: Univ. California Press.

Jin, Jiashun, and Cai, T. Tony. 2007. Estimating the null and the proportional of non-null effects in large-scale multiple comparisons. *J. Amer. Statist. Assoc.*, **102**(478), 495–506.

Jin, Jiashun, and Cai, T. Tony. 2010. Optimal rates of convergence for estimating the null density and proportion of non-null effects in large-scale multiple testing. *Ann. Statist.*, **38**(1), 100–145.

Johnson, Norman L., and Kotz, Samuel. 1970. *Distributions in Statistics. Continuous Univariate Distributions. 1.* Boston, Mass.: Houghton Mifflin Co.

Johnstone, Iain M., and Silverman, Bernard W. 2004. Needles and straw in haystacks: Empirical Bayes estimates of possibly sparse sequences. *Ann. Statist.*, **32**(4), 1594–1649.

Kerr, M.K., Martin, M., and Churchill, G.A. 2000. Analysis of variance for gene expression microarray data. *J. Comput. Biology*, **7**(6), 819–837.

Lancaster, H.O. 1958. The structure of bivariate distributions. *Ann. Math. Statist.*, **29**, 719–736.

Lehmann, Erich L., and Romano, Joseph P. 2005a. Generalizations of the familywise error rate. *Ann. Statist.*, **33**(3), 1138–1154.

Lehmann, Erich L., and Romano, Joseph P. 2005b. *Testing Statistical Hypotheses.* 3rd edn. Springer Texts in Statistics. New York: Springer.

Lemley, Kevin V., Lafayette, Richard A., Derby, Geraldine, Blouch, Kristina L., Anderson, Linda, Efron, Bradley, and Myers, Bryan D. 2008. Prediction of early progression in recently diagnosed IgA nephropathy. *Nephrol. Dialysis Transplant.*, **23**(1), 213–222.

Liao, J.G., Lin, Y., Selvanayagam, Z.E., and Shih, W.C.J. 2004. A mixture model for estimating the local false discovery rate in DNA microarray analysis. *BMC Bioinformatics*, **20**(16), 2694–2701.

Lindsey, J.K. 1974. Construction and comparison of statistical models. *J. Roy. Statist. Soc. Ser. B*, **36**, 418–425.

Marcus, Ruth, Peritz, Eric, and Gabriel, K.R. 1976. On closed testing procedures with special reference to ordered analysis of variance. *Biometrika*, **63**(3), 655–660.

McCullagh, P., and Nelder, J.A. 1989. *Generalized Linear Models.* 2nd edn. Monographs on Statistics and Applied Probability. London: Chapman & Hall.

Miller, Jr., Rupert G. 1981. *Simultaneous Statistical Inference.* 2nd edn. Springer Series in Statistics. New York: Springer-Verlag.

Morris, Carl N. 1983. Parametric empirical Bayes inference: Theory and applications. *J. Amer. Statist. Assoc.*, **78**(381), 47–65. With discussion.

Muralidharan, Omkar. 2009. *High dimensional exponential family estimation via empirical Bayes.* Unpublished.

Muralidharan, Omkar. 2010. An empirical Bayes mixture method for effect size and false discovery rate estimation. *Ann. Appl. Statist.*, **4**(1), 422–438.

Newton, M.A., Noueiry, A., Sarkar, D., and Ahlquist, P. 2004. Detecting differential gene expression with a semiparametric hierarchical mixture method. *Biostatistics*, **5**(2), 155–176.

Olshen, Richard, and Rajaratnam, Bala. 2009. *The effect of correlation in false discovery rate estimation*. Tech. rept. Stanford University Department of Statistics.

Owen, Art B. 2005. Variance of the number of false discoveries. *J. Roy. Statist. Soc. Ser. B*, **67**(3), 411–426.

Pan, Wei, Lin, Jizhen, and Le, Chap T. 2003. A mixture model approach to detecting differentially expressed genes with microarray data. *Funct. Integr. Genomics*, **3**(3), 117–124.

Pavlidis, Paul, Lewis, Darrin P., and Noble, William Stafford. 2002. Exploring gene expression data with class scores. *Pac. Symp. Biocomput.*, 474–485.

Pitman, E.J.G. 1937. Significance Tests Which May be Applied to Samples From Any Populations. *Suppl. J. Roy. Statist. Soc.*, **4**(1), 119–130.

Pounds, S., and Morris, S.W. 2003. Estimating the occurrence of false positives and false negatives in microarray studies by approximating and partitioning the empirical distribution of p-values. *BMC Bioinformatics*, **19**(10), 1236–1242.

Qiu, Xing, Klebanov, Lev, and Yakovlev, Andrei. 2005a. Correlation between gene expression levels and limitations of the empirical Bayes methodology for finding differentially expressed genes. *Stat. Appl. Genet. Mol. Biol.*, **4**, Art. 34, 32 pp. (electronic).

Qiu, Xing, Brooks, Andrew, Klebanov, Lev, and Yakovlev, Andrei. 2005b. The effects of normalization on the correlation structure of microarray data. *BMC Bioinformatics*, **6**(1), 120.

Rahnenführer, Jörg, Domingues, Francisco S., Maydt, Jochen, and Lengauer, Thomas. 2004. Calculating the statistical significance of changes in pathway activity from gene expression data. *Stat. Appl. Genet. Mol. Biol.*, **3**, Art. 16, 31 pp. (electronic).

Ridgeway, Greg, and MacDonald, John M. 2009. Doubly robust internal benchmarking and false discovery rates for detecting racial bias in police stops. *J. Amer. Statist. Assoc.*, **104**(486), 661–668.

Robbins, Herbert. 1956. An empirical Bayes approach to statistics. Pages 157–163 of: *Proceedings of the Third Berkeley Symposium on Mathematical Statistics and Probability, 1954–1955, vol. I*. Berkeley and Los Angeles: University of California Press.

Robbins, Herbert. 1968. Estimating the total probability of the unobserved outcomes of an experiment. *Ann. Math. Statist.*, **39**, 256–257.

Romano, Joseph P., Shaikh, Azeem M., and Wolf, Michael. 2008. Control of the false discovery rate under dependence using the bootstrap and subsampling. *TEST*, **17**(3), 417–442.

Schwartzman, A., and Lin, X. 2009. *The effect of correlation in false discovery rate estimation*. Biostatistics Working Paper Series number 106. Harvard University.

Schwartzman, A., Dougherty, R.F., and Taylor, J.E. 2005. Cross-subject comparison of principal diffusion direction maps. *Magn. Reson. Med.*, **53**(6), 1423–1431.

Simes, R.J. 1986. An improved Bonferroni procedure for multiple tests of significance. *Biometrika*, **73**(3), 751–754.

Singh, Dinesh, Febbo, Phillip G., Ross, Kenneth, Jackson, Donald G., Manola, Judith, Ladd, Christine, Tamayo, Pablo, Renshaw, Andrew A., D'Amico, Anthony V., Richie, Jerome P., Lander, Eric S., Loda, Massimo, Kantoff, Philip W., Golub, Todd R., and Sellers, William R. 2002. Gene expression correlates of clinical prostate cancer behavior. *Cancer Cell*, **1**(2), 203–209.

Singh, R.S. 1979. Empirical Bayes estimation in Lebesgue-exponential families with rates near the best possible rate. *Ann. Statist.*, **7**(4), 890–902.

Smyth, Gordon K. 2004. Linear models and empirical Bayes methods for assessing differential expression in microarray experiments. *Stat. Appl. Genet. Mol. Biol.*, **3**, Art. 3, 29 pp. (electronic).

Soric, Branko. 1989. Statistical "discoveries" and effect-size estimation. *J. Amer. Statist. Assoc.*, **84**(406), 608–610.

Spevack, Marvin. 1968. *A Complete and Systematic Concordance to the Works of Shakespeare, Vols 1–6*. Hildesheim: George Olms.

Stein, Charles M. 1956. Inadmissibility of the usual estimator for the mean of a multivariate normal distribution. Pages 197–206 of: *Proceedings of the Third Berkeley Symposium on Mathematical Statistics and Probability, 1954–1955, vol. I*. Berkeley and Los Angeles: University of California Press.

Stein, Charles M. 1981. Estimation of the mean of a multivariate normal distribution. *Ann. Statist.*, **9**(6), 1135–1151.

Storey, John D. 2002. A direct approach to false discovery rates. *J. Roy. Statist. Soc. Ser. B*, **64**(3), 479–498.

Storey, John D. 2003. The positive false discovery rate: A Bayesian interpretation and the *q*-value. *Ann. Statist.*, **31**(6), 2013–2035.

Storey, John D. 2007. The optimal discovery procedure: A new approach to simultaneous significance testing. *J. Roy. Statist. Soc. Ser. B*, **69**(3), 347–368.

Storey, John D., Taylor, Jonathan E., and Siegmund, David. 2004. Strong control, conservative point estimation and simultaneous conservative consistency of false discovery rates: A unified approach. *J. Roy. Statist. Soc. Ser. B*, **66**(1), 187–205.

Subramanian, A., Tamayo, P., Mootha, V. K., Mukherjee, S., Ebert, B.L., Gillette, M.A., Paulovich, A., Pomeroy, S.L., Golub, T.R., Lander, E.S., and Mesirov, J.P. 2005. Gene set enrichment analysis: A knowledge-based approach for interpreting genome-wide expression profiles. *Proc. Natl. Acad. Sci.*, **102**(43), 15545–15550.

Thisted, Ronald, and Efron, Bradley. 1987. Did Shakespeare write a newly-discovered poem? *Biometrika*, **74**(3), 445–455.

Tibshirani, Robert, and Efron, Brad. 2002. Pre-validation and inference in microarrays. *Stat. Appl. Genet. Mol. Biol.*, **1**, Art. 1, 20 pp. (electronic).

Tibshirani, Robert, and Hastie, Trevor. 2007. Outlier sums for differential gene expression analysis. *Biostatistics*, **8**(1), 2–8.

Tibshirani, Robert, Hastie, Trevor, Narasimhan, Balasubramanian, and Chu, Gilbert. 2002. Diagnosis of multiple cancer types by shrunken centroids of gene expression. *Proc. Natl. Acad. Sci. USA*, **99**(10), 6567–6572.

Tibshirani, Ryan, and Tibshirani, Robert. 2009. A bias-correction for the minimum error rate in cross-validation. *Ann. Appl. Statist.*, 822–829.

Tomlins, S.A., Rhodes, D.R., Perner, S., Dhanasekaran, S.M., Mehra, R., Sun, X.W., Varambally, S., Cao, X.H., Tchinda, J., Kuefer, R., Lee, C., Montie, J.E., Shah, R.B.,

Pienta, K.J., Rubin, M.A., and Chinnaiyan, A.M. 2005. Recurrent fusion of TM-PRSS2 and ETS transcription factor genes in prostate cancer. *Science*, **310**(5748), 644–648.

van't Wout, A.B., Lehrman, G.K., Mikheeva, S.A., O'Keefe, G.C., Katze, M.G., Bumgarner, R.E., Geiss, G.K., and Mullins, J.I. 2003. Cellular gene expression upon human immunodeficiency virus type 1 infection of CD4(+)-T-cell lines. *J. Virol*, **77**(2), 1392–1402.

Westfall, P.H., and Young, S.S. 1993. *Resampling-based Multiple Testing: Examples and Methods for p-Value Adjustment.* Wiley Series in Probability and Statistics. New York, NY: Wiley-Interscience.

Wu, Baolin. 2007. Cancer outlier differential gene expression detection. *Biostatistics*, **8**(3), 566–575.

Index